Remote Sensing
Models and Methods for Image Processing
Second Edition

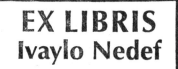

Front cover:

In early June 1986, the advancing face of Hubbard Glacier, located in Southeastern Alaska, closed off the entrance to Russell Fiord with an ice dam that created the world's largest known modern glacier-dammed lake. This Landsat TM CIR composite is from September 11, 1986. Over 70 miles long, the Hubbard is North America's largest tidewater glacier. The Valerie Glacier, a tributary to the Hubbard, moved at rates of up to 130 feet per day causing the Hubbard to advance more than a mile during the summer of 1986. The current advance is part of a long-term cycle of advance and retreat. About 800 years ago, the glacier extended all the way to the mouth of Yakutat Bay. Because of the ice dam, Russell Fiord became a lake and water level rose at a very rapid rate during the summer months. Marine mammals (porpoise and seals) and other marine life were trapped. If the lake had continued to rise, it was predicted that it would overflow to the south within a year. The dam formed by the surging glacier broke on October 8, 1986, restoring the dammed-up lake to a fiord, and recovering the vital sea-water environment for the marine life. It is expected that the glacier will continue to surge and form another dam in the future as the cycle continues. (Image data and description courtesy U.S. Geological Survey [Meridian Data, 1991 #528])

Remote Sensing
Models and Methods for Image Processing

Second Edition

Robert A. Schowengerdt

Department of Electrical and
Computer Engineering
University of Arizona
Tucson, Arizona

ACADEMIC PRESS
An Imprint of Elsevier
San Diego   London   Boston
New York   Sydney   Tokyo   Toronto

Academic Press
An Imprint of Elsevier
525 B Street, Suite 1900, San Diego, California 92101-4495, USA
http://www.academicpress.com

Academic Press
84 Theobalds Road, London WC1X 8RR, UK
http://www.academicpress.com

Library of Congress Cataloging-in-Publication Data

Schowengerdt, Robert A.
    Remote sensing, models and methods for image processing / Robert A. Schowengerdt.–2nd ed.
        p.   cm.
        Rev. ed. of: Techniques for image processing and classification in remote Sensing.
        Includes bibliographical references and index.
        ISBN 0-12-628981-6 (alk. paper)
        1.   Image processing. 2. Remote-sensing images. I. Schowengerdt, Robert A. Techniques for image processing and classification in Remote sensing. II. Title.
    TA1632.S3   1997   97-8078
    621.36'7–dc21   CIP

PRINTED IN THE UNITED STATES OF AMERICA
            05   06   07   GS   9   8   7   6

To my best friend, Amy

Contents

CHAPTER 7 Correction and Calibration 287

CHAPTER 9 *Thematic Classification* *389*

Image Gallery

Plate 1-1

Remote sensing imagery can provide a long-term record of man's impact on the environment. This series of Landsat MSS images of the Twin Buttes copper mining complex south of Tucson, Arizona, also includes irrigated pecan orchards along the right side, and the growing retirement community of Green Valley between the orchards and the mines. The large polygonal features are tailings waste ponds from the open pit mining, and the presence of blue-black color indicates standing water on the surface. Note the noisy data in the 1974 image and the cloud and its shadow in the 1977 image. The expansion of the mining activity during this 13 year period is evident, as is growth in the irrigated agricultural areas and in Green Valley (white-grey area at the lower right).

Plate 1-2

This July 23, 1990 AVIRIS image of Palo Alto, California, illustrates hyperspectral data. The CIR composite is made from bands 51, 27 and 17 (arrows in the graph) displayed in red, green and blue, respectively, and represents only about 1.5% of the total data in the image. Five pixels were selected for extraction of spectral profiles and are located at the centers of the 5x5 pixel yellow squares. From left to right, they represent: dry grassland, building, Palo Alto golf course, San Francisco Bay and salt evaporator pond. The spectral profiles are plotted below, color coded to match the CIR composite colors. The AVIRIS data are uncorrected for atmospheric effects; note the similarity between the golf course (grass) profile and the simulated signature of Fig. 2-9 in Chapter 2. Atmospheric scattering increasing towards the blue spectral region is evident in all the signatures, as are the H_2O and CO_2 absorption bands. (Image courtesy of NASA/JPL)

Plate 1-3

This TM CIR composite from June 9, 1984, shows the All-American Canal used to irrigate agriculture in the desert near Yuma, Arizona. Crops such as lettuce and melons are typically grown in this area. The degree of "redness" indicates the growth stage of the crop in a particular field; grey fields are fallow and black fields are flooded with water. The Algodones Dunes at the top show periodic sand dune patterns. An image acquired at a different date would likely show changes in these patterns, as well as in the agricultural area.

Plate 1-4

The potential for remote sensing in global applications is illustrated in these two images. They show the Normalized Difference Vegetation Index (NDVI) calculated from AVHRR satellite data (an example of Level 3 processing, see Chapter 1) reduced to about 16km GIFOV. At the top is a 10 day composite (used to increase the potential for cloud-free coverage at every pixel) from May 1–10, 1992 and at the bottom is a composite from June 1–10, 1992. The degree of "greeness" indicates the amount of green vegetation at the time. Note the increase in vegetation between the two periods in the Northern Hemisphere, corresponding to spring growth. The geometric projection used here is known as the Interrupted Goode Homolosine equal-area projection (Goode, 1925; Steinwand, 1994). (NDVI data courtesy U.S. Geological Survey; image coloring by the author)

Plate 1-5

This declassified photograph, showing a major Soviet strategic bomber base near Dolon, Kazakhstan, was taken on August 20, 1966 with a KH-4A camera during mission 1036 of the CORONA reconnaissance satellite. The original photographic resolution of about 2m allowed analysts to distinguish between heavy bombers and transport aircraft on the airfield's main parking apron. CORONA thus provided critical quantitative information to dispel the so-called "bomber gap," as well as the "missile gap," of the 1960s. Military surveillance satellites have become a vital component in national policy decision-making. CORONA imagery was photographic and returned to earth in a re-entry capsule that was retrieved in the lower atmosphere by hooking its parachute with a specially-equipped airplane. This digital image was created by scanning a 30X enlarged photograph at 300 pixels per inch. (Image and description courtesy of U. S. Geological Survey)

August 22, 1972   *November 15, 1973*   *June 1, 1974*

November 14, 1975   *May 30, 1976*   *June 30, 1977*

October 11, 1978   *July 13, 1983*   *June 14, 1985*

Plate 1-1

Plate 1-2

Plate 1-3

May 1 - 10, 1992

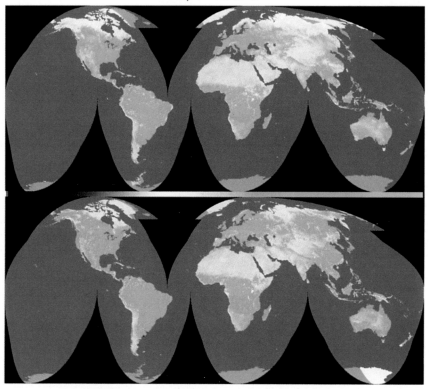

June 1 - 10, 1992

Plate 1-4

Plate 1-5

Plate 1-6

Plate 1-7

TMS3　　　*TMS4*　　　*TMS5*

TMS7　　　*TMS6*　　　*TMS6 (high gain)*

Plate 1-8

TM1 TM2 TM3

TM4 TM5 TM6

TM7

Plate 1-9

Cuprite: TM bands 3,2,1   *TM bands 7,5,4*   *Yuma: TM bands 4,3,2*

original bands

color space transforms

original

min-max stretch

50% saturation

normalization stretch

50% intensity

decorrelation stretch

cycle hue

Plate 5-1

original

min-max stretch

histogram equalization stretch

normalization stretch

decorrelation stretch

HSI intensity stretch

Plate 5-2

Cuprite, Nevada *Palo Alto, California*

at-sensor radiance (true color composite: bands 27, 17, 7)

scene radiance (dark object radiance subtracted)

scene "reflectance" (divided by bright target radiance)

Plate 7-1

TM bands 4,3,2

SPOT-P

PCT composite

IHS composite

HFM composite (3x3)

HFM composite (PSF)

Plate 8-1

TM bands 4,3,2

SPOT-P

PCT composite

IHS composite

HFM composite (3x3)

HFM composite (PSF)

Plate 8-2

*Training sites for dark soil, light soil, and crop in the
Marana, Arizona, TM image*

DN$_4$

DN$_3$

maximum-likelihood

artificial neural network

*Classification decision regions; each color represents a different class
(© 1995 IEEE).*

Plate 9-1

AVIRIS 2.3, 2.2, 2.1 μm   *residual image*

Residual image processing (Cuprite, NV)

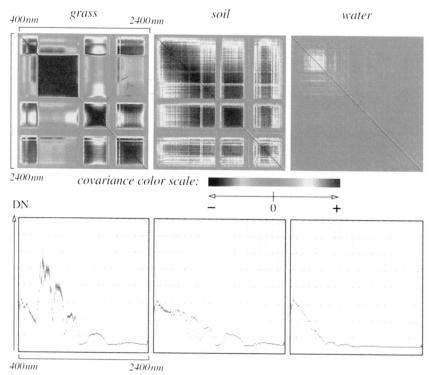

Statistics image visualization of hyperspectral signatures

Plate 9-2

Plate 1-6

Spaceborne imaging radar image of Isla Isabella in the western Galapagos Islands taken by the L-band radar in HH polarization from the Spaceborne Imaging Radar C/X-Band Synthetic Aperture Radar on the 40th orbit of the space shuttle Endeavour on April 15, 1994. The image is centered at about 0.5 degree south latitude and 91 degrees west longitude. The radar incidence angle at the center of the image is about 20 degrees. The western Galapagos Islands, which lie about 1,200 km west of Ecuador in the eastern Pacific, have six active volcanoes, and since the time of Charles Darwin's visit to the area in 1835, there have been over 60 recorded eruptions. This SIR-C/X-SAR image of Alcedo and Sierra Negra volcanoes shows the rougher lava flows as bright features, while ash deposits and smooth pahoehoe lava flows appear dark. (Image and description courtesy of NASA/JPL)

Plate 1-7

Environmental monitoring with thermal imagery is illustrated by this HCMM image of the northeastern US acquired at 3A.M. on June 11, 1978. Cheasepeake Bay extends northward from the right center of the image, with Washington D.C. on the Potomac River at the end. Water is generally warmer than land at night, and cooler Appalachian valleys can be seen in the upper left. HCMM was a unique mission that acquired 25,000 images from April 26, 1978 to September 30, 1980 (Short and Stuart, 1982). The purpose was to acquire day and night image pairs that could be used for thermal inertia mapping of geologic materials. The GSD was about 500×500 m.

Plate 1-8

In 1988, there was a devastating wildfire in Yellowstone National Park, Wyoming. This airborne Thematic Mapper Simulator (TMS) image was acquired on September 2, 1988. The TMS bands are the same as those of TM. In the VNIR bands, TMS3 and TMS4, only the smoke from the fire is visible. The fire itself begins to be visible in TMS5 (1.55–1.75mm) and is clearly visible in TMS7 (2.08–2.35mm) and TMS6 (8.5–14mm). The high gain setting in the lower right image provides higher signal level in the TIR. (Imagery courtesy of Jeffrey Myers, Aircraft Data Facility, NASA/Ames Research Center)

Plate 1-9

All seven TM bands are shown for the same area as in Plate 1-3. Although the data are uncalibrated, they show some particular properties of TM data in general. Note, for example, the increase in brightness of some agricultural fields in band 4 relative to band 3, indicating the vegetation spectral "edge." Band 6 is the thermal IR channel and exhibits the 120m GIFOV compared to 30m for the other six bands (band 6 data are resampled in post-acquisition processing to 30m GSD). It also shows characteristically low contrast. Band 5 is saturated in the desert area at the top (the image is from June 9, 1984, and so has a high solar elevation angle). Also, variable stripes can be seen in one detector in this band. Such noise is usually scene-dependent and in this case is probably instigated by the high radiance levels in this band.

Plate 5-1

Examples of color contrast enhancement of three band composites of TM imagery are shown at the left. This desert area has relatively little spectral information in the visible region (bands 1, 2, and 3); the decorrelation stretch manages to enhance the small differences, but results in a noisy image. In the NIR and SWIR spectral region (bands 4, 5 and 7), however, the varied mineralization is clearly evident and easily enhanced. The CIR Yuma, Arizona, TM image at the top right is used to illustrate various spectral transforms in the text. Below are the RGB images resulting from the processing in HSI space described in Fig. 5–33.

Plate 5-2

More examples of color contrast enhancement are shown, in this case for a CIR TM composite of an urban area in the desert, with some vegetation. All methods produce substantial enhancement; as with all contrast enhancement processing, the "best" result is in the eye of the beholder!

Plate 7-1

The effects of atmospheric scattering correction by DOS, followed by in-scene flat-field calibration using a bright target are shown for a true color composite of two AVIRIS images, one of a dry desert area (Cuprite) and the other of a coastal area (Palo Alto). The bottom images represent surface reflectance except for topographic shading. The same display LUT is used for all images to preserve color differences.

Plate 8-1

Fusion of multispectral TM and panchromatic SPOT imagery over downtown San Francisco using three methods. This area has little spectral contrast, and the

TM and SPOT bands have reasonable spectral correlation. The major source for a lack of correlation is different shadow patterns from buildings on the two dates.

Plate 8-2

Fusion of multispectral TM and panchromatic SPOT imagery over Golden Gate Park in San Francisco using three methods. Half of this area is vegetated and hilly. The TM4 and SPOT P bands are poorly correlated because of the vegetation signature, and all TM bands are poorly correlated to the SPOT P band because of the different topographic shading on the two dates. We see here a large difference between the results of the spectral domain methods and those of the HFM method, because the former are strongly affected by the large proportion of vegetation in the scene.

Plate 9-1

Supervised training sites are shown at the top for the three class, two band (TM3 and TM4) supervised classification examples in the text. The image is a CIR composite. Below are two-band decision regions obtained from 12 supervised training sites for an urban land use and land cover classification (Paola and Schowengerdt, 1995b). The maximum-likelihood decision boundaries are constrained by the Gaussian distribution assumption, unlike the neural network decision boundaries.

Plate 9-2

Techniques for hyperspectral image analysis. The image residual method provides visual discrimination among different minerals. Red corresponds to alunite, dark blue to kaolinite and yellow-orange (the two sites in the center) to the rare mineral, buddingtonite. Image statistics visualization provides a visual tool for class differentiation. The covariance matrices for three training classes are pseudo-colored in the top row and the class mean spectral signatures are shown in the bottom row. Note the regions of negative covariance and the high NIR covariance in the grass covariance matrix.

Figures

CHAPTER 1 *The Nature of Remote Sensing*

CHAPTER 2 Optical Radiation Models

CHAPTER 3 Sensor Models

CHAPTER 4 Data Models

CHAPTER 5 Spectral Transforms

CHAPTER 6 Spatial Transforms

CHAPTER 7 Correction and Calibration

CHAPTER 8 Image Registration and Fusion

CHAPTER 9 *Thematic Classification*

Tables

CHAPTER 1 The Nature of Remote Sensing

CHAPTER 3 Sensor Models

CHAPTER 4 Data Models

CHAPTER 5 Spectral Transforms

CHAPTER 6 Spatial Transforms

CHAPTER 9 *Thematic Classification*

APPENDIX A *Sensor Acronyms*

Preface

This book began as a rather conservative revision of my earlier textbook, *Techniques for Image Processing and Classification in Remote Sensing*. Like many "limited" endeavors, however, it soon grew to be a much larger project! When it became clear that simply a revision would not suffice, I gave considerable thought on a new way to present the subject of image processing in the context of remote sensing. After much mental wandering about, it became clear that there was a unifying theme through many of the image processing methods used in remote sensing, namely that they are based, directly or indirectly, on *models* of physical processes. In some cases there is a direct dependence, for example, on physical models that describe orbital geometry or the reflectance of radiation. In other cases, the dependence is indirect. For example, the common assumption of data similarity implies that neighboring pixels in the space or spectral domains are likely to have similar values. The origin of this similarity is in the physical processes leading up to the acquisition of the data, and in the acquisition itself. In nearly all cases, the motivation and rationale for remote sensing image processing algorithms can be traced to an assumption of one or more such models. Thus, I settled on this viewpoint for the book.

It was obvious from the beginning that the book should be an entirely digital production. The computer tools currently available for desktop publishing easily support this, and given the subject matter, seem almost obligatory. Therefore, extensive use is made of computer-generated graphics and image processing. Nearly all figures are entirely new and produced especially for this edition. Three-dimensional graphing programs were used to visualize multidimensional data, and an assortment of desktop image processing programs was used to produce the images. These include in particular, IPT, a development version of the MacSADIE image processing software from my laboratory, and MultiSpec, a multispectral classification program from David Landgrebe's laboratory at Purdue University.

To enhance the use of the book in the classroom, exercises are included for each chapter. They range from conceptual, "gedanken" experiments to mathematical derivations. The exercises are intended to promote an understanding of the material presented in the chapter. Extensive bibliographies of many of the subjects covered are provided in the form of tables to conserve space and provide a compact source for further information. In the references, I've emphasized archival journal papers, because they are generally easiest for the reader to acquire.

Chapter 1 provides an overview of remote-sensing science and technology as of 1996. The basic parameters for optical remote sensing are established here, and the main types of scanning sensors are described. In Chapter 2, the most important optical radiation processes in remote-sensing are described mathematically. These include solar radiation, atmospheric scattering, absorption and transmission, and surface reflectance. The wavelength region from 400nm wavelength to the thermal infrared is analyzed. Sensor models for radiometric and spatial response are explained in Chapter 3. Satellite imaging geometry is also included because of its importance for image rectification and geocoding and for extraction of elevation information from stereo images.

In Chapter 4, data models provide a transition between the physical models of Chapters 2 and 3 and the image processing methods of later chapters. Spectral and spatial statistical models for remote sensing data are described. A series of imaging simulations illustrate and explain the influence of the sensor's characteristics on the data acquired by remote-sensing systems.

Chapter 5 begins the discussion of image processing methods and covers spectral transforms, including various vegetation indicies, principal components and contrast enhancement. Chapter 6 includes convolution and Fourier filtering, multiresolution image pyramids and scale-space techniques such as wavelets. The latter types of image analyses appear to have considerable potential for efficient and effective spatial information extraction. The concept of image spatial decomposition into two or more components is used here to provide a link among the different spatial transforms. In Chapter 7, several examples of the use of image processing for image radiometric and geometric calibration are given. The importance of image calibration for high spectral resolution imagery ("hyperspectral") data is also discussed. The topic of multiimage fusion is addressed in Chapter 8, with reference to the spatial decomposition concept of Chapter 6. The various approaches are explained and illustrated with Landsat TM multispectral and SPOT panchromatic image fusion. An image pyramid-based scheme for digital elevation model (DEM) extraction from a stereo image pair is also described in detail. Chapter 9 is devoted to thematic classification of remote-sensing images, including the traditional statistical approaches and newer neural network and fuzzy classification methods. Techniques specifically developed for hyperspectral imagery are also described.

Some topics that one usually finds in a remote sensing image processing book, such as classification map error analysis, were deliberately excluded. This was done not only for space reasons, but also because I felt they departed too far from the main theme of the relation of image processing methods to remote sensing physical models. Likewise, classification methods such as rule-based systems that rely on higher level abstractions of the data, although effective and promising in many cases, are not described. I also view Geographic Information Systems (GIS) as being outside the scope of this work.

I'm indebted to many colleagues for their advice and assistance. In some cases, their contributions were substantial: Phil Slater and Kurt Thome (Optical Sciences Center, University of Arizona) provided knowledgeable guidance to keep me on the right path in Chapters 2 and 3, as did Jennifer Dungan (NASA/ Ames Research Center) with respect to Chapter 4. Others provided reviews of selected portions, including Chuck Hutchinson and Stuart Marsh (Office of Arid Lands Studies, University of Arizona) and Chris Hlavka (NASA/Ames Research Center). I also wish to acknowledge the insight provided by Eric Crist (Environmental Research Institute of Michigan) on the tasseled cap transformation. Many of my former and current students provided invaluable data and examples, including Dan Filiberti (Science Applications International Corporation), Steve Goisman (University of Arizona), Per Lysne (University of Arizona), Justin Paola (Oasis Research Center), and Ho-Yuen Pang (University of Arizona). Gerhard Mehldau (Photogrammetrie GMBH) supported my efforts with updated versions of IPT. Colleagues in the U. S. Geological Survey, including William Acevedo, Susan Benjamin, Brian Bennett, Rick Champion, Len Gaydos and George Lee, and Jeff Meyers of the NASA/Ames Research Center, kindly supplied image and digital elevation data that provided important examples throughout the book. I would also like to thank Peter B. Keenan, a longtime friend and colleague, for helping me collect ground truth by bicycle on a typically beautiful day in the San Francisco Bay area!

I am grateful for technical and management contributions by several editors and professional staff at Academic Press, including Lori Asbury, Sandra Lee and Bruce Washburn of the San Diego office, and Diane Grossman, Abby Heim and Zvi Ruder of the Chestnut Hill office.

Finally, I must thank my family, Amy, Andrea and Jennifer, for they sacrificed more than I did during the long hours devoted to this book. Hopefully, the result is to some degree worthy of their support.

CHAPTER 1

The Nature of Remote Sensing

1.1 Introduction

Compared to 1972, when the first Landsat system was launched, remote-sensing systems now exhibit an extraordinary diversity. There are operational satellite systems that sample nearly all available parts of the electromagnetic spectrum, with spatial resolutions from 5 to 1000m, complemented by a number of aerial systems with 10nm spectral resolution. The general characteristics of electro-optical remote-sensing instruments and images are described in this chapter to demonstrate this diversity and to establish the context for processing images of the earth's surface.

1.2   *Remote Sensing*

Remote sensing is defined, for our purposes, as the measurement of object properties on the earth's surface using data acquired from aircraft and satellites. It is therefore an attempt to measure something *at a distance*, rather than *in situ*. While remote-sensing data can consist of discrete, point measurements or a profile along a flight path, we are most interested here in measurements over a two-dimensional spatial grid, i.e. *images*. Remote-sensing systems, particularly those deployed on satellites, provide a repetitive and consistent view of the earth that is invaluable to monitoring the earth system and the effect of human activities on the earth. Some of the important applications of remote-sensing technology are:

- environmental assessment and monitoring (urban growth, hazardous waste)
- global change detection and monitoring (atmospheric ozone depletion, deforestation, global warming)
- agriculture (crop condition, yield prediction, soil erosion)
- nonrenewable resource exploration (minerals, oil, natural gas)
- renewable natural resources (wetlands, soils, forests, oceans)
- meteorology (atmosphere dynamics, weather prediction)
- mapping (topography, land use, civil engineering)
- military surveillance and reconnaissance (strategic policy, tactical assessment)
- news media (illustrations, analysis)

To meet the needs of different data users, there are many remote-sensing systems, offering a wide range of spatial, spectral and temporal parameters. Some users may require frequent, repetitive coverage with relatively low spatial resolution (meteorology).[1] Others may desire the highest possible spatial resolution with repeat coverage only infrequently (mapping); while some users need both high spatial resolution and frequent coverage, plus rapid image delivery (military surveillance). Remote-sensing data can be used to initialize and validate large computer models, such as Global Climate Models (GCMs), that attempt to simulate and predict the earth's environment. In this case, high spatial resolution may

---

[1]   "Resolution" is a term that can lead to much confusion. We take the common meaning in this chapter, namely the (linear) size of a single sensor detector element, geometrically projected to the earth's surface (Fig. 1-9). The subject is discussed in detail in Chapter 3.

be undesirable because of computational requirements, but accurate and consistent sensor calibration over time and space is essential.

The modern era of earth remote sensing began with the first Landsat Multispectral Scanner System (MSS) in 1972, which provided for the first time a consistent set of synoptic, high resolution earth images to the world scientific community. The characteristics of this new sensor were multiple spectral bands (sensing four regions of the electromagnetic spectrum, each about 100nm wide[2]—a coarse spectrometer, if you will), with reasonably high spatial resolution (80m), large area (185km by 185km) and repeating coverage (every 18 days). Moreover, the MSS provided general purpose, satellite image data directly in digital form for the first time. Much of the foundation of multispectral data processing was developed in the early 1970s by organizations such as the National Aeronautics and Space Administration (NASA), Jet Propulsion Laboratory (JPL), U. S. Geological Survey (USGS), Environmental Research Institute of Michigan (ERIM), and the Laboratory for Applications of Remote Sensing (LARS) at Purdue University.

Since 1972, we have seen four additional MSS systems, as well as the Landsat Thematic Mapper (TM) in 1982 and 1984 with 30m spatial resolution and 7 spectral bands, and the SPOT HRV in 1986, 1990 and 1993 with 20m spatial resolution in 3 spectral bands, and 10m resolution in a panchromatic band. There have also been several lower resolution AVHRR and GOES systems put into operation, as well as a wide variety of other multispectral sensors on aircraft and satellites. A depiction of many of these systems, present and future, in a performance space defined by two key sensor parameters, the number of spectral bands and the *Ground-projected Instantaneous Field of View (GIFOV),*[3] is shown in Fig. 1-1. A relatively new *hyperspectral* sensor class occupies the upper portion of the chart. The Advanced Visible/InfraRed Imaging Spectrometer (AVIRIS) and the HYperspectral Digital Image Collection Experiment (HYDICE) are airborne sensors that produce over 200 images of the same area on the ground in spectral bands about 10nm wide over the solar reflective portion of the spectrum from 400 to 2400nm. The HyperSpectral Imager (HSI) will be carried on NASA's Lewis spacecraft as the first hyperspectral satellite system. The separation of spectral bands in these systems is achieved with a continuously dispersive optical element, such as a grating or prism. Discrete, filter-based systems, such as the MODerate Imaging Spectroradiometer (MODIS), part of the NASA Earth Observing System (EOS) in the late 1990s, will also provide images in numerous spectral bands over the range 0.4 to 14μm from orbit. Such sensors promise large improvements in the quality of information that can be gathered about the earth's

---

[2] The spectral range of sensitivity is referred to as the *bandwidth* and can be defined in a number of ways (Chapter 3). It determines the spectral resolution of the sensor.

[3] The *GIFOV* is synonymous with the simple meaning of spatial resolution.

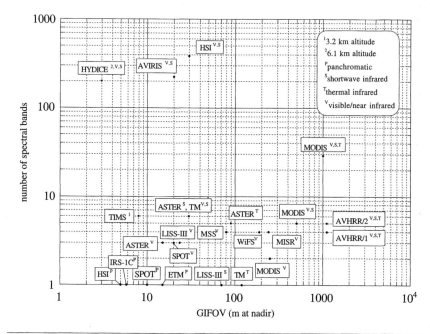

FIGURE 1-1. A plot of some remote-sensing systems in a two-dimensional parameter space. The sensor acronyms are defined in Appendix A. Three of the systems are flown on aircraft: AVIRIS, HYDICE and TIMS. The remainder are satellite systems. There are a number of airborne simulators of satellite systems which are not shown, e.g. the Modis Airborne Simulator (MAS) and the Thematic Mapper Simulator (TMS). For a thorough summary of sensor data, the book by Kramer is recommended (Kramer, 1996).

surface and near environment (Table 1-1). They also present a huge challenge to modern data storage and computing systems. For example, the anticipated rate of data production for all EOS systems, including both raw and derived products, is about one terabyte (10^{12} bytes) per day, with an accumulated database of about 10 petabytes (10^{16} bytes) expected over 15 years (Vetter *et al.*, 1995).

Although electro-optical imaging sensors and digital images dominate earth remote sensing today, earlier technologies remain viable. For instance, aerial photography, although the first remote-sensing technology, is still an important source of data because of its high spatial resolution and flexible coverage. Photographic imagery also plays a role in remote sensing from space. The EOSAT Corporation markets photography from the Russian KVR-1000 panchromatic film camera, with a ground resolution of about 2 to 3m, and the TK-350 camera, which provides stereo coverage with a ground resolution between 5 and 10m.

TABLE 1-1. Primary geophysical variables measurable with each spectral band of the EOS MODIS system (Salomonson et al., 1995). Note the units of spectral range are nanometers (nm) for bands 1–19 and micrometers (µm) for bands 20–36.

geophysical variables		band	spectral range	GIFOV (m)
general	specific			
land/cloud boundaries	vegetation chlorophyll	1	620–670nm	250
	cloud and vegetation	2	841–876	
land/cloud properties	soil, vegetation differences	3	459–479	500
	green vegetation	4	545–565	
	leaf/canopy properties	5	1230–1250	
	snow/cloud differences	6	1628–1652	
	land and cloud properties	7	2105–2155	
ocean color	chlorophyll observations	8	405–420	
	chlorophyll observations	9	438–448	
	chlorophyll observations	10	483–493	
	chlorophyll observations	11	526–536	
	sediments	12	546–556	
	sediments, atmosphere	13	662–672	
	cholorophyll flourescence	14	673–683	
	aerosol properties	15	743–753	
	aerosol/atmosphere properties	16	862–877	
atmosphere/clouds	cloud/atmosphere properties	17	890–920	
	cloud/atmosphere properties	18	931–941	
	cloud/atmosphere properties	19	915–965	
thermal	sea surface temperatures	20	3.66–3.84µm	1000
	forest fires/volcanoes	21	3.929–3.989	
	cloud/surface temperature	22	3.929–3.989	
	cloud/ surface temperature	23	4.02–4.08	
	troposphere temp/cloud fraction	24	4.433–4.498	
	troposphere temp/cloud fraction	25	4.482–4.549	
atmosphere/clouds	cirrus clouds	26	1.36–1.39	
thermal	mid-troposphere humidity	27	6.535–6.895	
	upper-troposphere humidity	28	7.175–7.475	
	surface temperature	29	8.4–8.7	
	total ozone	30	9.58–9.88	
	cloud/surface temperature	31	10.78–11.28	
	cloud/surface temperature	32	11.77–12.27	
	cloud height and fraction	33	13.185–13.485	
	cloud height and fraction	34	13.485–13.785	
	cloud height and fraction	35	13.785–14.085	
	cloud height and fraction	36	14.085–14.385	

Imagery from both systems can be provided in scanned digital form (800 to 1000 MB per frame).

Two actions by the United States government will make an unprecedented volume of high spatial resolution satellite imagery available to scientists. The first is declassification of photography from the early national surveillance satellite systems, CORONA, ARGON and LANYARD (McDonald, 1995a; McDonald, 1995b). This collection consists of hundreds of thousands of photographs (some scanned and digitized), mostly black and white, but some in color and stereo, over large portions of the earth at resolutions of 2 to 8 m. The imagery covers the period 1960 through 1972 and, although it was less systematically acquired over the whole globe than Landsat data, provides a previously unavailable, 12-year historical record that is an invaluable baseline for environmental studies.

The second policy change is authorization for commercial development of high-performance sensors, with resolutions as high as 1 m in panchromatic mode and 3 to 4 m in multispectral mode. The routine availability of 1 m resolution imagery opens many new commercial and public service opportunities for satellite imagery, such as real estate marketing, design of cellular telephone and wireless Personal Communications System (PCS) coverage areas (which depend on topography and building structures), urban and transportation planning, and natural and man-made disaster mapping and management. These systems will undoubtedly have equal value to military intelligence and environmental remote sensing. At this time, three consortia are planning such systems (Fritz, 1996).

1.2.1 Information Extraction from Remote-Sensing Images

One can view the use of remote-sensing data in two ways. The traditional approach might be called *image-centered*. Here the primary interest is in the spatial relationships among features on the ground, which follows naturally from the similarity between an aerial or satellite image and a cartographic map. In fact, the common goal of image-centered analyses is creation of a map. Historically, aerial photographs were analyzed by *photointerpretation*. This involves a skilled and experienced human analyst who locates and identifies features of interest. For example, rivers, geologic structures and vegetation may be mapped for environmental applications, or airports, troop convoys and missile sites for military purposes. The analysis is done by examination of the photograph, sometimes under magnification or with a stereo viewer (when two overlapping photos are available), and transfer of the spatial coordinates and identifying attributes of ground features to a map of the area. Special instruments like the *stereoplotter* are used to extract elevation points and contours from stereo imagery. Examples of photointerpretation are provided in many textbooks on remote sensing

(Colwell, 1983; Lillesand and Kiefer, 1987; Sabins, 1987; Avery and Berlin, 1992; Campbell, 1996).

With most remote-sensing imagery now available in digital form, the use of computers for information extraction is standard practice. For example, images can be *enhanced* to facilitate visual interpretation or *classified* to produce a digital thematic map (Swain and Davis, 1978; Moik, 1980; Schowengerdt, 1983; Niblack, 1986; Mather, 1987; Richards, 1993; Jensen, 1996). In recent years, the process of creating feature and elevation maps from remote-sensing images has been partially automated by *softcopy photogrammetry*. Although these computer tools speed and improve analysis, the end result is still a map; and in most cases, visual interpretation cannot be supplanted completely by computer techniques (Fig. 1-2).

The second view of remote sensing might be called *data-centered*. In this case, the scientist is primarily interested in the data dimension itself, rather than the spatial relationships among ground features. For example, specialized algorithms are used with hyperspectral data to measure *spectral absorption features* (Rast *et al.*, 1991; Rubin, 1993) and estimate *fractional abundances* of surface materials for each pixel (Goetz *et al.*, 1985; Vane and Goetz, 1988).[4] Atmospheric and ocean parameters can be obtained with *profile retrieval* algorithms that invert the integrated signal along the view path of the sensor. Accurate absolute or relative radiometric calibration is generally more important for data-centered analysis than for image-centered analysis. Even in data-centered analysis, however, the results and products should be presented in the context of a spatial map in order to be fully understood.

Interest in global change and in long-term monitoring of the environment and man's effect on it naturally leads to the use of remote-sensing data (Townshend *et al.*, 1991). Here the two views, image-centered and data-centered, converge. The science required for global change monitoring means that we must not only extract information from the spectral and temporal data dimensions, but also must integrate it into a spatial framework that can be understood in a global sense. It is particularly important in this context to ensure that the data are spatially and radiometrically calibrated and consistent over time and from one sensor to another. For example, imagery is *georeferenced* to a fixed spatial grid relative to the earth (geodetic coordinates) to facilitate analysis of data from different sensors acquired at different times. The data can then be "inverted" by algorithms capable of modeling the physics of remote sensing to derive *sensor-independent* geophysical variables.

---

[4] A *pixel* is one element of a two-dimensional digital image. It is the smallest sample unit available for processing in the original image.

line map   aerial photo

photo registered to map   composite

FIGURE 1-2. An example of how maps and imagery complement each other. The line
map of an area in Phoenix, Arizona, produced manually from a stereo pair of aerial
photographs and scanned into the digital raster form shown here, is an abstraction of
the real world; it contains only the information that the cartographer intended to
convey: an irrigation canal (the Grand Canal across the top), roads, elevation
contours (the curved line through the center), and large public or commercial
buildings. An aerial photograph can be registered to the map using image processing
techniques and the two superimposed to see the differences. The aerial photo contains
information about land use which is missing from the map. For example, an apartment
complex that may not have existed (or was purposely ignored) when the map was
made can be seen as a group of large white buildings to the left-center. Also,
agricultural fields are apparent in the aerial photo to the right of the apartment
complex, but are not indicated on the line map. In the lower half, the aerial photo
shows many individual houses that also are not documented on the map.

1.2.2 Spectral Factors in Remote Sensing

The major spectral regions used for earth remote sensing are shown in Table 1-2. These particular spectral regions are of interest because they contain relatively transparent atmospheric "windows," through which (barring clouds in the non-microwave regions) the ground can be seen from above, and because there are effective radiation detectors in these regions. Between these windows, various constituents in the atmosphere absorb radiation, e.g. water vapor and carbon dioxide absorb from 2.5–3 μm and 5–8 μm. In the microwave region given in Table 1-2, there is a minor water absorption band near 22 GHz frequency (about 1.36 cm wavelength)[5] with a transmittance of about 0.85 (Curlander and McDonough, 1991). Above 50 GHz (below 0.6 cm wavelength), there is a major oxygen absorption region to about 80 GHz (Elachi, 1988). At the frequencies of high atmospheric transmittance, microwave and radar sensors are noted for their ability to penetrate clouds, fog and rain, as well as an ability to provide nighttime reflected imaging by virtue of their own active illumination.

TABLE 1-2. The primary spectral regions used in earth remote sensing. The TIR region is also known as the Long Wave InfraRed (LWIR). The boundaries of some atmospheric windows are not distinct and one will find small variations in these values in different references.

name	wavelength range	radiation source	surface property of interest
Visible (V)	0.4 – 0.7 μm	solar	reflectance
Near InfraRed (NIR)	0.7 – 1.1 μm	solar	reflectance
Short Wave InfraRed (SWIR)	1.1 – 1.35 μm 1.4 – 1.8 μm 2 – 2.5 μm	solar	reflectance
Mid Wave InfraRed (MWIR)	3 – 4 μm 4.5 – 5 μm	solar, thermal	reflectance, temperature
Thermal InfraRed (TIR)	8 – 9.5 μm 10 – 14 μm	thermal	temperature
microwave, radar	1 mm – 1 m	thermal (passive) artificial (active)	temperature (passive) roughness (active)

---

[5] For all electromagnetic waves, the frequency in Hertz (Hz; cycles/second) is given by $\nu = c/\lambda$, where c is the speed of light (2.998×10^8 m/sec in vacuum) and λ is the wavelength in meters (Slater, 1980; Schott, 1996). A useful nomograph is formed by a log-log plot of λ versus ν (Fig. 1-3).

Passive remote sensing in all of these regions employs sensors that measure radiation naturally reflected or emitted from the ground, atmosphere and clouds. The V, NIR and SWIR regions (from $0.4\mu m$ to about $3\mu m$) are the *solar-reflective* spectral range because the energy supplied by the sun at the earth's surface exceeds that emitted by the earth itself. The MWIR region is a transition zone from solar-reflective to thermal radiation. Above $5\mu m$, *self-emitted thermal radiation* from the earth generally dominates. Since this phenomenon does not depend directly on the sun as a source, TIR images can be acquired at night, as well as in the daytime. This self-emitted radiation can be sensed even in the microwave region as *microwave brightness temperature,* by passive systems such as the Special Sensor Microwave/Imager (SSM/I) (Hollinger *et al.,* 1987; Hollinger *et al.,* 1990).

Active remote-sensing techniques employ an artificial source of radiation as a probe. The resulting signal that scatters back to the sensor characterizes the atmosphere or earth. For example, the radiation scattered and absorbed at a particular wavelength from a laser beam probe into the atmosphere can provide information on molecular constitutents such as ozone. In the microwave spectral region, *Synthetic Aperture Radar (SAR)* is an imaging technology in which radiation is emitted in a beam from a moving sensor, and the backscattered component returned to the sensor from the ground is measured. The motion of the sensor platform creates an effectively larger antenna, thereby increasing the spatial resolution. An image of the backscatter spatial distribution can be reconstructed by computer processing of the amplitude and phase of the returned signal. Wavelengths used for microwave remote sensing, active and passive, are given in Table 1-3 and graphically related to frequencies in Fig. 1-3.

Figure 1-4 shows the solar energy spectrum received at the earth (above the atmosphere), with an overlaying plot of the daylight response of the human eye. Notice that what we see with our eyes actually occupies only a small part of the total solar spectrum, which in turn is only a small part of the total electromagnetic spectrum. Much remote-sensing data is therefore "non-visible," although we can of course display the digital imagery from any spectral region on a monitor. Visual interpretation of TIR and microwave imagery is often considered difficult, simply because we are not innately familiar with what the sensor "sees" outside the visible region.

As indicated, most optical remote-sensing systems are multispectral, acquiring images in several spectral bands, more or less simultaneously. They provide multiple "snapshots" of spectral properties which are often much more valuable than a single spectral band or broad band (a so-called "panchromatic") image. Microwave systems, on the other hand, tend to be single frequency, with the exception of the passive SSM/I. SAR systems emit radiation in two polarization planes, horizontal (H) and vertical (V), and sense the return in either the same planes (HH, VV modes) or in the orthogonal planes (HV, VH modes) (Avery and

TABLE 1-3. *Microwave wavelengths and frequencies used in remote sensing. Compiled from (Sabins, 1987), (Hollinger et al., 1990), (Way and Smith, 1991), and (Curlander and McDonough, 1991).*

band	frequency (GHz)	wavelength (cm)	examples (frequency in GHz)
Ka	26.5 – 40	0.8 – 1.1	SSM/I (37.0)
K	18 – 26.5	1.1 – 1.7	SSM/I (19.35, 22.235)
Ku	12.5 – 18	1.7 – 2.4	Cassini (13.8)
X	8 – 12.5	2.4 – 3.8	X-SAR (9.6)
C	4 – 8	3.8 – 7.5	SIR-C (5.3), ERS-1 (5.25), RADARSAT (5.3)
S	2 – 4	7.5 – 15	Magellan (2.385)
L	1 – 2	15 – 30	Seasat (1.275), SIR-A (1.278), SIR-B (1.282), SIR-C (1.25), JERS-1 (1.275)
P	0.3 – 1	30 – 100	NASA/JPL DC-8 (0.44)

Berlin, 1992; Richards, 1993). More and more, images from different spectral regions, different sensors or different polarizations are combined for improved interpretation and analysis. Examples include composites of thermal and visible imagery (Haydn *et al.*, 1982), radar and visible imagery (Wong and Orth, 1980; Welch and Ehlers, 1988) and aerial photography and hyperspectral imagery (Filiberti *et al.*, 1994).

1.3 Spectral Signatures

The spatial resolution of satellite remote-sensing systems is too low to identify many objects by their shape or spatial detail. In some cases, it is possible to identify such objects by spectral measurements. There has therefore been great interest in measuring the *spectral signatures* of surface materials, such as vegetation, soil, and rock, over the spectral range in Fig. 1-4. The spectral signature of a material may be defined in the solar-reflective region by its reflectance as a function of wavelength, measured at an appropriate spectral resolution. In other spectral regions, signatures of interest are temperature and emissivity (TIR) and surface roughness (radar). The motivation of multispectral remote sensing is that different types of materials can be distinguished on the basis of differences in their spectral signatures. Although this desirable situation is often reached in practice, it is also often foiled by any number of factors, including

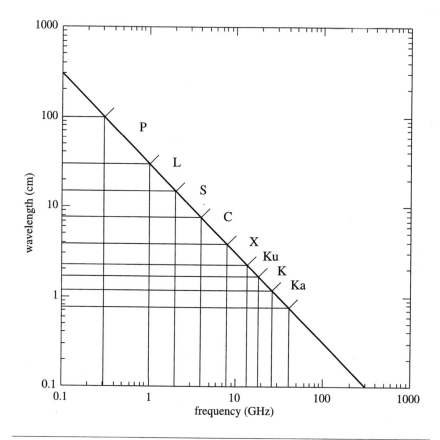

FIGURE 1-3. A nomograph for finding wavelength given frequency, or vice versa, in the microwave spectral region. The major radar bands are indicated. A similar nomograph can be drawn for any region of the electromagnetic spectrum.

- natural variability for a given material type
- coarse spectral quantization of many remote-sensing systems
- modification of signatures by the atmosphere

Therefore, even though we may wish to apply different labels to different materials, *there is no guarantee that they will exhibit measurably different signatures in the natural environment.*

Figure 1-5 shows spectral reflectance curves for different types of grasses and agricultural crop types. Note that all of these vegetation "signatures" exhibit similar general characteristics, namely a low reflectance in the green–red

FIGURE 1-4. *Exo-atmospheric (i.e., arriving at the top of the atmosphere) solar spectral irradiance and the daylight-adapted response of the human eye.*

spectrum,[6] a sharp increase in reflectance near 700 nm[7] and strong dips in reflectance near 1400 nm and 1900 nm caused by liquid water absorption in the plant leaves. The spectral signature for vegetation is perhaps the most variable in nature since it changes completely during the seasonal life cycle of many plants, acquiring a "yellow" characteristic in senescence, with a corresponding increase in the red region reflectance caused by a loss of photosynthetic chlorophyll.

Spectral reflectance data for some geologic materials are shown in Fig. 1-6. The dry and wet clay example illustrates the overall decrease in reflectance that results from an increase in water content in the sample material. Note also the characteristic SWIR water absorption bands, similar to those seen in vegetation. The curves for the alteration minerals are high spectral resolution laboratory measurements. Each mineral shows distinguishing absorption features, in some cases "doublets." Such features will not be seen with broadband sensors, such as Landsat TM and SPOT, but can be measured with a narrowband hyperspectral sensor, such as AVIRIS, with a 10 nm spectral bandwidth.

All spectral reflectance data are unique to the sample and the environment in which they are measured. Mineral signatures, for example, will vary from sample

---

6   The small peak in the green near 550 nm is due to low chlorophyll absorption relative to the blue and red spectral regions on either side. This peak is the reason healthy plants appear green to the human eye.

7   This so-called vegetation "red edge" is caused by the cellular structure within plant leaves.

FIGURE 1-5. *Example vegetation spectral reflectance curves (Bowker et al., 1985). The curves in the upper graph show variability among three types of grasses; even under relatively well-controlled laboratory conditions, the reflectance of corn leaves has been found to vary as much as ±17% near the reflectance edge at 0.67µm (Landgrebe, 1978).*

FIGURE 1-6. *Example mineral spectral reflectance curves for clay (top) and several alteration minerals (bottom) (Bowker et al., 1985; Clark et al., 1993). The presence of liquid water in any material generally lowers the reflectance; for a theory of this effect, see (Twomey et al., 1986).*

to sample. Vegetation is even more variable, being dependent on growth stage, plant health and moisture content. Complicating matters further, it is impossible to duplicate field reflectance measurement conditions in the laboratory. Even if the reference reflectance data are taken in the field, aerial and satellite imagery suffers from atmospheric, topographic and calibration influences (Chapters 2 and 3) that alter the signature as imaged and recorded by the sensor (Marsh and Lyon, 1980). Therefore, the use of laboratory or field reflectance data should be tempered by the fact that they only approximate the "signature" in the real world, and that remote sensor data need careful calibration for comparison to laboratory or field measurements. What saves image-based analysis in many cases is that we can compare relative signatures, one material versus another within a single image acquisition, rather than absolute signatures.

1.4   Remote-Sensing Systems

The details of sensor construction and materials vary with the wavelengths of interest, and the dimensions of optical systems and detectors depend on the engineering limitations in particular spectral regions. However, all passive, scanning optical sensors (visible through thermal spectral regions) operate on the same principles of optical radiation transfer, image formation, and photon detection. Microwave sensors, active and passive, are decidedly different in their nature, and will not be described here.

1.4.1 Optical Sensors

Every pixel represents an average in each of three dimensions: space, wavelength and time. The average over time is usually very small (on the order of microseconds for a whiskbroom scanner such as TM and milliseconds for a pushbroom scanner such as SPOT) and is inconsequential in most applications. The averages over space and wavelength, however, define the characteristics of the data in those critical dimensions.

If we imagine a three-dimensional continuous parameter space (x, y, λ), defined over spatial coordinates (x, y) and spectral wavelength (λ), we can visualize each pixel of a given image as representing an integration over a relatively small volume element in that continuous space (Fig. 1-7). We will see in Chapter 3 that the (x, y, λ) space is not quite as neatly divided as Fig. 1-7 indicates. Specifically, the volume of integration represented by each pixel is not a well-defined box, but overlaps in both the spatial and spectral dimensions with the integration volumes of neighboring pixels. For now, however, we will assume this convenient subdivision.

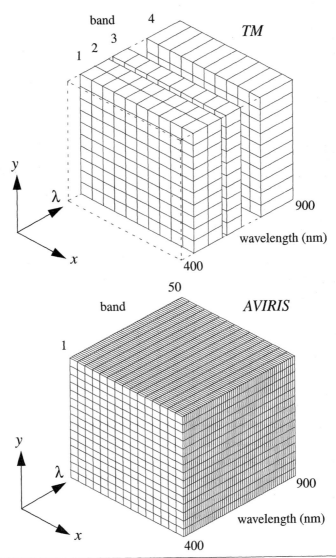

FIGURE 1-7. *Visualization of the spatial and spectral resolutions of the Landsat TM and AVIRIS in the VNIR spectral range. The relative proportions between the two sensors are correct along each axis, and each small rectangular box represents one image pixel. The TM samples the spectral dimension incompletely and with relatively broad spectral bands. In comparison, AVIRIS represents almost a continuous spectral sampling. It also has a somewhat smaller GIFOV. This volume visualization is called an "image cube."*

The grid of pixels that constitutes a digital image is achieved by a combination of scanning in the *cross-track* direction (orthogonal to the motion of the sensor platform) and by the platform motion along the *in-track* direction (Fig. 1-8) (Slater, 1980). A pixel is created whenever the sensor system electronically samples the continuous data stream provided by the scanning. A line scanner uses a single detector element to scan the entire scene. Whiskbroom scanners, such as the Landsat TM, use several detector elements, aligned in-track, to achieve parallel scanning. Pushbroom scanners, such as SPOT, have a linear array of thousands of detector elements, aligned cross-track, which scan the full width of the scene in parallel.

The spacing between pixels on the ground is the *Ground Sample Interval (GSI)*. The cross-track and in-track *GSI*s are determined by the cross-track and in-track sampling rates, respectively. It is common practice to design the sample rates so that the *GSI* equals the *Ground-projected Instantaneous Field Of View (GIFOV)*,[8] the geometric projection of a single detector *width, w,* onto the earth's surface (Fig. 1-9 and Fig. 1-10). Thus, the *GIFOV*s of neighboring pixels will abut, both in-track and cross-track. The in-track *GSI* is fixed by the necessary combination of satellite velocity and data sample timing to match the cross-track *GSI* at nadir. One notable exception is the Landsat MSS, which has an 80m *GSI* and *GIFOV* in-track, but a 57m *GSI* and 80m *GIFOV* cross-track. The latter "oversampling" results in some improvement in the data quality.

The *GSI* is determined by the altitude of the sensor system H, the sensor's focal length, f, and the inter-detector spacing (or spatial sample rate as explained previously). If the sample rate is equal to one pixel per inter-detector spacing, the relation for the *GSI* at *nadir*, i.e. directly below the sensor, is simply,

$$GSI = \text{inter-detector spacing} \times \frac{H}{f} = \frac{\text{inter-detector spacing}}{m} \qquad (1-1)$$

where f/H is the *geometric magnification*, m, from the ground to the sensor focal plane.[9] As we mentioned, the inter-detector spacing is usually equal to the detector width, w.

The *GIFOV* depends in a similar fashion on H, f and w. System design engineers prefer to use the *Instantaneous Field of View (IFOV)*, defined as the angle subtended by a single detector element on the axis of the optical system (Fig. 1-9),

$$IFOV = 2\operatorname{atan}\left(\frac{w}{2f}\right) \cong \frac{w}{f} . \qquad (1-2)$$

---

[8]   Also called the *Ground Sample Distance (GSD)*.

[9]   Since $f \ll H$, m is much less than one.

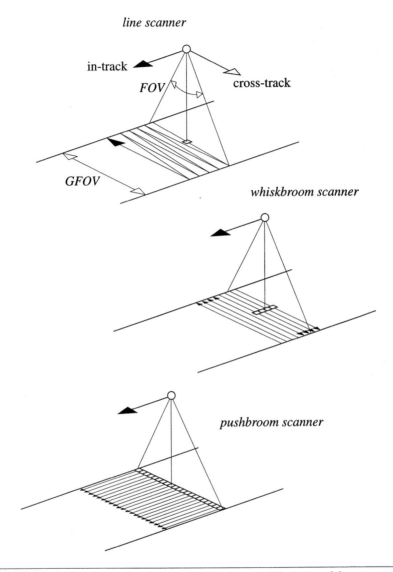

FIGURE 1-8. *Definition of basic scanner parameters and depiction of three types of scanning methods. The solid arrows represent motion relative to a stationary earth. In reality, the earth is rotating during the scanning process, approximately in the cross-track direction since most satellite remote-sensing systems are in a near-polar orbit. This results in a east-west skew in the surface coverage over the full scene.*

one detector element
width = w

focal length f *image space*

optics

altitude H *object space*

IFOV

earth's surface

GIFOV

FIGURE 1-9. Simple geometric description of a single detector element in an optical sensor. The sizes of w and f are greatly exaggerated relative to H for clarity. Likewise for the optics (which, more often than not, would be a series of curved mirrors, possibly with multiple, folded optical paths). Angular dimensions in image and object space are equal in this model, but linear dimensions are related by the magnification between the two spaces, f/H. Everything in this diagram is assumed stationary; with scan, platform and earth motion, the GIFOV moves during the integration time of the detector, resulting an effective projected area somewhat larger than shown (Chapter 3).

The *IFOV* is independent of sensor operating altitude, H, and is the same in the image and the object space. It is a convenient parameter for airborne systems where the operating altitude may vary. For the *GIFOV*, we therefore have,

$$GIFOV = 2H\tan\left(\frac{IFOV}{2}\right) = w \times \frac{H}{f} = \frac{w}{m} . \qquad (1-3)$$

The *GSI* and *GIFOV* are found by scaling the inter-detector spacing and width, respectively, by the geometric magnification, m. Users of satellite and aerial

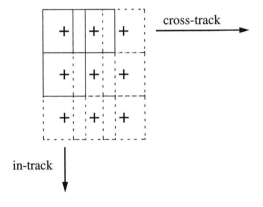

FIGURE 1-10. *The relationship between GIFOV and GSI for most scanning sensors (top) and the particular relationship for the Landsat MSS (bottom). Each cross is a pixel. In the MSS case, neighboring pixels have overlapping GIFOVs in the cross-track direction.*

remote-sensing data generally (and justifiably) prefer to use the *GIFOV* in their analysis, rather than the *IFOV*.

At every pixel location, the radiance[10] of a relatively small area of the total scene is measured by the sensor. This area is approximated by the *GIFOV* squared, but is actually larger (Chapter 3). The radiance seen by the detector at each pixel is converted to an electrical signal and then quantized to a discrete integer value, the *Digital Number (DN)*. As with all digital data, a finite number of bits, Q, is used to code the continuous data measurements as binary numbers. The number of discrete *DN*s is given by,

$$N_{DN} = 2^Q \qquad\qquad (1-4)$$

and the *DN* can be any integer in the range,

$$DN_{range} = [0, 2^Q - 1] . \qquad\qquad (1-5)$$

The larger the value of Q, the more closely the quantized data approximates the original continuous data, and the higher the *radiometric resolution* of the sensor. Both SPOT and TM have 8 bits per pixel, while AVHRR has 10 bits per pixel. To achieve high radiometric precision in a number of demanding applications, the EOS MODIS is designed with 12 bits per pixel.

In summary, a pixel is characterized, to the first order, by three quantities: the *GSI*, *GIFOV* and *Q*. These parameters are always associated with the term "pixel." Whenever there may be confusion as to what is meant between the *GSI* and *GIFOV*, "pixel" should be reserved to refer to the *GSI*.

Multispectral channels are typically created in an optical sensor by splitting the optical beam into multiple paths and inserting different spectral filters in each path. Some hyperspectral sensors, such as HYDICE, use a two-dimensional array of detectors in the focal plane (Fig. 1-11). For HYDICE, the detector has 320 elements cross-track and 210 in-track. The cross-track dimension serves as a line of pixels in a pushbroom mode, while the optical beam is *dispersed over wavelength* by a prism along the other direction (in-track) of the array. Therefore, as the sensor platform (an aircraft in the case of HYDICE) moves in-track, a full line of cross-track pixels is acquired *simultaneously* in all 210 spectral bands, a total of 67,200 values. These data are read out of the detector array in time for a new line to be acquired, contiguous to the previous line by virtue of the platform's ground velocity.

---

[10] Radiance is a precise scientific term used to describe the power density of radiation; it has units of $W\text{-}m^{-2}\text{-}sr^{-1}\text{-}\mu m^{-1}$, i.e. watts per unit source area, per unit solid angle, and per unit wavelength. For a thorough discussion of the role of radiometry in remote sensing, see (Slater, 1980) and (Schott, 1996).

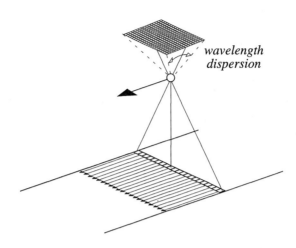

wavelength dispersion

FIGURE 1-11. *The pushbroom 2D-array concept used in the HYDICE and Lewis HSI hyperspectral imagers. Each cross-track line of pixels is simultaneously dispersed in wavelength along the in-track direction of the array. Therefore, the 2D-array measures cross-track spatial variation in a large number of spectral bands simultaneously. The number of spectral bands is equal to the number of detector elements in the in-track direction. All of this data must be read out of the array before the sensor progresses to the next line.*

The *Field Of View (FOV)* of a sensor is the angular extent of data acquisition cross-track (Fig. 1-8). The corresponding cross-track ground distance is given by,[11]

$$GFOV = 2H \tan\left(\frac{FOV}{2}\right). \qquad (1-6)$$

The *GFOV* is also called the *swath width*. While the *GSI* and *GIFOV* are specified for nadir viewing, i.e. directly below the aircraft or spacecraft, they increase towards either end of the *GFOV* as described in Chapter 3.

The in-track *FOV* is not naturally defined, since data acquisition in-track is controlled by the continuous aircraft or satellite motion. Instead, it is often determined by ground processing limitations, data rate constraints and a desire to have approximately equal coverage cross-track and in-track.

---

[11] We're ignoring the curvature of the earth, which increases this distance. The error in this approximation will increase as the *FOV* or *H* increases; a more accurate description that includes earth curvature is provided in Chapter 3.

The Landsat MSS and TM are whiskbroom scanners, while the AVHRR is a line scanner, with a single detector in each band. The Landsat MSS has 6 detectors in each of 4 bands, or 24 total, and the TM has 16 detectors in each of the 6 non-TIR bands plus 4 in band 6, for a total of 100 detectors. Each detector is a discrete electronic element with its own particular characteristics, so relative radiometric calibration among detectors is particularly important for whiskbroom systems. Errors in calibration lead to cross-track "striping" and "banding" noise in these systems, which can be particularly noticeable artifacts.

Line and whiskbroom scanners clearly have many dynamic motions occuring during scene acquisition (mirror rotation, earth rotation, satellite or aircraft roll/pitch/yaw) and consequently require complex post-processing to achieve accurate geometry. This can be done to a high level, however, as exemplified by the high quality of Landsat TM imagery. The geometry of a pushbroom scanner such as SPOT is comparatively simple, since the detector motion is in only one direction. However, factors such as earth rotation and satellite pointing during image acquisition still affect the image geometry. Also, a pushbroom system has a very large number of cross-track detector elements (6000 in the SPOT panchromatic mode, averaged to 3000 in the multispectral mode), requiring proportionally more calibration effort.

The total time required to "build" an image of a given in-track length depends on the satellite ground velocity, which is about 7 km/sec for low altitude, earth-orbiting satellites. A full MSS or TM scene therefore requires about 26 seconds to acquire, and a full SPOT scene about 9 seconds. The number of detectors scanning in parallel directly affects the amount of time available to integrate the incoming optical signal at each pixel. Pushbroom systems therefore have an advantage because all pixels are measured at once cross-track. If there were no satellite motion, the radiometric quality of the image would increase with increased integration time, because the signal would increase relative to detector noise. With platform motion, however, longer integration times also imply greater smearing of the image as it is being sensed, leading to reduced spatial resolution.

1.4.2 Temporal Characteristics

One of the most valuable aspects of unmanned, near polar-orbiting satellite remote-sensing systems is their inherent repeating coverage of the same area on the earth. Landsat, NOAA AVHRR and SPOT are in sun-synchronous orbits which means the local time over the same spot on the ground is always the same. The interval between "revisits" depends solely on the particulars of the satellite orbit for sensors that have a fixed view direction (nadir), such as the Landsat MSS and TM. If more than one system is in orbit at the same time, it is possible to increase the revisit frequency. The SPOT systems can be programmed to point cross-track to a maximum of ±26° from nadir, which allows a single satellite to view the same area more frequently from different orbits. Some example revisit times for a single system are given in Table 1-4 (Slater, 1980; Jensen, 1996).

TABLE 1-4. Revisit intervals and equatorial crossing times for several satellite remote-sensing systems. The specifications assume only one system in operation, with the exception of the AVHRR, which normally operates in pairs, with a 7 hour spacing, allowing morning and afternoon viewing of the same area in one day. The GOES is in a geostationary orbit and is always pointed at the same area of the earth. Equatorial crossing times are only approximate, as they change continually by small amounts and orbit adjustments are needed periodically.

system	revisit interval (days)	daylight equatorial crossing time
Landsat 1, 2, 3	18	9:30–10:00 A.M.
Landsat 4, 5	16	9:45 A.M.
AVHRR	1 7 hours (two systems)	7:30 A.M. 2:30 P.M.
SPOT	26 (nadir only); 1, or 4–5 (pointing)	10:30 A.M.
IRS-1A, B	22	10:30 A.M.
MODIS	2	10:30 A.M. or 1:30 P.M.
GOES	30 minutes	NA

Regular revisits are important for monitoring agricultural crops; in fact, *temporal signatures* can be defined for certain crop types in different regions of the world and used to classify them from multitemporal images (Haralick *et al.*, 1980; Badhwar *et al.*, 1982). Manned systems such as the Space Shuttle, which has carried several experimental SAR and other remote-sensing systems, have non-polar orbits which allow less regular revisit cycles. Their operating life is also limited, until the advent of a long-term space station.

1.5   Image Display Systems

Remote-sensing images are stored on disk or tape in one of three formats: Band SeQuential (BSQ), Band-Interleaved-by-Line (BIL), or Band-Interleaved-by-Sample (BIS). These formats are determined by different ordering of the three data dimensions (Fig. 1-12). From an access time viewpoint, the BSQ format is preferred if one is mainly interested in working with individual spectral bands, and BIS is preferred if one is working with all spectral bands from a relatively small image area. The BIL format is a commonly-used compromise.[12]

Computer image displays convert the digital image data to a continuous, analog image for viewing. They are usually preset to display 8 bits/pixel in grey-scale, or 24 bits/pixel in additive color, achieved with red, green and blue primary screen colors. Three bands of a multispectral image are processed by three hardware *Look-Up Tables (LUTs)* to convert the integer *DN*s of the digital image to integer *Grey Levels (GLs)* in each band,

$$GL = LUT_{DN}. \qquad (1-7)$$

The *DN* serves as an integer index in the *LUT*, and the *GL* is an integer index in the video memory of the display (Fig. 1-13). The range in image *DN*s is given by Eq. (1 – 5), while GL typically has a range,

$$GL_{range} = [0, 255] \qquad (1-8)$$

in each color. The hardware *LUT* can be used to apply a "stretch" transformation to the image *DN*s to improve the displayed image's contrast or, if the *DN* range of the original image is greater than the *GL* range, the *LUT* can be used to "compress" the range for display. The output of the *LUT* will always be limited according to Eq. (1 – 8).

Color images are formed from composites of the triplet of *GLs* corresponding to any three bands of a multispectral image. With a 24 bit display, each band is assigned to one of three 8 bit integers corresponding to the display colors: red (R), green (G) or blue (B). There are therefore 256 *GLs* available for each band. Every displayed pixel has a color defined by a triplet of *GLs*, which we may consider a three-dimensional column vector *RGB*,[13]

$$RGB = [GL_R, GL_G, GL_B]^T \qquad (1-9)$$

There are 256^3 possible *RGB* vectors, but fewer, distinguishable display colors because there are no monitors that can display all of the colors in the color cube. The exact color displayed for a given *RGB* data vector depends on the phosphor characteristics and control settings of the monitor.

The *GL* space of a color video memory system is depicted in Fig. 1-14. Computer monitors create color in an *additive* way; that is, a pixel with equal *GLs* in red, green and blue will appear as grey on the screen (if the monitor is properly

---

[12] The BIS format is also called *band-interleaved-by-pixel (BIP)*.

[13] The triplet is conveniently written as a row vector in Eq. (1 – 9). The superscript *T* converts it to a column vector by a transpose operation.

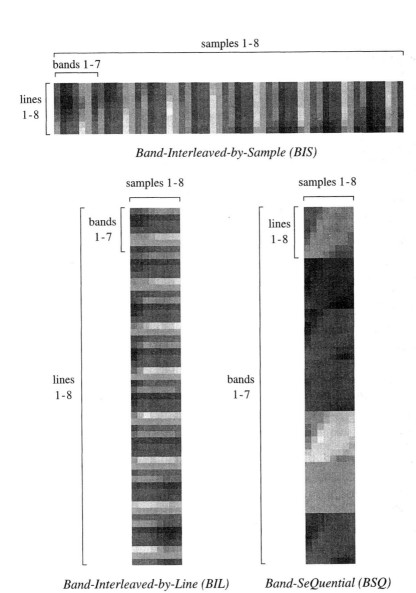

Band-Interleaved-by-Sample (BIS)

Band-Interleaved-by-Line (BIL)   *Band-SeQuential (BSQ)*

FIGURE 1-12. The three most common multispectral image formats: BIS, BIL and BSQ, illustrated with an 8 sample-by-8 line-by-7 band TM image. Note the very low contrast in the TIR band 6 relative to the other bands.

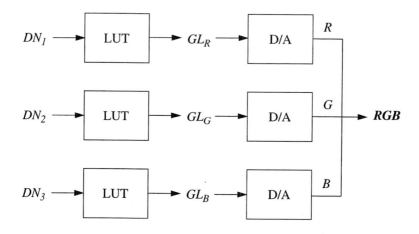

FIGURE 1-13. The conversion from DN to GL to color in a 24 bits/pixel digital video display. Three images are converted by individual LUTs to the three display GLs that determine the amplitude of the primary display colors, red, green and blue. The last step is a digital-to-analog (D/A) conversion and combination of the three channels to achieve the color seen on the monitor.

adjusted), a pixel with equal amounts of red and green, and with no blue, appears as yellow, and so forth. Certain color combinations are widely used in remote sensing (Table 1-5). However, since many commonly-used bands are not even in the visible spectrum, color assignment in the display image is arbitrary. The "best" colors to use are those that enhance the data of interest. The popularity of the Color IR (CIR) type of display derives from its emulation of color IR photography, in which vegetation appears as red because of its relatively high reflectance in the NIR and low reflectance in the visible (Fig. 1-5). Anyone with photointerpretation experience is usually accustomed to interpreting such images. The sensor-specific spectral bands used to achieve the "standard" color composites are given in Table 1-6. Example color composites of remote-sensing images are shown in the accompanying plates.

Single bands of a multispectral image can be displayed as a greyscale image or *pseudo-colored* by converting each *DN* or range of *DN*s to a color using the display's *LUT*s. Pseudo-coloring makes it easier to see small differences in *DN*.

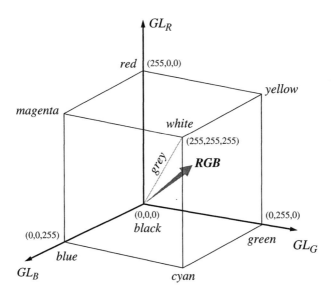

FIGURE 1-14. *Color cube for a 24 bits/pixel display. The vector* **RGB** *specifies any triplet of GLs within the cube.*

TABLE 1-5. *Sensor band color assignments for standard color composites.*

display color	sensor spectral band		
	TrueColor (TC)	Color IR (CIR)	False Color (FC)
red (R)	red	NIR	any
green (G)	green	red	any
blue (B)	blue	green	any

TABLE 1-6. *Specific sensor band color assignments for TC and CIR composites.*

composite	bands assigned to RGB			
	MSS	TM	SPOT	AVIRIS
TC	NA	3,2,1	NA	27,17,7
CIR	4,2,1	4,3,2	3,2,1	51,27,17

1.6   Data Systems

The data volume and technical complexity of modern remote sensing systems dictate that some *preprocessing* of the data be performed before it is supplied to the science community. The aim of preprocessing is to create a consistent and reliable image database by

- calibrating the image radiometry
- correcting geometric distortions
- removing some types of sensor noise
- formatting to a standard prescription

The particular types of preprocessing required obviously depends on the sensor's characteristics, since the goal is to remove any undesirable image characteristics caused by the sensor. Since not all data users need, or can afford, the highest degree of correction, various levels of preprocessing are provided as options.

A taxonomy of data products has evolved for major remote-sensing systems along the lines shown in Tables 1-7, 1-8 and 1-9. A generic hierarchy that encompasses these examples is given by 4 major processing levels

- reformatted raw data
- sensor-corrected data

    geometric corrections
    radiometric corrections
- scene-corrected data

    geometric corrections
    radiometric corrections
- geophysical data

Each level normally requires more ancillary data and processing than the preceding level. Typically, only a small fraction of the total data collected is processed to the higher levels, and the cost to the data customer generally increases with the level of processing. NASA was responsible for Landsat TM data production until that task was assumed by the commercial firm, EOSAT, on September 27, 1985. The levels of processing produced by NASA during the pre-commercialization period are documented (Clark, 1990). The engineering characteristics of the Landsat TM also were described thoroughly in two special journal issues (Salomonson, 1984; Markham and Barker, 1985).

Information extraction from the preprocessed images is generally done by earth science researchers. But, since not all data is processed to the highest level

TABLE 1-7. Processing levels for NASA Earth Observing System (EOS) imagery (Asrar and Greenstone, 1995). For some instruments, there will be no Level 1B product that is distinct from the Level 1A product.

processing level	description
0	Reconstructed unprocessed instrument/payload data at full resolution; any and all communications artifacts (e.g., synchronization frames, communications headers) removed
1A	Reconstructed unprocessed instrument data at full resolution, time-referenced, and annotated with ancillary information, including radiometric and geometric calibration coefficients and georeferencing parameters (i.e., platform ephemeris) computed and appended, but not applied, to the Level 0 data (or if applied, applied in a manner in which Level 0 data can be fully recovered)
1B	Level 1A data that have been processed to calibrated sensor units (not all instruments will have a Level 1B equivalent; Level 0 data is not recoverable from Level 1B data)
2	Derived geophysical variables at the same resolution and location as the Level 1 source data
3	Level 1 or 2 data or other variables mapped on uniform space-time grid scales, usually with some completeness and consistency
4	Model output or results from analyses of lower level data, e.g., variables derived from multiple measurements

TABLE 1-8. Processing levels for SPOT imagery (SPOTImage, 1991)

processing level	description
1A	Radiometric corrections for detector normalization
1B	Geometric correction for earth rotation and panoramic distortion in oblique views
SPOTView Precision	Map projection using maps, Geographic Positioning System (GPS) points or satellite ephemeris data
SPOTView Ortho	Orthographic projection using 3 arc-second Digital Elevation Models (DEMs) or other models

TABLE 1-9. Processing levels for EOSAT TM imagery (EOSAT, 1993)

processing level	description
system	Geometric correction and projection to one of 20 map projections and rotated to north, using ephemeris data and systematic error models. Relative detector calibration non-thermal bands. UTM, SOM or polar stereographic projection with residual distortions of 200 – 300m (orbit-oriented product).
precision	Map projection using Ground Control Points (GCPs).
terrain	Orthographic projection using DEMs.

at a central site, the earth scientist often must do equivalent processing before information extraction can begin, e.g. co-registration of scenes from different dates or sensors. It is therefore important that ancillary calibration data (primarily radiometric and geometric) be included with data supplied to users.

With the rapidly changing capabilities of computers and high-speed computer networks, many sets of data can now be "browsed" and ordered electronically. Sometimes, reduced resolution versions of the imagery can be viewed before ordering. The *EOS Data Information System* (EOSDIS) is evolving at this time into a distributed source for EOS data stored at different locations, while central sites for Landsat, SPOT and ERS data are maintained. Landsat MSS and TM data that are more than a few years old are archived and available for a nominal fee from the U. S. Geological Survey's EROS Data Center.

1.7   Summary

We have surveyed the field of remote sensing and the types of systems used for imaging the earth. Several points can be observed:

- Remote sensing of the earth's surface is limited to spectral transmission windows in the atmosphere.

- Materials on the earth's surface may *potentially* be identified by their spectral–temporal optical reflectance signatures.

- Therefore, a key component of remote sensing is repetitive multispectral or hyperspectral imagery.

In the next three chapters, we will look in detail at the influence of the atmosphere and terrain relief on spectral "signatures" and the impact of sensor characteristics on remote-sensing measurements. These interactions affect the quality and characteristics of remote-sensing data and the design and performance of image processing algorithms discussed in later chapters.

1.8 Exercises

1-1. Construct a nomograph like that of Fig. 1-3 for the full remote-sensing spectrum from 0.4µm to 1m and label each of the major atmospheric windows from Table 1-2.

1-2. A Landsat scene covers 185km by 185km. How many scenes are required to cover the entire earth, assuming an average 10% endlap (top and bottom) and 30% sidelap (adjacent orbital paths)? How many pixels does this correspond to? How many bytes?

1-3. We are able to visually interpret the aerial photograph of Fig. 1-2 to describe an apartment complex, roads and agricultural fields. Think about how this interpretation might possibly be automated using computer analysis of the photograph. What physical (spatial, spectral, temporal) characteristics of these features could be used? What general algorithm characteristics would be required?

1-4. The data formats in Fig. 1-12 can be obtained by planar slices from different directions of the multispectral image cube in Fig. 1-7. Determine the slices and the order in which the data are extracted for each format.

1-5. Identify four important characteristics of a multispectral imaging system. Select one of the remote sensing applications discussed at the beginning of this chapter and explain how the sensor characteristics you identified affect the quality of the data for that application.

CHAPTER 2

Optical Radiation Models

2.1 Introduction

Passive remote sensing in the optical regime (visible through thermal) depends on two sources of radiation. In the visible to shortwave infrared, the radiation collected by a remote sensing system originates with the sun. Part of the radiation received by a sensor has been reflected at the earth's surface and part has been scattered by the atmosphere, without ever reaching the earth. In the thermal infrared, thermal radiation is emitted directly by materials on the earth and combines with self-emitted thermal radiation in the atmosphere as it propagates upward. In this chapter, we outline the basic models appropriate to the optical region from the visible through the thermal infrared. The science of radiometry is applied here as a means to an end, i.e. to define the major processes involved in remote sensing in this spectral region. For thorough treatments of optical radiometry in remote sensing, the books by Slater (Slater, 1980) and Schott (Schott, 1996) are recommended.

2.2   *Visible to Shortwave Infrared Region*

All materials on the earth's surface passively absorb and reflect solar radiation in the 0.4 to 3 μm spectral range. Some materials also transmit solar radiation; for example, water bodies and plant canopies. At longer wavelengths, materials at normal tempertures begin to actively emit thermal radiation, which is discussed later in this chapter. A description of how solar radiation propagates and is modified prior to sensing by an optical system follows.

2.2.1 Solar Radiation

The source of energy for remote sensing in the *solar-reflective* spectral region (visible to shortwave IR) is the sun. The sun is a near-perfect *blackbody* radiator; that is, it emits radiation at nearly the maximum efficiency possible for a body at its effective temperature. The *spectral radiant exitance*, M_λ, from the sun can therefore be modeled by *Planck's blackbody equation*,

$$M_\lambda = \frac{C_1}{\lambda^5 [e^{C_2/(\lambda T)} - 1]} \qquad (2-1)$$

where

T is the blackbody's temperature in Kelvin (K),
$C_1 = 3.74151 \times 10^8$ W-m^{-2}-μm^4, and
$C_2 = 1.43879 \times 10^4$ μm-K.

Using these values and specifying the wavelength, λ, in μm and the temperature in Kelvin results in units for spectral radiant exitance of W-m^{-2}-μm^{-1} (Slater, 1980), i.e. power (or flux) per unit area of the sun's surface, per unit wavelength interval.

The blackbody function peaks at a wavelength given by *Wien's displacement law*,

$$\lambda|_{max} = 2898/T. \qquad (2-2)$$

where the blackbody's temperature is in Kelvin and the wavelength at which the radiation is a maximum is in μm. Thus, as the temperature of a blackbody increases, the wavelength of maximum radiant exitance decreases.

We, of course, are interested in the radiation that reaches the earth. To calculate that, the spectral radiant exitance of Eq. (2 − 1) must first be propagated to the top of the earth's atmosphere and then converted to *spectral irradiance*, E_λ^0. This transformation is accomplished with the following equation,

$$\textit{top-of-the atmosphere:} \quad E_\lambda^0 = \frac{M_\lambda}{\pi} \times \frac{\text{area solar disk}}{(\text{distance-to-earth})^2} \qquad (2-3)$$

The units of spectral irradiance are the same as those of spectral radiant exitance, namely those of *spectral flux density*, $\text{W-m}^{-2}\text{-}\mu\text{m}^{-1}$. The spectral content of the radiation does not change in the transit through space. The magnitude of solar irradiance at the earth changes by a few percent as the sun-earth distance changes throughout the year. The exo-atmospheric flux density is plotted in Fig. 2-1 using a blackbody model and a more detailed, empirical function stored in the atmospheric modeling program MODTRAN. As can be seen from the figure, the blackbody model with a temperature of 5900K is a good approximation to measured solar radiation.

FIGURE 2-1. *Comparison of the exo-atmospheric (top-of-the-atmosphere) solar spectral irradiance as used in the atmospheric modeling code MODTRAN (Berk et al., 1989) to the blackbody curve for T = 5900K. Deviations from the blackbody model are mainly in narrow absorption lines present in the actual solar spectrum.*

As the wavelength increases to the shortwave IR, less radiation is available from the sun for signal detection by remote sensing. If we ignore atmospheric effects, the solar radiation arriving at the earth is matched by self-emitted thermal energy from the earth at about 4.5 μm (Fig. 2-2). The wavelength at which they are equal contributors to the *at-sensor radiance above the atmosphere* depends on earth surface reflectance and emissivity (see Sect. 2.3) and the atmosphere, and can range between 2.5 and 6μm (Slater, 1980).

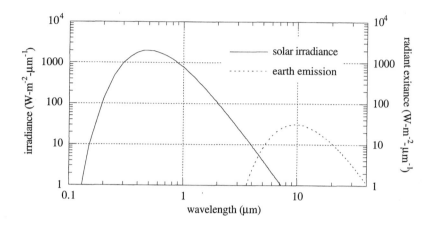

FIGURE 2-2. Spectral distributions at the top-of-the-atmosphere for the two radiation sources in the visible through thermal infrared spectral regions. The earth is assumed to be a blackbody at T = 300K and the sun a blackbody at T = 5900K, and atmospheric effects are ignored. Blackbody curves for different temperatures cannot cross each other; it only appears to occur here because the solar radiant exitance has been scaled as described in Eq. (2 – 3).

2.2.2 Radiation Components

The major radiation transfers of concern in the visible through SWIR spectral regions are shown in Fig. 2-3. There are generally three significant components in the upwelling at-sensor radiation:

- the unscattered, surface-reflected radiation, L_λ^{su}

- the down-scattered, surface-reflected skylight, L_λ^{sd}

- the up-scattered path radiance, L_λ^{sp}

We can therefore write for the total upwelling radiance at a high altitude or satellite sensor,[1]

$$L_\lambda^s = L_\lambda^{su} + L_\lambda^{sd} + L_\lambda^{sp} .\qquad(2-4)$$

In the following sections, we will look at appropriate models for each component.

---

[1] We use the superscript *s* to denote "solar," thereby distinguishing these terms from thermal emitted radiation, described later.

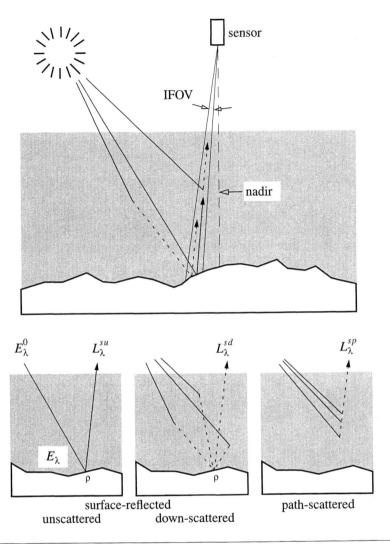

FIGURE 2-3. *Radiation components in solar reflective remote sensing. The most significant components seen by the sensor are generally surface-reflected, unscattered (direct) and surface-reflected, down-scattered (reflected "skylight"), and path-scattered. Other components can exist, such as an "adjacency component," consisting of a direct reflection from a neighboring GIFOV, followed by either an up-scatter directly to the sensor, or a down–scatter into the GIFOV of interest, where a second reflection occurs towards the sensor. The adjacency phenomenon increases local spatial correlation among pixels, but the magnitude of the effect is difficult to characterize. Multiple atmospheric scattering can also produce additional components, but their magnitude is reduced by each additional scattering event.*

Surface-reflected, unscattered component (L_λ^{su})

The atmosphere is an unavoidable influence in satellite and high altitude aerial remote sensing in the visible through shortwave IR. It scatters and absorbs radiation between the sun and the earth along the solar path, and again between the earth and the sensor along the view path. The fraction of radiation that arrives at the earth's surface is called the *solar path transmittance*, $\tau_s(\lambda)$, and is, by definition, between zero and one, and unitless. A typical spectral transmittance curve for the solar path, as generated from atmospheric models, is shown in Fig. 2-4.

FIGURE 2-4. *Transmittance of the atmosphere as calculated by the program MODTRAN. This is the transmittance along the solar path, i.e. between the sun and the earth's surface, for a solar elevation angle of 45°. The absorption bands are primarily associated with water vapor and carbon dioxide. This figure, and many of the others in this chapter, are produced with specific atmospheric parameters in the modeling program MODTRAN (Berk et al., 1989). The atmosphere model is "mid-latitude summer" with a "visibility" of 23 km (a relatively high value, implying that Mie scattering effects are small relative to Rayleigh scattering effects). The solar elevation angle is 45° and the azimuth angle from north is 135°. Special thanks to Dr. Kurt Thome, who performed the MODTRAN runs and provided assistance in data interpretation. The reader should realize that these curves are intended to be only examples; the specifics of any particular atmospheric condition will be different, and other atmospheric modeling programs, e.g. 5S (Tanre et al., 1990), may yield somewhat, but not vastly, different curves.*

The molecular absorption bands of water and carbon dioxide cause "deep" absorption features that, in two bands near 1.4μm and 1.9μm, completely block transmission of radiation. These spectral regions, therefore, are avoided for remote sensing of the earth's surface. They can be useful, however, for the detection of cirrus clouds, which are not easily distinguished from lower altitude clouds or surface features, at other wavelengths (Gao *et al.*, 1993). Since cirrus clouds are above most of the atmosphere's water vapor, they return a signal to a high-altitude sensor, while the signals from the earth's surface and lower altitude clouds are absorbed in the 1.4μm band. This application is the motivation for the MODIS band 26 (Table 1-1). The water absorption bands near 0.9μm and 1.1μm are much narrower and will block energy completely for narrowband remote sensors when the water vapor content of the atmosphere is high. They can also affect broadband sensors such as the Landsat MSS (band 4) by decreasing the signal received at the sensor.

The transmittance generally decreases towards the blue spectral region. This is largely the result of light being scattered out of the solar path by air molecules, whose diameter is less than λ/π (Slater, 1980). The magnitude of scattered light is approximated by pure *Rayleigh scattering* with a λ^{-4} dependence on wavelength. The short wavelength blue light is scattered out of the direct transmitting path more than the longer wavelength red light, which explains the red sky at sunrise and sunset, when the solar path through the atmosphere is greatest. In an atmosphere containing aerosols and particulates (smoke, smog, dust, haze, fog), *Mie scattering* is also present. For particle sizes larger than about $2\lambda/\pi$, the magnitude of Mie scattering does not depend on wavelength. Thick clouds are white from above because of Mie scattering by water droplets. For particle sizes in the range from about λ/π to about $2\lambda/\pi$, there is a wavelength dependence that is less than that of Rayleigh scattering. Real atmospheres with a full range of molecular, aerosol and particulate sizes exhibit a combination of Rayleigh and Mie scattering.

The spectral effect of atmospheric transmission on the solar radiation is shown in Fig. 2-5. The atmosphere significantly alters the spectral irradiance before it arrives at the earth. Mathematically, the irradiance E_λ on a plane perpendicular to the solar path and at the earth's surface is given by,

$$earth's\ surface:\quad E_\lambda = \tau_s(\lambda)E_\lambda^0 \qquad (2-5)$$

where τ_s is the solar path atmospheric transmittance. Note that, by the definition of transmittance, E_λ must be less than or equal to E_λ^0.

With the exception of cast shadows, E_λ^0 can be assumed to be constant across a scene. The irradiance at the surface depends on the incident angle, being a maximum if the surface is perpendicular to the incident angle, and less as the angle decreases. The decrease varies as the cosine of the angle, which can be calculated

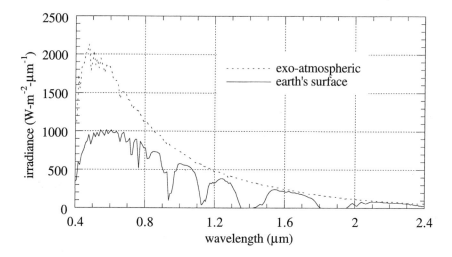

FIGURE 2-5. *Solar irradiance in the visible and shortwave IR regions (for a solar elevation angle of 45°), above the atmosphere and at the earth's surface. The ratio of these two curves is the path transmittance depicted in Fig. 2-4.*

by a dot product of two vectors (Fig. 2-6) (Horn, 1981). The incident irradiance E_λ in Eq. $(2-5)$ must then be modified to account for terrain shape as follows,

earth's surface:
$$E_\lambda(x, y) = \tau_s(\lambda)E_\lambda^0 \hat{n}(x, y) \cdot \hat{s}$$
$$= \tau_s(\lambda)E_\lambda^0 \cos[\theta(x, y)]$$
$(2-6)$

The next energy transfer occurs upon reflectance at the earth's surface. The irradiance downward onto a *Lambertian* surface is converted to the radiance leaving the surface with the aid of a geometric factor π and a *diffuse spectral reflectance* ρ,

earth's surface:
$$L_\lambda(x, y) = \rho(x, y, \lambda)\frac{E_\lambda(x, y)}{\pi}$$
$$= \rho(x, y, \lambda)\frac{\tau_s(\lambda)E_\lambda^0}{\pi}\cos[\theta(x, y)]$$
$(2-7)$

Like transmittance, reflectance is, by definition, unitless and between zero and one. Example spectral reflectance curves for some natural earth surface materials were shown in Chapter 1. Reflectance varies with wavelength and spatial location but does not depend on the view (sensor) direction for a truly

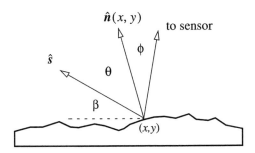

FIGURE 2-6. *The geometry of solar direct irradiance on the earth's surface. The unit vector* ŝ *points to the sun and the unit vector* $\hat{n}(x, y)$ *is normal to the surface. The solar elevation angle is β and the solar zenith angle is 90° − β. The solar incidence angle to the surface is θ and the exitance angle from the surface towards the sensor is φ. The cosine of the angle θ is given by the vector dot product* $\hat{n}(x, y) \bullet \hat{s}$ *. For simplicity in this diagram, the surface normal is assumed to lie in the vertical plane through the solar vector, but the dot product calculation is valid for any two vectors. Note this terrain-related effect does not involve the view angle of the sensor.*

Lambertian surface.[2] Deviation from this simple model by real materials is expressed in a *Bi-directional Reflectance Distribution Function (BRDF)*, which can be measured for a particular material by finding the ratio of outgoing radiance to incoming irradiance, as a function of incident and view angles. The quantity $\rho(x, y, \lambda)/\pi$ in Eq. (2 – 7) is then replaced by the BRDF (Schott, 1996).

The outgoing radiance from the earth's surface traverses the atmosphere to the sensor. The transmittances of the atmosphere along a nadir view path and along a view path at 40° from nadir are shown in Fig. 2-7. These curves differ from the one in Fig. 2-4 only by virtue of different path lengths through the atmosphere. Sensors such as Landsat TM with a small FOV show little within-scene atmospheric variation (other than localized clouds, smoke and haze) over areas with moderate terrain relief, since the path through the atmosphere is nearly constant for all pixels. In areas of high terrain relief, even Landsat images can show pixel-to-pixel atmospheric variation due to pixel-to-pixel altitude

---

[2] A Lambertian surface exhibits equal radiance in all directions. Visually, we say it appears equally bright at any view angle. Such a surface is also termed *perfectly diffuse*, with no mirror-like specular reflection. Many natural surfaces are approximately Lambertian within a limited range of view angles, typically 20°– 40°; as the view angle increases beyond that, most materials become non-Lambertian and show unequal reflectance in different directions. This property can be measured by sensors such as the Muli-angle Imaging SpectroRadiometer (MISR) to better characterize the surface radiative characteristics (Diner *et al.*, 1989).

differences (Proy *et al.*, 1989). Wide FOV sensors, such as the AVHRR, can show considerable scan-angle effects due to changes in the atmospheric path in the cross-scan direction. Note that the differences among the curves in Fig. 2-4 and Fig. 2-7 are relatively greater at shorter wavelengths. Visible and NIR imagery will therefore be particularly sensitive to changes in view angle across the FOV.

FIGURE 2-7. *Atmospheric path transmittance as viewed by a satellite sensor. View angle is defined as the angle from nadir. Note the similarity in shape, but differences in magnitude, compared to Fig. 2-4. The transmittance for a view angle of 40° is less because the path through the atmosphere is longer, leading to more scattering and absorption.*

We now must modify Eq. $(2-7)$ according to the *view path* transmittance, $\tau_v(\lambda)$, to obtain the at-sensor radiance,

$$L_\lambda^{su} = \tau_v(\lambda)L_\lambda$$

at-sensor:

$$= \rho(x, y, \lambda)\frac{\tau_v(\lambda)\tau_s(\lambda)E_\lambda^0}{\pi}\cos[\theta(x, y)] \qquad (2-8)$$

This component carries the signal of interest, namely the spatial-spectral reflectance distribution, $\rho(x, y, \lambda)$.

Surface-reflected, atmosphere-scattered component (L_λ^{sd})

The sensor also sees radiance arising from radiation that is scattered *downward* by the atmosphere ("skylight") and then reflected at the earth upward into the IFOV of the pixel of interest. This term, $L_\lambda^{sd}(x, y)$, is responsible for the commonly-observed fact that shadows are not totally dark. The reflected-skylight term is proportional to the diffuse surface reflectance, ρ, and the irradiance at the surface due to skylight, E_λ^d. This quantity is used because it is directly measurable with ground-based instruments. We accomodate the possibility that the sky may not be entirely visible from the pixel of interest due to intervening topography with a factor, $F(x, y)$,[3]

$$\textit{at-sensor:} \quad L_\lambda^{sd} = F(x, y)\rho(x, y, \lambda)\frac{\tau_v(\lambda)E_\lambda^d}{\pi} \qquad (2-9)$$

Path-scattered component (L_λ^{sp})

The path radiance term is a combination of molecular *Rayleigh scattering*, which varies with wavelength as λ^{-4}, and aerosol and particulate *Mie scattering,* which depends less strongly, or not at all, on wavelength. The combined effect of Rayleigh and Mie scattering in a clear atmosphere results in a net wavelength dependence of between λ^{-2} and $\lambda^{-0.7}$ (Curcio, 1961).

Path radiance can vary within a scene, for example, between a rural and an urban area, or in the presence of smoke plumes from fires. It also will vary with view angle, which is particularly important for a wide FOV sensor (such as AVHRR) or a sensor pointed off-nadir (such as SPOT). For scenes of homogeneous landscapes and for nadir-viewing sensors with relatively small FOVs (such as TM or SPOT), the path radiance is reasonably assumed to be a constant over the entire scene, and we simply write this term as L_λ^{sp} .

Total at-sensor, solar radiance (L_λ^s)

The total at-sensor, solar radiation is the sum of the three components described previously,

---

[3] F is the *fraction* of the sky hemisphere that is visible from the pixel of interest. For completely flat terrain, F equals one. A detailed description and calculated examples are given in (Schott, 1996).

at-sensor:

$$L_\lambda^s(x, y) = L_\lambda^{su}(x, y) + L_\lambda^{sd}(x, y) + L_\lambda^{sp}$$

$$= \rho(x, y, \lambda)\frac{\tau_v(\lambda)\tau_s(\lambda)E_\lambda^0}{\pi}\cos[\theta(x, y)] +$$

$$F(x, y)\rho(x, y, \lambda)\frac{\tau_v(\lambda)E_\lambda^d}{\pi} + L_\lambda^{sp} \qquad (2-10)$$

$$= \rho(x, y, \lambda)\frac{\tau_v(\lambda)}{\pi}\{\tau_s(\lambda)E_\lambda^0\cos[\theta(x, y)] + F(x, y)E_\lambda^d\} + L_\lambda^{sp}$$

The essence of Eq. (2 – 10) is that:

- the total spectral radiance received by the sensor is linearly proportional to the surface diffuse reflectance, modified by

- a multiplicative, spatially- and spectrally-variant factor that depends on the terrain shape, and

- an additive, spatially-invariant, spectrally-dependent term due to view path scattering.

As an example, Fig. 2-8 shows how the at-sensor radiance components vary with wavelength for a reflecting surface described by Kentucky Bluegrass (Fig. 1-5). The total radiance seen by a satellite or high altitude aircraft sensor is the sum of the components in Fig. 2-8, which is plotted in Fig. 2-9. Compare this curve to the original spectral reflectance of Kentucky bluegrass (Fig. 1-5)—they are dramatically different! How then is it possible to even hope to use remote sensing to recognize different terrestrial materials? The major reason why this drastic alteration of spectral signatures is not, in fact, a disaster, is that many image processing algorithms rely only on *relative spectral differences* among pixels. It is clear, however, that *remote sensing data must be corrected for atmospheric and solar effects if they are to be compared to a library of spectral reflectance curves.* Furthermore, *relative* atmospheric correction is needed if data signatures from one image date are to be compared to those from another date.

In Fig. 2-9, we also show a spectral profile for a pixel containing grass and trees in the Palo Alto AVIRIS scene. Even though we have made no attempt to emulate the atmospheric conditions, surface reflectance magnitude, or solar angles of this AVIRIS scene, the data are remarkably similar in shape to the results of the MODTRAN simulation. The MODTRAN atmospheric models are evidently realistic, and if *actual* atmosphere values are used in predicting the radiances of a particular image with a program such as MODTRAN, good agreement is possible. The problem, of course, is that such calibration data is almost never available for a given scene. Measurements of the atmosphere and ground

FIGURE 2-8. *The path-scattered and ground-reflected components of the total upwelling radiance seen by a satellite sensor for a surface reflectance of Kentucky Bluegrass. These components, as defined in MODTRAN, are related to the terms of Eq. (2 – 10) as follows. The path-scattered component is L_λ^{sp}, plus radiation that is reflected by the surface in a direction other than towards the sensor (remember, we assume the surface is perfectly diffuse and reflects equally in all directions), and is then scattered into the IFOV (we have not included this term in our discussion). The strong increase in the path-scattered component below 0.7µm is due to molecular scattering and is primarily the L_λ^{sp} term, since the surface reflectance here is relatively low. Above 0.7µm, the influence of the reflected and then scattered component is apparent. The ground-reflected component is the sum of L_λ^{su} and L_λ^{sd}. In the ground-reflected component, little information about the grass signature is seen until above 0.7µm, where the reflectance becomes relatively high. The ground-reflected component only exceeds the path-scattered component above 0.7µm, but both contain information about the signal (grass reflectance). Note the atmospheric water vapor absorption bands near 0.9, 1.1, 1.4 and 1.9µm (compare to Fig. 2-4).*

FIGURE 2-9. The spectral response of Kentucky Bluegrass as predicted by the MODTRAN model and a plot of a mixed grass and trees response from the AVIRIS image of Palo Alto (Plate 1-2). The shapes of the curves are quite similar, even though the MODTRAN model parameters have no direct relation to this particular AVIRIS data.

reflectance in conjunction with aircraft or satellite overpasses require substantial planning and effort, and are subject to the vagaries of weather and equipment failure. Therefore, there has been a large degree of interest in *image-based* atmospheric calibration techniques (Teillet and Fedosejevs, 1995).

2.2.3 Image Examples in the Solar Region

Some image examples are used in this section to illustrate the theory and models of optical radiation transfer in the solar reflective portion of the spectrum.

Terrain shading

The cosine factor arising from terrain relief (Eq. $(2 - 10)$) can be important to the spatial content of images as demonstrated in Fig. 2-10. The *Digital Elevation Model (DEM)*[4] has a GSI of 30m, matching the TM GSI. A *shaded relief* image is created from this DEM by calculating the $\cos[\theta(x,y)]$ term at each pixel and

---

[4] A DEM is a regular spatial grid of elevation values. It can be created in a variety of ways, for example indirectly via analysis of stereo imagery (see Chapter 8) or interferometric SAR data, or directly from laser altimeter measurements.

assuming that the surface reflectance is diffuse (i.e. the surface is Lambertian) and constant everywhere. In this case, we have set the solar elevation and azimuth angles to correspond to those for the particular TM image in Fig. 2-10. The shading caused by the terrain relief is clearly predictable with the DEM data.

The differences that remain between the contrast-stretched TM image and the shaded relief image of Fig. 2-10 are partially due to the reflectance spatial variations that are not modeled in the relief image. The modeling of vegetation reflectance from plant canopies is quite complex and is a matter of considerable research interest. The reader is referred to an extensive review (Goel, 1988) as a starting point for this subject.

The influence of terrain geometry on remote sensing measurements has been discussed in detail in (Sjoberg and Horn, 1983; Proy *et al.*, 1989), including the possibility of reflectance by neighboring ground elements in mountainous terrain into the IFOV at a pixel of interest. Looking at Eq. (2 – 10), we see that the image spatial content results from the product of two spatially-varying terms, the reflectance and the terrain-dependent cosine factor (assuming the down-scattered term is small). Either may be considered to *spatially modulate* the other. This relationship is useful in certain algorithms for image fusion, as described in Chapter 8.

Shadowing

Analysis of the DEM in Fig. 2-10 provides information on shadowing in the TM image (Dubayah and Dozier, 1986; Giles *et al.*, 1994). For example, the *self-shadowed* points can be found directly from the shaded relief image (see Exercise 2-1); these are terrain segments that face away from the solar irradiance direction. With a different algorithm, one can also find the pixels that lie within *projected shadows* (Fig. 2-11). Because of the mild terrain relief and high solar elevation for the TM image, there are relatively few projected shadows. Both types of shadow regions may be used to estimate atmospheric path radiance if a reasonable estimate is made of the reflectance within the shadowed area.

An image's appearance can vary dramatically with different solar angles. This effect is illustrated in Fig. 2-12. The difference in acquisition dates of four months creates a dramatically different image because the changes in solar elevation and, to a lesser extent, azimuth cause changes in the $\cos[\theta(x,y)]$ term of Eq. (2 – 10) and in shadowing.

Atmospheric correction

A portion of a Landsat TM scene near Oakland, California, is shown for bands 1 through 4 in Fig. 2-13. These data, although not calibrated for sensor gain and offset (Chapter 3), nevertheless indicate the general properties of atmospheric influence as a function of wavelength. A common technique for atmospheric digital elevation model (DEM)   shaded relief image

TM image   contrast-stretched TM image

FIGURE 2-10. The influence of terrain relief on image structure is depicted with a co-registered DEM and TM band 4 image near Berkeley, California. Elevation is coded as brightness in the DEM image. The solar elevation angle of 35° and azimuth angle of 151° for the October 25, 1984, TM image are used to create a shaded relief image from the DEM. The contrast of the TM image is significantly lower than that of the shaded relief image because of atmospheric contrast reduction and the fact that the displayed image contrast is controlled by the high reflectance man-made feature in the lower left. If the contrast of the TM image is adjusted to match that of the shaded relief image, the similarity between the two is evident. The residual differences are due to surface cover reflectance variation in the TM image and reflected skylight within shadows, neither of which are modeled in the shaded relief image. (The DEM is courtesy of William Acevedo and Len Gaydos of the U. S. Geological Survey.)

self-shadowed pixels   *projected shadows*

FIGURE 2-11. *Maps of the self-shadowed pixels and projected shadows for the solar angles and DEM of Fig. 2-10. (The projected shadows map was produced by Justin Paola, Oasis Research Center.)*

June 11, 1981   *October 20, 1980*

FIGURE 2-12. *Landsat MSS images of the Grand Canyon, Arizona, acquired on two dates. The lower sun elevation of 38° for the October image dramatically increases the shadowing in the Canyon, compared to the June image with a sun elevation of 65°.*

correction of multispectral imagery uses a "dark object" as a calibration target (Chavez, 1988). The dark object may be a region of cast shadow or, as in this case, a deep body of water. It is *assumed* that the dark object has uniformly zero radiance for all bands, and that any non-zero measured radiance must be due to atmospheric scattering into the object's pixels.[5] For the images in Fig. 2-13, the Briones Reservoir is a good choice for a dark object (it is significantly darker than the San Pablo Reservoir at shorter wavelengths). Averaging the darkest pixels in the Reservoir yields DNs of 53, 20, 11, and 14 for bands 1 through 4. It is tempting at this point to remove the atmospheric path scattering bias by simply subtracting these DN values from every pixel in the corresponding band, given that the at-sensor radiance-to-DN calibration is linear. However, that would not correct the data for the sensor gain. In Chapter 3, we will calibrate the raw image DNs to at-sensor radiances in order to compare them to the models discussed in this chapter.

2.3   Midwave to Thermal Infrared Region

At longer wavelengths, beyond the SWIR and into the MWIR spectral region, the importance of solar radiation declines and that of emitted thermal radiation increases for objects that are Lambertian reflectors (Fig. 2-2). At the longer wavelengths of the TIR, direct solar radiation is not a factor compared to self-emitted thermal radiation, other than solar-induced heating of the surface. The only exception to these statements is if the objects of interest are specular reflectors; in that case the solar reflected component may exceed the emitted component, even in the TIR (Slater, 1996). We will ignore this relatively rare circumstance in the remainder of this section.

2.3.1 Thermal Radiation

Every object at a temperature above absolute zero (0K) emits thermal radiation due to kinetic energy of molecules within the object. The radiation obeys Planck's equation (Eq. $(2-1)$) *if* the object is a perfect emitter and absorber, i.e. a blackbody. Real objects are not perfect emitters or absorbers, and Eq. $(2-1)$ is modified by an emission efficiency factor, the *emissivity,* which is generally a function of wavelength. Real materials are sometimes referred to as *greybodies,*

---

[5]   If any data are available on the actual reflectance of the "dark object," then it should be used to obtain a better estimate of the at-sensor path radiance for this correction (Teillet and Fedosejevs, 1995).

TM band 1 *TM band 2*

TM band 3 *TM band 4*

FIGURE 2-13. TM band 1 through band 4 images of the San Pablo (left) and Briones Reservoirs (right) north of Berkeley, California (part of the same TM scene used in Fig. 2-10). The individual bands are uncalibrated and shown with their original relative brightness and contrast. Atmospheric scattering reduces the contrast in band 1, while bands 2 and 3 are dark, due to low vegetation reflectance and lower sensor gain than band 1. Band 4 shows high contrast between the water-filled reservoirs and surrounding vegetated and bare soil terrain. Note the Briones Reservoir is relatively darker in the shorter wavelength spectral bands than the San Pablo Reservoir. This indicates the latter may have suspended sediments and particulates in the water, which is particularly likely since it is at a lower altitude and subjected to more runoff from the surrounding terrain. In band 4, both reservoirs have little radiance because of the near zero reflectance of water in the NIR.

but their spectral radiant exitance may not follow the blackbody curve because of the wavelength dependence of emissivity.

2.3.2 Radiation Components

The three emitted thermal components considered here arise from (Fig. 2-14):

- the surface-emitted radiation from the earth, L_λ^{eu},
- the down-emitted, surface-reflected radiation from the atmosphere, L_λ^{ed}, and
- the path-emitted radiance, L_λ^{ep}.

The total at-sensor radiance contribution from emission is,

$$L_\lambda^e = L_\lambda^{eu} + L_\lambda^{ed} + L_\lambda^{ep} \qquad (2-11)$$

In the MWIR, we write for the total radiance,[6]

$$\textit{at-sensor (MWIR):} \quad L_\lambda^{MWIR} = L_\lambda^s + L_\lambda^e \qquad (2-12)$$

where L_λ^s is given by Eq. $(2-10)$. In the $8-15\,\mu m$ region, however, the contribution of solar energy is negligible compared to that of the self-emitted thermal component, so we can just write,

$$\textit{at-sensor (TIR):} \quad L_\lambda^{TIR} = L_\lambda^e \qquad (2-13)$$

Surface-emitted component (L_λ^{eu})

The primary source of energy for thermal imaging is the earth itself, which has a typical temperature of 300 K. Different materials on the earth, however, can emit different amounts of thermal energy even if they are at the same temperature. Most materials are not ideal blackbodies with 100 percent radiative efficiency. The efficiency with which real materials emit thermal radiation at different wavelengths is determined by their *emissivity, ε.* Emissivity plays a proportionality role in the thermal region much like that of reflectance in the visible; it is defined as the ratio of the spectral radiant exitance of a greybody to that emitted by a blackbody (M_λ in Eq. $(2-1)$) at the same temperature, and is therefore unitless and between zero and one. The emitted radiance of the earth is therefore,

---

[6] We will use an extra superscript, *e*, on some radiance quantities in this section to distinguish emission-related terms from solar reflectance terms.

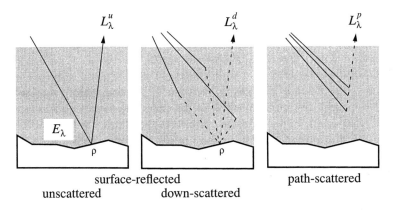

(a) reflected and scattered components

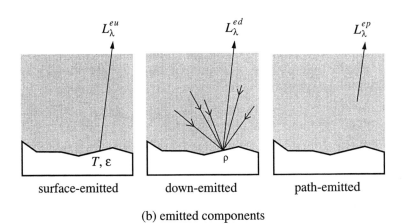

(b) emitted components

FIGURE 2-14. The reflected and scattered components of the at-sensor radiance (Fig. 2-3) and the analogous emitted components. In the 2.5 to 6μm spectral region, both must generally be considered; in the thermal IR region (8 to 15μm) only the emitted components are important.

$$earth's\ surface:\quad L_\lambda(x, y) = \varepsilon(x, y, \lambda)\frac{M_\lambda[T(x, y)]}{\pi} \qquad (2-14)$$

It is implied in this equation that different objects or materials on the earth's surface can have different temperatures, and hence different spectral radiant exitances, as well as different emissivities. Note the similarity between this relation and that for the solar reflective region, Eq. (2 – 7).

To separate the effects of emissivity and temperature, scientists usually assume one or the other is spatially constant. In thermal studies, such as aerial thermal scanning for building heat loss, the emissivity of various roof materials might be assumed equal in order to estimate temperatures. On the other hand, in some geologic applications, temperature might be assumed constant in order to estimate emissivity. The fact that both affect the radiance, however, means that one should be able to justify ignoring the variation of either parameter. The situation is particularly complicated because emissivity can vary with wavelength, and even temperature.

The relation between emitted radiance and the source temperature is not obvious from Eq. (2 – 1) and Eq. (2 – 14). To get a better feeling for that, we plot in Fig. 2-15 the spectral radiance as a function of temperature for three fixed wavelengths in the TIR, assuming constant emissivity. The temperature range, 250 K to 320 K, includes normal daytime and nighttime temperatures on the earth. We see that *spectral radiance is approximately linear with temperature over this range*, and for any smaller range, as might actually be encountered in a thermal image, a linear approximation is even better. Thus, for our purposes, we can approximate Eq. (2 – 14) by,

$$earth's\ surface:\quad L_\lambda(x, y) \approx \varepsilon(x, y, \lambda)\frac{[a_\lambda T(x, y) + b_\lambda]}{\pi} \qquad (2-15)$$

where a_λ and b_λ are relatively weak functions of wavelength λ. In practice, these coefficients would be given by their average over the spectral passband of the sensor (Chapter 3). Our intent with Eq. (2 – 15) is to provide a simple, easily-visualized relationship between radiance and temperature; in effect, a complex system has been approximated by a linear relationship that applies over a limited temperature range. If one's goal is to actually calculate temperatures from remote-sensing imagery, then the accurate form of Eq. (2 – 14) should be used.

The radiation emitted from the earth is transmitted by the atmosphere along the view path to the sensor,

$$L_\lambda^{eu}(x, y) = \tau_v(\lambda)L_\lambda(x, y)$$

$$at\text{-}sensor: \qquad\qquad\qquad\qquad\qquad\qquad\qquad\qquad (2-16)$$

$$= \varepsilon(x, y, \lambda)\frac{\tau_v(\lambda)[a_\lambda T(x, y) + b_\lambda]}{\pi}.$$

FIGURE 2-15. The dependence of radiant exitance from a blackbody on its temperature, at three wavelengths. Emissivity is held constant at one, whereas it actually can vary with temperature and wavelength for a greybody. The temperature range depicted is that for normal temperatures at the earth's surface.

The atmosphere's transmittance from 2.5 μm through the thermal IR is shown in (Fig. 2-16). There are four distinct spectral windows available for remote sensing in this spectral region, as determined by molecular absorption bands. As discussed earlier, the solar energy contribution in this spectral region is small; the rapid fall-off in solar irradiance above 2.5 μm is shown in Fig. 2-17.

Surface-reflected, atmosphere-emitted component (L_λ^{ed})

The atmosphere also emits thermal radiation downward, which is then reflected at the earth's surface, and transmitted through the atmosphere to the sensor. This term is analogous to the skylight component arising from scattering in the visible spectral region.

FIGURE 2-16. Atmospheric transmittance (solar elevation = 45°) in the midwave IR and thermal IR spectral regions. This curve, like those in the visible and shortwave IR, is obtained from the atmospheric modeling program MODTRAN. The curve appears more detailed than Fig. 2-4 because the abscissa covers six times the range in wavelength with about the same spectral resolution.

$$\text{at-sensor:} \quad L_\lambda^{ed} = F(x, y, \lambda)\rho(x, y, \lambda)\frac{\tau_\nu(\lambda)M_\lambda^a}{\pi} \qquad (2-17)$$

where M_λ^a denotes the atmospheric-emitted radiance. The factor F is the fraction of the sky hemisphere that is seen from the surface at (x,y) and is the same function that was used for reflected skylight (compare to Eq. $(2-9)$).

A useful relation between an object's emissivity and its reflectance (assuming that the object is thick and does not transmit radiation) is *Kirchhoff's Law*,

$$\rho(x, y, \lambda) = 1 - \varepsilon(x, y, \lambda) \qquad (2-18)$$

which is valid at any wavelength. It implies that any object which is a good emitter of thermal energy ($\varepsilon \cong 1$) is also a poor reflector, and vice versa. This equation may be substituted in Eq. $(2-17)$, if desired.

Path-emitted component (L_λ^{ep})

The atmosphere also emits radiation upward (according to Planck's blackbody law) as a function of the temperature at different altitudes. The total energy that

FIGURE 2-17. Solar irradiance in the midwave and thermal IR regions (solar elevation = 45°). The ratio of these two curves is the path transmittance depicted in Fig. 2-16.

arrives at the sensor is integrated over the view path from the contributions at each altitude. We will call this the *path-emitted* component, L_λ^{ep} . The resulting spectral distribution does not particularly resemble that of a blackbody at a single temperature; it is a mixture of blackbodies over a range of temperatures. Moreover, radiation from lower altitudes is absorbed and re-emitted at higher altitudes, making the situation quite complicated.

It is reasonable to assume that this component does not vary significantly over a scene; the only exceptions are for large angles from nadir (above about ±20°), where the path-emitted term tends to increase (Schott, 1996), and in areas where the surface temperature has significant spatial variation (for example, areas of burning vegetation) that may influence the near-surface atmospheric temperature.

Total at-sensor, emitted radiance (L_λ^e)

We write the total at-sensor, emitted radiance as the sum of the three components described previously,

$$L_\lambda^e(x, y) = L_\lambda^{eu} + L_\lambda^{ed} + L_\lambda^{ep}$$

at-sensor: $$= \varepsilon(x, y, \lambda)\frac{\tau_v(\lambda)}{\pi}[a_\lambda T(x, y) + b_\lambda] +$$ $$\qquad\qquad (2-19)$$

$$F(x, y, \lambda)\rho(x, y, \lambda)\frac{\tau_v(\lambda)M_\lambda^a}{\pi} + L_\lambda^{ep}.$$

Similarly to the solar reflectance region (Eq. (2 – 10)), we note that:

- the total spectral thermal radiance received by the sensor is approximately linearly proportional to the surface temperature, modified by

- a multiplicative, spatially- and spectrally-variant emissivity factor, and

- an additive, spatially-invariant, spectrally-dependent term due to view path emission.

2.3.3 Total Solar and Thermal Upwelling Radiance

The sum of the non-thermal and thermal radiation contributions is,

at-sensor: $$\quad L_\lambda(x, y) = L_\lambda^s(x, y) + L_\lambda^e(x, y) \qquad\qquad (2-20)$$

where L_λ^s is given by Eq. (2 – 10) and L_λ^e is given by Eq. (2 – 19). As discussed earlier, the second term is negligible in the visible and shortwave IR, while the first term is negligible in the thermal IR. In the midwave IR, both can be important, depending on the surface reflectance, emissivity and temperature.

The at-sensor radiance that would be seen by a satellite thermal sensor is depicted in Fig. 2-18. In this MODTRAN simulation, the reflectance and emittance are assumed to be spectrally uniform. The direct component due to solar irradiance at the earth is small, but contributes the majority of energy between 2.5 and 5 µm. Above 5 µm, the upwelling radiance is due to the three components discussed previously. The path-emitted component shifts to longer wavelengths as the atmosphere becomes colder with increasing altitude, and the result seen by a downward-looking sensor is a mixture from all levels. For these reasons, the total at-sensor radiance of Fig. 2-18 does not resemble a blackbody curve for a single temperature source (Fig. 2-1).

2.3.4 Image Examples in the Thermal Region

Although the atmosphere emits thermal radiation, only the layer near the earth is near 300 K. As altitude above the surface increases, the temperature falls off

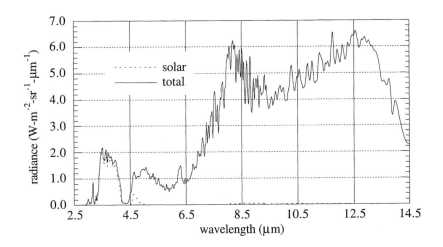

FIGURE 2-18. *The at-sensor radiance above the atmosphere in the middle and thermal IR regions. Note how the two sources of radiation, solar and thermal emission, exchange relative importance from the MWIR to the TIR. The satellite view angle is zero degrees from nadir and the surface emissivity is assumed to be one. The spectral reflectance is also assumed to be uniform in wavelength.*

rapidly. Clouds appear cooler than land in daytime TIR imagery since they are at some altitude above the earth (Fig. 2-19).

Nighttime thermal imagery is particularly useful because solar heating is not present. In Fig. 2-20, a nighttime HCMM image is compared to a co-registered Landsat MSS image. Since the emissivity can safely be assumed constant across the lake, the HCMM-measured radiance is a good representation of the lake's temperature. The large GIFOV of the HCMM sensor, however, mixes the radiance coming from the lake and surrounding terrain, which is mostly vegetated. A detailed analysis of mixing proportions for water and land at each pixel is required to estimate lake temperatures (Schowengerdt, 1982).

Urban areas expectedly show a "heat island" effect in thermal imagery. In Fig. 2-21, New Orleans appears to be warmer than the surrounding undeveloped areas and water. One must know the emissivities of the different surface materials represented, however, to correctly find the temperature differences.

Thermal images do not show topographic shading for the same reasons as visible imagery. Since thermal radiation is self-emitted by the surface, there is no directional dependence due to external irradiance. However, one sees so-called "thermal shadows" caused by preferential heating of the surface depending on its orientation to the sun. The example in Fig. 2-22 clearly shows the effect of solar

TM band 4   TM band 6

FIGURE 2-19. *TM band 4 and band 6 images of the San Francisco area. The bright clouds visible in band 4 appear darker than the ground in band 6, implying they are cooler than the ground in this daytime image. The cloud shadows, visible to the upper left of each cloud in band 4, are not visible or appear only slightly darker than the ground in band 6. The band 6 GIFOV is 120m compared to 30m in band 4. It is interesting, and somewhat disheartening to a remote sensing scientist, to note that weather satellite TIR images shown on television often have an inverted greyscale in order to make cold clouds appear bright!*

heating. It lends a different appearance to the topography compared to the visible image; note the brighter appearance of the southeast facing slopes in the lower half of the mountain range.

MSS band 4

HCMM TIR

FIGURE 2-20. Lake Anna, Virginia, viewed by the Landsat MSS band 4 (June 12, 1978, at 9:30A.M.) and the HCMM thermal band (June 11, 1978, at 3:30A.M.). This lake is used to cool a nuclear power plant, located about midway up the left side. Cool water from the center of the lake is pumped through the power plant's reactor, and then, via a series of canals, into the lake fingers at the bottom. Even though the HCMM GIFOV was only about 600m, it is clear that the water in these cooling ponds is warmer than that in the main body of the lake. (Imagery courtesy of Dr. Alden P. Colvocoresses, U. S. Geological Survey.)

TM band 2

TM band 6

FIGURE 2-21. Landsat TM band 2 and band 6 images of New Orleans, Louisiana, including Lake Pontchartrain and the Mississippi River (September 16, 1982). The urban area, particularly the denser core, appears warmer than surrounding vegetation, although their respective emissivities may be a factor. The darker, rectangular feature above the city center and adjoining the lake is a city park.

TM band 2

TM band 6

FIGURE 2-22. The Santa Rita Mountains, south of Tucson, Arizona, viewed by Landsat TM on January 8, 1983. The elevation ranges from about 1000 meters to 2880 meters at Mt. Wrightson, and the mountains are heavily vegetated at the higher elevations. Note the thermal "shadows" on slopes facing away from the direction of solar irradiance (from the lower right); these valleys are cooler than the solar-facing slopes in this mid-morning, winter image.

2.4   Summary

The basic models for optical radiation transfer at the earth's surface and through the atmosphere have been described. The most important components for remote sensing were covered and illustrated with image examples. The key elements of this chapter are:

- The energy collected by a remote sensing system is proportional to the surface reflectance (solar reflective region) and to the surface emissivity and temperature (thermal region).

- A spatially-invariant, but spectrally-dependent, constant bias term arising from atmospheric scattering (solar reflective region) and atmospheric emission (thermal region) is present in the sensed signal.

- A "coupling" exists between the surface and the atmosphere; they interact as a function of the surface reflectance, emission and topography.

In the next chapter, we will look at the influence of the sensor on remote sensing measurements. It's primary impact is a local spatial and spectral integration of the signal, by which the "resolution" of the image data is determined.

2.5   Exercises

2-1.   Suppose you have a DEM and wish to calculate a shaded relief image. The formula in Eq. $(2 - 6)$ can be used, but it does not explicitly account for surface points that are self-shadowed, i.e. points that face away from the sun. What mathematical constraint needs to be imposed on Eq. $(2 - 6)$ for such points so that they are correctly depicted? How would you determine points that are shadowed by other points, i.e. points that lie in a cast shadow? (This requires a rather involved mathematical solution—in your response, just indicate what needs to be done.) For simplicity, assume a 1–D problem in the vertical plane defined by the solar vector.

2-2.   Suppose we want to measure the amount of liquid water contained in the leaves of an agricultural crop by remote sensing in the absorption bands at 920nm and 1120nm. What effect will the atmosphere have on our ability to obtain accurate measurements?

2-3.   What physical factors may contribute to the differences between the two curves in Fig. 2-9?

CHAPTER 3

Sensor Models

3.1 Introduction

The sensor converts the upwelling radiance (reflected and/or emitted) into an image of the radiance spatial distribution. Several important transformations of the radiometric, spatial and geometric properties of the radiance occur at this stage. Generally, the sensor degrades the signal of interest, i.e. the portion of the total radiance that contains information about the earth's surface. It is important to understand the nature of this degradation to properly design image processing algorithms and interpret their results.

67

3.2   Overall Sensor Model

An electro-optical sensor may be modeled by the processes shown in Fig. 3-1. The radiation carrying the remotely-sensed signal is optical in nature up to the detectors, where it is transduced to an electronic signal. The scanning operation (Chapter 1) converts the spatial information to a continuous, temporal electronic signal, which is further processed up to the Analog/Digital (A/D) converter. There it is quantized into discrete *DN* values representing the image pixels.

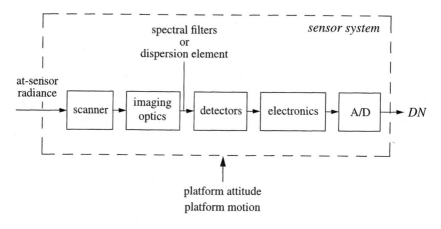

FIGURE 3-1. *The primary components in an electro-optical remote-sensing system. Although the platform geometry is external to the sensor per se, it has important effects on the final image characteristics and quality.*

3.3   Resolution

No property of images is more widely quoted, and simultaneously misused, than *resolution*. It is a term that conveys a strong intuitive meaning, but is difficult to define quantitatively. Remote sensing systems have "resolution" in the spectral, spatial and temporal measurement domains. Furthermore, there is a numerical resolution associated with the data itself by virtue of radiometric quantization. We will discuss the notion of resolution in this section, and in the process, introduce important concepts such as convolution, mixed pixels and sampling.

3.3.1 The Instrument Response

No instrument, remote-sensing systems included, can measure a physical signal with infinite precision. If the signal varies in time, the instrument must average over a non-zero response time; if the signal varies in wavelength, the instrument must average over a non-zero spectral bandwidth; or if the signal varies in space, the instrument must average over a non-zero spatial response. In general, we can write the output of most instruments as

$$o(z_0) = \int_W i(\alpha) r(z_0 - \alpha) d\alpha \qquad (3-1)$$

where

$i(\alpha)$ = input signal

$r(z_0 - \alpha)$ = instrument response (unit area), inverted and shifted by z_0

$o(z_0)$ = output signal at $z = z_0$

W = range over which the instrument response is significant

The physical interpretation of Eq. (3 – 1) is that the instrument weights the input signal in the vicinity (W) of z_0 and integrates the result. If we allow z_0 to have any continuous value, say z, then this relation is known as a *convolution*. A convenient, commonly-used shorthand notation for Eq. (3 – 1) is,

$$o(z) = i(z) * r(z) \qquad (3-2)$$

which reads "the output signal equals the input signal convolved with the response function." This mathematical description can be applied to a wide range of instruments.[1] In the following sections, we use it to describe the spatial and spectral response characteristics of a remote-sensing imaging system.

3.3.2 Spatial Resolution

While the spatial "resolution" of a sensor (or its image) is often quoted as the *GIFOV*, it is well known that it is possible to *detect* objects that are considerably smaller than the *GIFOV* if their contrast to the surrounding background is sufficiently high. Even though such objects may be detectable, they are not necessarily *recognizable*, except by the general context of the image. Thus, in TM imagery one frequently "sees" roads or bridges over water that are 10m (one third of a pixel) or less wide (Fig. 3-2).

---

[1] Specifically, systems that are linear (the output for multiple inputs is the sum of their individual outputs) and invariant (the form of $r(z_0 - \alpha)$ does not depend on z_0).

1X   *2X*

FIGURE 3-2. Example of sub-pixel object detection. This is part of a TM band 3 image of San Francisco showing the Berkeley Pier. The pier is 7m wide and made of concrete. An older extension of the same width, but made of wood with a lower reflectance than concrete, is barely visible. (Acknowledgments to Joseph Paola for providing details on the size and construction of the pier.)

Suppose we have a sensor that produces a linear *DN* output versus scene reflectance.[2] A ground area that is larger than *GIFOV*-squared[3] and has zero reflectance produces a zero *DN*, and an area with a reflectance of one produces a maximum *DN*, say 255. Now, suppose the area within one *GIFOV* contains two types of materials, with reflectances of zero and one. Surrounding pixels contain only the dark material and represent the background against which the pixel of interest is compared. At that pixel, the signal produced by the sensor will be the integrated effect of the mixture of the two components, target and background, within the *GIFOV*. If their relative proportion within the *GIFOV* is 50 percent, then the *DN* will be 128 (rounded to the nearest integer *DN*). If the brighter material occupies only ten percent of the *GIFOV* area, the *DN* will be 26. Since the sensor presumably can reliably distinguish two *DN*s separated by only one unit, we could theoretically (assuming no image noise) detect a bright object against a

---

[2]   We simplify the discussion by assuming remote-sensing measurements are a direct function of reflectance, without intervening influences.

[3]   The product of the in-track and cross-track *GIFOV*s. For brevity, this two-dimensional area will simply be called a *GIFOV* in the rest of this section

dark background even if it occupied only 0.4 percent of the *GIFOV* area (Fig. 3-3).

Now, we will make the object reflectances more realistic. The darker background might have a reflectance of 4% and the lighter target a reflectance of 8%, yielding a *contrast ratio* of 2:1. If the target covers 50% of the *GIFOV*, the *DN* of that pixel will be 15, while the *DN* of pure background pixels will be 10. The target is still detectable since it differs from the background by 5 *DN*s. If the size of the target is less than 10% of a pixel, however, it falls below the threshold (one *DN*) for detection. We see, therefore, that the radiometric quantization, coupled with the target and background reflectances, and the sensor *GIFOV*, all conspire to determine the "resolution" of the image. If there is any noise in the image, the threshold for detection will be higher, i.e. a greater target-to-background contrast will be required for reliable detection.

An often neglected influence on image resolution is the *sample-scene phase*, i.e. the relative location of the pixels and the target (Park and Schowengerdt, 1982; Schowengerdt *et al.*, 1984). This relative spatial phase is unpredictable (and almost always unknown) for any given image acquisition, and varies from acquisition–to–acquisition with a uniform probability distribution between ±1/2 pixel. Two of an infinite number of possible sample-scene phases are shown in Fig. 3-4 for the low contrast case described previously. On the left, the target appears equally in each of four *GIFOV*s, and on the right, unequally among the four. The resulting *DN*s of the four pixels containing part of the target are shown below the diagram. In the case of equal subdivision of the target, the four pixels would have a *DN* of 11 and the target would be detectable in each. If the image pixel grid happened to partition the target unequally as shown to the right in the figure, the percent coverage by the target in each *GIFOV* would be 30, 5, 2, and 13 (clockwise from the upper left). The four pixels would thus have *DN*s of 13, 11, 10, and 12, respectively. The target would remain detectable in three of the pixels (again, if there were no noise), but with little margin for error.

Sample-scene phase is important for more complex targets, as well. For example, suppose the target consists of a series of equally-spaced bright bars against a dark background. If the sensor *GIFOV* is equal to the width of one bar, the two extreme sample-scene phases result in a spatial signal that either has maximum contrast, or no contrast! If the sensor *GIFOV* is exactly twice the width of one bar, the spatial signal will have zero contrast for *any* sample-scene phase.

An enlargement of the TM image in Fig. 3-2 is shown in Fig. 3-5. It can be seen that the *DN* profile along each scan line across the pier is different. This is the effect of sample-scene phase variation because the pier is not oriented at 90° to the scan. Thus, in one particular scan the pier may appear to be one pixel wide, while in another it appears to be two pixels wide. To estimate the true width of this sub-pixel object, an interleaved composite of many lines should be made, with careful attention to phasing them correctly to fractions of a pixel. The

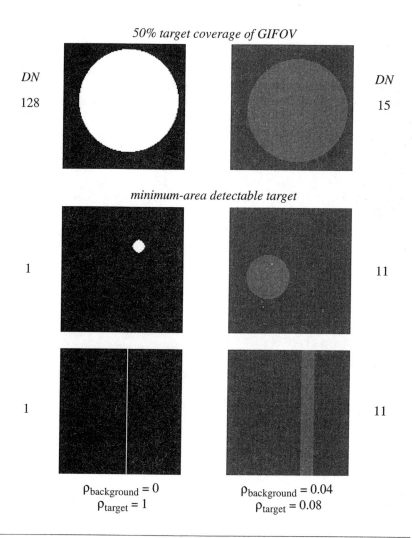

FIGURE 3-3. Detectability analysis for a single target at two different contrasts to the surrounding background and an idealized sensor. The squares represent the area sampled by a single GIFOV. The minimum-area detectable target results in a one DN difference from the background. Note that the target does not have to be centered in the GIFOV (if the sensor response is uniform across the GIFOV); the spatially-integrated signal will be the same for any internal location. For the same reason, the shape of the target cannot be discerned from a single pixel. If the target is linear, such as a road or bridge, then the shape may be inferred by the context of several pixels. A case in point is Fig. 3-2.

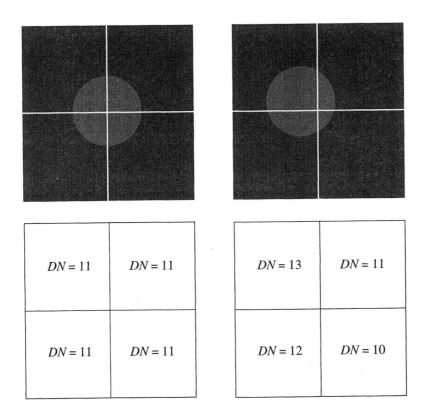

FIGURE 3-4. The effect of spatial phasing between the pixel grid and the ground target. Four adjacent GIFOVs are shown and the target area is 50% of the GIFOV area. On the left, the target fills 12.5% of each GIFOV. On the right, the target occupies 30%, 5%, 2%, and 13% of the four GIFOVs. The location of the grid is unpredictable prior to imaging any given scene; it could be anywhere within ± 1/2 pixel interval with equal probability. The imaging of long linear features can sometimes allow precise measurement of the phase to a small fraction of a pixel.

composite would represent the convolution, Eq. (3 – 2), of the sensor spatial response function and the pier radiance profile, sampled at an interval much finer than the pixel *GSI*.

With the inclusion of factors that come into play in real images, namely sensor noise, non-uniform targets and backgrounds, and variable solar angle and topography, it becomes clear that the situation with respect to image resolution is

FIGURE 3-5. A contrast-enhanced enlargement of Fig. 3-2 and DN profile plots along three adjacent scanlines near the center of the pier, illustrating the sample-scene phase effect. The pier's profile is different in each line. (The linear interpolation between individual pixels is only to aid visualization of the graph)

not simple! A common statement such as "the resolution of this image is 30 meters," probably refers to the sensor *GIFOV* or *GSI*, but it is not a precise statement as we have seen.

3.3.3 Spectral Resolution

The total energy measured in each spectral band of a sensor is a spectrally-weighted sum of the image irradiance over the spectral passband, Eq. (3 – 5). This weighting by wavelength is the primary determinant of the sensor's capability to *resolve* details in the spectral signal. To see why this is so, we will take the reflectance data for the mineral alunite (Chapter 1) in the vicinity of the OH absorption doublet between 1350 and 1550nm wavelength, and simulate it's measurement by a multispectral sensor. Each half of the doublet is only about 10 to 20nm wide, and they are separated by about 50nm. Now, imagine these data are measured with a sensor having many spectral bands, each 10 nm wide (at 50% of the peak responsivity) and spaced at 10nm intervals in wavelength. Each band sees an effective reflectance which is the weighted reflectance over the band (Fig. 3-6); the weighting function is the spectral response of the sensor in each band. We will use a bell-shaped function which approximates actual spectral band responsivities. Since the spectral passband is comparable in width to the details in the signal, they are preserved. The exact reflectance minima are not found, however, because of the positioning of the bands along the wavelength scale.[4]

If we do a similar exercise with 50nm-wide spectral bands, the result shows complete loss of information about the doublet (Fig. 3-6); it is literally "averaged away" by the broad spectral bands. Even if the spectral band locations were shifted, the doublet would not be "resolved." If the spectral bands do not overlap as they do in Fig. 3-6, (for example, bands 3, 4, 5, and 7 of TM), the sensor's ability to resolve even coarser features is seriously hampered. The trade-off, of course, is the increased data burden of finely-sampled spectra.

To illustrate how the actual TM spectral response modifies the at-sensor radiance, we will use the Kentucky Bluegrass spectral reflectance shown in Chapter 1 and the atmospheric propagation model of Chapter 2. The net at-sensor radiance is shown in Fig. 3-7. This is multiplied by the spectral response of each of the four VNIR TM bands (Fig. 3-8), to yield the weighted spectral distribution seen by each band. The integral of this function over wavelength then provides the total *effective* radiance in each band.

Now, in fact, the spectral reflectances of many natural materials are fairly smooth, without fine absorption lines (see the reflectance curves for soil and vegetation in Chapter 1). So the high spectral resolution of hyperspectral sensors at first seems to be of little value in those cases. With better understanding of the absorption properties and biochemical interactions among natural material components, for example cellulose, lignin and protein in vegetation, the additional resolution afforded by imaging spectroscopy can be fully utilized (Verdebout *et al.*, 1994; Wessman, 1994). As seen in the prior examples, the *placement* of spectral bands is as important as the spectral *bandwidth* to a sensor's ability to resolve spectral features. Hyperspectral sensors offer generally contiguous band placement over a wide spectral range, and are therefore superior to multispectral sensors with a few, albeit judiciously-placed, spectral bands.

---

[4]   To simplify the illustration, we are not including solar irradiance or atmospheric propagation as described in Chapter 2. These factors would change the input function, but not alter the concept.

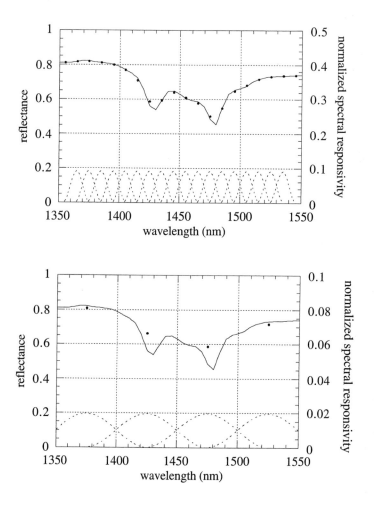

FIGURE 3-6. The effective reflectance of alunite as measured by a multispectral sensor. The solid line is the original reflectance sampled at 1nm and the individual band spectral responsivities are shown as dashed lines. Each solid dot is the output of the corresponding band. The upper graph corresponds to a sensor with 10nm-wide spectral bands, which approximates a hyperspectral sensor such as AVIRIS or HYDICE. The lower graph represents a sensor with 50nm-wide spectral bands, such as TM (although TM does not actually have any spectral bands in this part of the spectrum).

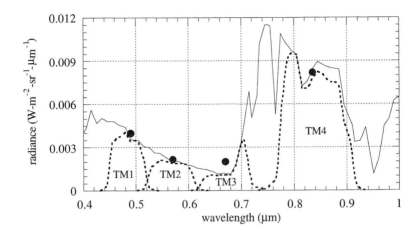

FIGURE 3-7. The at-sensor radiance for Kentucky Bluegrass (solid curve), the weighted spectral distribution seen by each TM band (dotted curves), and the total (integrated over the weighted spectral distribution) effective radiance (solid circles). Note how the broad bandwidth of bands 3 and 4, in particular, averages spectral detail present in the original radiance.

3.4   Spectral Response

The radiance arriving at the sensor, as described in Chapter 2, is transferred by the sensor optics to the detector focal plane where an image is formed. The spectral irradiance on a detector located on the optical axis, is related to the scene radiance L_λ by the *camera equation* (Slater, 1980)

$$at\text{-}image\ plane: \quad E_\lambda^i(x, y) = \frac{\pi \tau_o(\lambda)}{4N^2} L_\lambda(x, y) \quad (\text{W-m}^{-2}\text{-}\mu\text{m}^{-1})$$

$$(3-3)$$

where N is the optics *f-number*, given by the ratio of the optical focal length divided by the aperture stop diameter.[5] For notational simplicity, we are assuming that the geometric magnification between the ground and the image plane (Chapter 1),

$$m = f/H ,$$

$$(3-4)$$

---

[5]   Actually, the *entrance pupil* of the system, which is an image of the aperture stop. Space does not permit us to elaborate further on this topic from geometrical optics.

is one, and we therefore use the same (x,y) coordinate system for both the scene and image. The optics transmittance $\tau_o(\lambda)$, excluding any filters, is reasonably high (often 90% or more) for most optical systems and is nearly spectrally flat, so little, if any, signature modification is caused by the optics *per se*.

At this point, multispectral filters or wavelength dispersion elements, such as prisms, are introduced to separate the energy into different wavelength bands. For example, in the MSS and TM the optical path is split into multiple paths, each with a different spectral band filter. If we denote the product of the spectral filter transmittance and detector spectral sensitivity as the *spectral responsivity* $R_b(\lambda)$, the signal s_b measured by the sensor in band b is,

$$s_b(x, y) = \int_{\lambda_{min}}^{\lambda_{max}} R_b(\lambda)E_\lambda^i(x, y)d\lambda \quad \text{(W-m}^{-2}) \tag{3-5}$$

where λ_{min} and λ_{max} define the range of sensitivity in the band.

Equation $(3-5)$ defines the spectral integration of the received radiance and subsequent conversion to detector current, the units of R being amps-W^{-1} or volts-W^{-1} (Dereniak and Boreman, 1996). The spectral responsivities for some current multispectral sensors are shown in Fig. 3-8. The spectral characteristics of the Landsat MSS have been documented in detail; in particular, it has been shown that small differences in the responsivities among individual detectors (recall the MSS has 6 in each band and the TM has 16) can explain some striping noise in the imagery (Slater, 1979; Markham and Barker, 1983). Hyperspectral sensors have a narrow spectral sensitivity in each band that can usually be considered Gaussian in shape, but with a bandwidth that varies across the full wavelength sensitivity range. The individual bands are also approximately linearly spaced over wavelength, as shown in Fig. 3-9 for the AVIRIS system.

3.5   Spatial Response

The sensor modifies the spatial properties of the scene in two ways: (1) blurring due to the sensor's optics, detectors and electronics and (2) distortion of the geometry. In this section, we discuss spatial blurring, which generally occurs on a smaller spatial scale (a few pixels) than distortion.

The image of the scene viewed by the sensor is not a completely faithful reproduction. Small details are blurred relative to larger features; this blurring is characterized by the net (total) sensor *Point Spread Function (PSF_net)*, which we can accurately view as the *spatial responsivity* of the sensor (just as we described a spectral responsivity in the previous section). The PSF is the weighting function

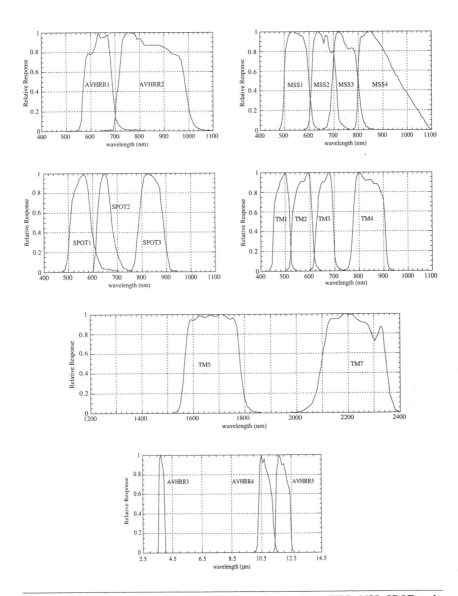

FIGURE 3-8. *Normalized spectral response functions for the AVHRR, MSS, SPOT, and TM sensors. These data are only meant to be representative; individual detectors within a band can vary by several percent. Note the passbands are not rectangular and neighboring bands often overlap. An absolute vertical scale is determined by the responsivity of the detectors measured in electron units, e.g. amps or volts per unit irradiance.*

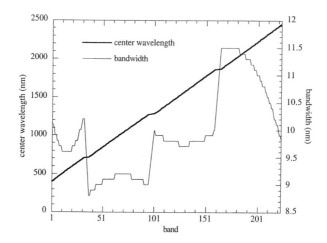

FIGURE 3-9. Spectral characteristics of the AVIRIS sensor. The full wavelength sensitivity range is from 396.9nm to 2454.9nm. The three "steps" in the center wavelength curve are where the bands in the four spectrometers overlap. These characteristics have changed slightly over the years of AVIRIS operation because of engineering changes in the spectrometers; every set of data contains the appropriate sensor spectral characteristics for that data.

for a spatial convolution (refer to Eq. (3 – 1)), resulting in the electronic signal, e_b, out of the detectors,

$$e_b(x, y)= \int_{\alpha_{min}}^{\alpha_{max}} \int_{\beta_{min}}^{\beta_{max}} s_b(\alpha,\beta)PSF_{net}(x-\alpha,y-\beta)d\alpha d\beta \qquad (3-6)$$

The spatial integration of Eq. (3 – 6) is much like the spectral integration of Eq. (3 – 5). In both cases, the sensor response function weights the physical signal, and the value measured is the integrated, weighted physical signal. In the present case, the limits of the integral define the *spatial extent* of the PSF about the coordinate (*x,y*). Note that the left side of Eq. (3 – 6) still depends on the continuous spatial coordinates (*x,y*); we do not convert this to discrete pixel coordinates until the signal is sampled.

The PSF consists of several components. First, the optics induce blurring by the *optical PSF*. The image formed by the optics on the detectors may in some cases move during the integration time for each pixel; this introduces an *image motion PSF*. Then the detector adds additional blurring due to the *detector PSF*. The detected signal is further degraded by the *electronics PSF*. The process is outlined in the flow diagram of Fig. 3-1. The following type of analysis has been described for the Landsat MSS (Park *et al.*, 1984) and TM (Markham, 1985). The appropriate sensor parameters for the models are generally found in contractor reports, e.g. (SBRC, 1984), although such data are normally pre-flight laboratory measurements which may not be appropriate after launch and possible sensor changes.

An important assumption in this analysis is that the net 2-D sensor PSF is given by a product of two 1–D PSFs in the cross-track and in-track directions,

$$PSF_{net}(x, y) = PSF_c(x)PSF_i(y) \qquad (3-7)$$

The net PSF is termed *separable* if this equation holds. Separability is generally valid for the types of scanners of interest, and allows a simpler analysis with 1-D functions (see Appendix B for an elaboration on separable functions).

3.5.1 Optical PSF_{opt}

The optical PSF is defined as the spatial energy distribution in the image of a point source, such as a star, or in the laboratory, an illuminated pinhole. An optical system is never "perfect" in the sense of forming a point image of a point source. The energy from the source is spread over a small area in the focal plane. The extent of spreading depends on many factors, including optical diffraction, aberrations and mechanical assembly quality.

A common model for the optical PSF is the 2–D Gaussian function,

$$PSF_{opt}(x, y) = \frac{1}{2\pi ab} e^{-x^2/2a^2} e^{-y^2/2b^2} . \qquad (3-8)$$

The parameters a and b determine the width of the optics PSF in the cross- and in-track directions, respectively. For well-designed optics, a equals b. Notice that the Gaussian is separable, consistent with Eq. (3 – 7).

3.5.2 Image Motion PSF_{IM}

If the image moves across the detectors during the time it takes to integrate the signal at each pixel, a blur will result.[6] For a whiskbroom scanner, the image motion is modeled by a rectangular PSF (Appendix B) in the cross-track direction only,

$$whiskbroom\ scanner: \quad PSF_{IM}(x, y) = rect(x/S) \qquad (3-9)$$

---

[6] In a conventional camera photograph, this smearing of the image is all too visible if the camera is moved during the exposure, or the object being photographed moves.

where S is the equivalent spatial "smear" of the image in the focal plane and,

$$S = \text{scan velocity} \times \text{integration time} . \qquad (3-10)$$

In whiskbroom scanners, such as AVHRR and TM, the scanning is so fast that the integration time is negligible compared to the sample interval (the pixel spacing, or the *GSI*). Typically, the integration time causes a spatial blurring on the order of 1/10 pixel. In a pushbroom scanner, such as SPOT, the integration time causes in-track blurring. In this case,

$$\text{pushbroom scanner:} \quad PSF_{IM}(x, y) = rect(y/S) \qquad (3-11)$$

where S equals,

$$S = \text{platform velocity} \times \text{integration time} . \qquad (3-12)$$

For SPOT, the equivalent spatial movement during the integration time is one *GIFOV*, and is therefore a significant part of the sensor PSF.

3.5.3 Detector PSF_{det}

This is the spatial blurring caused by the non-zero spatial area of each detector in the sensor. Although there can be small non-uniformities in detector response across its dimension, it is usually a good assumption to let,

$$PSF_{det}(x, y) = rect(x/GIFOV)rect(y/GIFOV) , \qquad (3-13)$$

which is a separable, uniform square function. If the *GIFOV* is different in the two directions, then Eq. (3 – 13) should be changed accordingly.

3.5.4 Electronics PSF_{el}

The signal from the detectors is filtered electronically to reduce noise. The electronic components operate in the time domain of the signal as it is scanned and read from the detectors. The time dependence can be converted to an equivalent spatial dependence by,

$$\text{whiskbroom scanner:} \quad x = \text{scan velocity} \times \text{sample time interval} \qquad (3-14)$$

or

$$\text{pushbroom scanner:} \quad y = \text{platform velocity} \times \text{sample time interval} . \qquad (3-15)$$

In the MSS and TM, the electronic filter is a low-pass, Butterworth-type filter operating in the cross-track direction (Park *et al.*, 1984; Markham, 1985). The effect of this filter is to smooth the data cross-track. In the SPOT sensor, there is no in-track filter *per se*, but the data is integrated as the sensor moves one $10\,\text{m}$ pixel in-track, as described previously, which is equivalent to filtering along the in-track direction.[7]

3.5.5 Net PSF_{net}

By a simple theorem (Chapter 6), the net PSF is the convolution of the component PSFs,

$$PSF_{net}(x, y) = PSF_{opt} * PSF_{IM} * PSF_{det} * PSF_{el} \, . \qquad (3-16)$$

The width of the net PSF is the sum of the widths of each of the component PSFs.

3.5.6 Comparison of Sensor PSFs

The total PSF, modeled as described previously, is shown for four satellite sensors in Fig. 3-10. Note that for the AVHRR and MSS, the amount of sensor blur is about twice as great in the cross-track direction than in the in-track direction. All four systems, however, show a cross-track response that is *considerably broader than that of the detector.* This fact has important implications in all aspects of information extraction from remote sensing imagery, from calculations of small target signatures to spatial-spectral signature mixing at each pixel. Because of this blurring, the *effective GIFOV* of remote sensing systems is larger than the oft-quoted geometric *GIFOV.* For example, the effective TM *GIFOV* was found to be 40 to $45\,\text{m}$, rather than $30\,\text{m}$ (Anuta *et al.*, 1984; Schowengerdt *et al.*, 1985).

PSF summary for TM

The component PSFs involved in the image formation process of the TM sensor are shown in Fig. 3-11. A higher greyscale value indicates a higher response, and the black surround indicates a zero response.

3.5.7 Imaging System Simulation

Simulation of imaging systems is useful in visualizing the effects of various image sensor parameters. For example, Fig. 3-12 shows an aerial image with a small *GSI* that is successively subsampled (taking every n^{th} pixel along rows and columns). It becomes increasingly difficult to recognize the man-made structures such as buildings and roads as the *GSI* increases. Magnification of the image does

---

[7]   See Chapter 6 for a discussion of spatial filtering and low-pass filters.

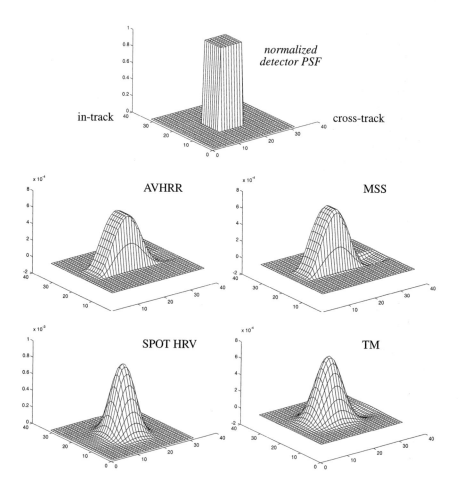

FIGURE 3-10. Perspective views of the total system PSF for four remote sensing systems. The PSFs are normalized to the same scale—eight grid points per in-track pixel; the PSF for the detector is shown at the top. Several interesting comparisons can be made: (1) All system PSFs are considerably broader cross-track than the detector PSF. (2) The three whiskbroom scanners (AVHRR, MSS and TM) have asymmetric spatial responses, caused by the low-pass electronic filter applied in the cross-track direction. The same filter also causes a small negative response to one side of the PSF (which leads to the vertical offsets in the plots). (3) The AVHRR and MSS in-track responses are nearly that of the detector PSF alone, because their optical PSFs are relatively small compared to the detector. The SPOT and TM have smaller detectors and the optical system PSF causes blurring comparable to that of the detector in both directions.

cross-track

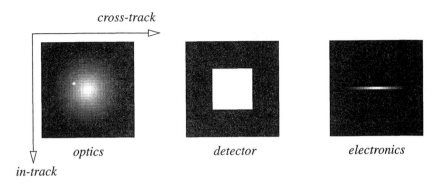

in-track

FIGURE 3-11. *The individual component PSFs for TM. All are displayed at the same scale; the detector width corresponds to a 30m GIFOV. These arrays were used for digital convolution of the lower image in Fig. 3-14 to simulate TM imaging.*

not help because the scene detail is not present in the image to begin with (analogous to so-called "empty magnification" in microscopy).

The *GIFOV* also affects the level of detail within the image, as seen in Fig. 3-13. Here the original aerial image is not only subsampled, but is also averaged over *n*-by-*n* pixels at each level to simulate a sensor with the *GIFOV* equal to the *GSI* (in Fig. 3-12, the *GIFOV* remained fixed at 10m). As the *GIFOV* and *GSI* increase, more scene detail is lost in the digital representation, although, curiously, the overall visual effect does not seem as great as that incurred by the *GSI* alone, as in Fig. 3-12. This is because the increased *GIFOV* reduces "aliasing" caused by the subsampling.

To illustrate the cascading of sensor components, we will do a simulation of TM imagery starting with the scanned aerial photograph of Fig. 3-14. The *GSI* and *GIFOV* of this image are each about 2m. First, the image is rotated to align with the in-track and cross-track directions of the Landsat orbit. This now represents the radiance spatial distribution seen by the TM. The first image in Fig. 3-15 depicts the effect of the optical PSF; it represents the irradiance spatial distribution on the detectors in the TM focal plane. Next, the blurring induced by the scanning detector IFOV is shown. If we were also simulating detector noise (which is quite small for TM), it would be added at this stage. Finally, the electronic filter is applied in the cross-track direction. The original 2m *GSI* has been maintained to this point for simulation accuracy, but is now reduced to the 30m of the TM in the final image of Fig. 3-15.

In Fig. 3-16, an enlargement of the final, simulated TM image is shown with an actual TM band 3 image acquired about four months later than the date of the aerial photograph used in the simulation. The simulation has clearly produced an

GIFOV = 10m
GSI = 10m

GIFOV = 10m
GSI = 20m

2X

GIFOV = 10m
GSI = 40m

4X

FIGURE 3-12. *The effect of GSI on reproduction of scene details. Here different GSIs are simulated by subsampling the original image at the top with the GIFOV fixed at 10m. This results in severe undersampling ("aliasing") of the scene at 20m and 40m GSI, which is particularly evident in the street patterns in the lower half of the image. One would not design a sensor with the latter two sets of parameters.*

$GIFOV = 10\text{m}$
$GSI = 10\text{m}$

$GIFOV = 20\text{m}$
$GSI = 20\text{m}$

$GIFOV = 40\text{m}$
$GSI = 40\text{m}$

*FIGURE 3-13. The effect of equal GSI and GIFOV on reproduction of scene details.
Note how the fine detail is better preserved than in Fig. 3-12 as the GSI is increased.
This is because the spatial averaging provided by the larger GIFOV helps reduce
aliasing effects. The sensor GSI is typically specified to be equal to the GIFOV.*

*rotated to align
with TM track and
scan directions*

FIGURE 3-14. *Scanned aerial photograph of an area in Tucson, Arizona, used in spatial simulation of TM imagery. The scanned photograph has a GSI and GIFOV of about 2m and is panchromatic, with a spectral response similar to the average of TM bands 2 and 3. The dark feature is a focal plane fiducial mark used for distortion calibration in aerial cameras. The bottom image is rotated to align its rows and columns with the TM scan and orbit directions, respectively, and then windowed to the final size used in the simulation.*

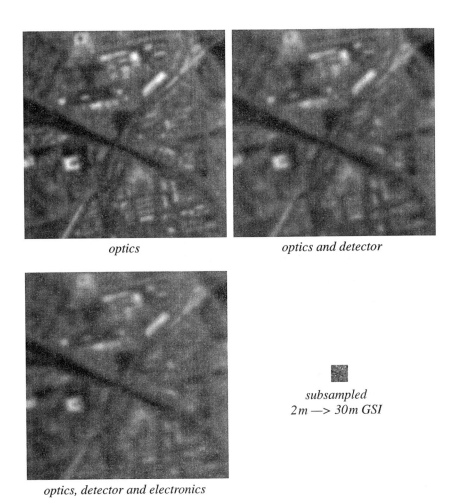

optics   *optics and detector*

subsampled
2m —> 30m GSI

optics, detector and electronics

FIGURE 3-15. Sequential series of images produced by the components of the TM spatial response. Each component introduces additional blurring.

image with a similar spatial resolution to that of the TM. If the TM image's contrast is adjusted to account for unspecified radiometric differences, they are in even closer agreement. No attempt was made to simulate radiometric properties, including different sun angles, shadows, atmospheric conditions and spectral response, which probably accounts for the residual differences.

A spatial degradation process similar to that used above has been described in (Justice *et al.*, 1989) for the simulation of lower resolution data from Landsat

MSS imagery. In that work, the original transfer function of the MSS was taken as the "source" for further degradation to a "target" transfer function representing a sensor with lower spatial resolution, such as the AVHRR.

simulated TM   *real TM*

real TM
(contrast-adjusted)

FIGURE 3-16. *Enlargement of the final, simulated TM image from Fig. 3-15 and an actual TM image, acquired about four months later. The TM image is the average of bands 2 and 3, which approximates the spectral response of the panchromatic film image used in the simulation. Contrast adjustment of the TM image compensates for radiometric differences between it and the simulated TM image due to atmospheric effects and sensor gain and bias.*

3.6 Signal Amplification

The electronic signal, e_b, in band b is amplified electronically and filtered by the electronics PSF. We included the filtering in the previous section for convenience. The amplification stage is designed to provide sufficient signal level to the A/D for quantization, without incurring saturation. This is done at the sensor design stage by estimating the maximum scene radiance range and the corresponding detector output range. The electronics gain and offset values are then set to yield a full *DN* range out of the A/D converter (Fig. 3-17). Some saturation under relatively infrequent conditions (high solar elevation over sand, for example) may be accepted in order to achieve higher gain (and therefore higher radiometric resolution) for the majority of scenes. The radiometric resolution of the sensor is, in part, controlled by the gain setting. The amplified signal, a_b,, is given by,

$$a_b = gain_b \times e_b(x, y) + offset_b .$$ $(3 - 17)$

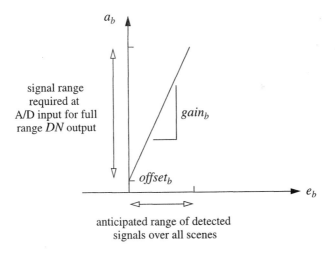

FIGURE 3-17. *Relation between the amplified signal, a_b, and the raw signal out of the detector, e_b. The gain and offset are set to provide sufficient signal amplitude to the A/D converter and are different for each band because of different sensitivities and anticipated irradiance levels. They are usually preset before flight in the sensor electronics and cannot be changed during flight. The SPOT system has in-flight selectable low- and high-gain settings, the latter being useful for imaging very dark targets (Chavez, 1989).*

3.7   Sampling and Quantization

Finally, the amplified and filtered signal is sampled and quantized into *DN*s, usually with a linear quantizer (Fig. 3-18). This can be expressed as an *int[ ]* operator that converts the output of the signal amplifier to the nearest integer value. The final *DN* at pixel p in band b is therefore,

$$DN_{pb} = int[a_b]$$
$$= int[gain_b \times e_b(x, y) + offset_b] \tag{3-18}$$

The spatial-temporal sampling that occurs at the A/D is implicit in this equation in the conversion from continuous to discrete spatial coordinates.[8] Although the *int[ ]* operation in the A/D converter makes this a non-linear relationship between *DN* and e_b, the non-linearity can be ignored for mid-range or higher signal levels because the quantization error is a small percentage of the total signal. At lower *DN*s, however, the quantization error is more important, and should be considered in a detailed analysis. The number of discrete *DN*s, determined by the number of bits/pixel Q, defines the *radiometric resolution* of the system. One can express the radiometric resolution as 2^{-Q} times the dynamic range in radiance.

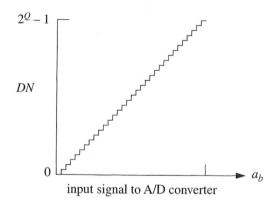

FIGURE 3-18. *Relation between the signal level at the A/D converter, a_b, and the DN output. The transform is usually linear in remote sensing systems, as shown, resulting in uniform quantization. Q is only six in the diagram for clarity.*

---

[8]   Spatial-temporal sampling can be made mathematically explicit, but we have chosen not to do so, in order to keep the discussion as short and direct as possible. For a mathematical treatment of sampling see, for example, (Park and Schowengerdt, 1982) and (Park *et al.*, 1984).

3.8   Simplified Sensor Model

Although Eq. (3 – 18) appears simple, it contains three integrations, one over the system spectral response in λ (Eq. (3 – 5) and two over the system spatial response in x and y (Eq. (3 – 6)). A simplification of Eq. (3 – 18) is often made in practice as follows. The spectral response of the sensor, $R_b(\lambda)$, is assumed to be an average constant over an *effective spectral band* (Palmer, 1984). Similarly, the spatial response of the sensor, *PSF(x,y)*, is assumed to be an average constant over an *effective GIFOV*. Thus, both functions can be removed from the integrals, and we can write,

$$DN_{pb} = int\left[K_b \iiint L_\lambda(x, y)d\lambda dx dy + \text{offset}_b\right] ,\qquad (3-19)$$

where the sensor response functions are combined with the various other constants into the single constant K_b. The integrals in Eq. (3 – 19) are over the effective spectral band and the effective *GIFOV*. We therefore have, notwithstanding quantization, a *linear* relationship between *DN* and at-sensor radiance. If we further simplify the notation by calling L_{pb} the *band- and space-integrated at-sensor radiance* at a particular pixel p in band b, we can write,

$$DN_{pb} = K_b L_{pb} + offset_b.\qquad (3-20)$$

This simplification relates the image *DN*s directly to the at-sensor radiances, integrated over the effective spectral passband and *GIFOV*. The inversion of Eq. (3 – 20) to obtain band radiance values from image *DN*s is known as *sensor calibration* or "calibration-to-radiance." If we go further, using the models and equations of Chapter 2, and convert radiance to reflectance, a *scene calibration,* or "calibration-to-reflectance," is achieved. Scene calibration is more difficult, of course, because it requires knowledge or assumptions about atmospheric conditions and surface terrain; a more detailed discussion appears in Chapter 7.

3.9   Geometric Distortion

The previous sections of this chapter dealt with sensor characteristics that affect the radiometric quality of the imagery, which is important in answering the question "what are we looking at?" Another relevant question is "where are we looking?" The answer to that question is determined by the geometric characteristics of the imagery, which in turn are set by the orbit, platform attitude, scanner properties, and earth rotation and shape.

As an ideal reference, consider conventional still frame imaging with a stationary scene. If the scene is a regular grid pattern and flat, and the camera optics have no distortion, the image will also be a regular grid, correct except for uniform scaling by virtue of the camera magnification. Now, imagine the camera is a pushbroom scanner, moving across the scene in a straight, constant altitude and velocity path. The resulting image will be geometrically identical to the still frame case. Any deviations from these conditions will result in *distortion* of the grid in the image. This non-ideal, but realistic, situation is the subject of this section.

3.9.1 Orbit Models

The orbits of most earth remote sensing satellites are nearly circular because a constant image scale is desired. For precise modeling, an elliptical orbit is assumed.[9]

The orbital velocity of satellites can be considered constant in time (e.g. 1.0153×10^{-3} radians/second for the Landsat-1 and -2 (Forrest, 1981)). Airborne sensors are another story, however, and variation in platform altitude and ground speed cannot be assumed negligible. In many cases, neither is recorded sufficiently often to permit post-flight correction.

3.9.2 Platform Attitude Models

Platform attitude is critical to geometric precision because of the long "moment arm" of high altitude aircraft or satellite pointing. A very small change in the pointing angle results in a large change in the viewed location on the ground. To see this, calculate the angle corresponding to the *GSI* between two neighboring pixels (Table 3-1). Any change in satellite attitude by this amount will result in a change of one pixel in the viewed location. The new generation of civilian high resolution sensors has attitude control requirements that are an order of magnitude higher than those for SPOT.

Attitude is expressed by three angles of platform rotation: *roll, pitch,* and *yaw.* These are shown for one coordinate system convention in Fig. 3-19. Various schemes are used to automatically monitor and control satellite attitude within specified limits, including horizon sensors and torque flywheels. The actual values of roll, pitch and yaw are sampled and recorded with the image data. Unfortunately, these data are not always available to the end user, and even if available, they are usually in complicated, non-standard formats. While any

---

[9]   Even smaller pertubations are possible by changes in gravitational field and terrain elevation under the satellite orbit; these are difficult to model and can safely be ignored here.

information on spacecraft attitude is useful, it is important that the precision and frequency of the reported values be sufficiently high to meet the geometric specifications for the imagery.

TABLE 3-1. The pointing angle between two adjacent pixels, calculated from nominal altitude and nadir in-track GSI values.

system	altitude (km)	in-track *GSI* (m)	pointing angle (mrad)
AVHRR	850	800	0.941
Landsat-4,-5 MSS	705	80	0.113
Landsat-4,-5 TM	705	30	0.0425
SPOT-XS (multispectral)	832	20	0.024
SPOT-P (panchromatic)	832	10	0.0123
Space Imaging (panchromatic) (Fritz, 1996)	680	1	0.00147

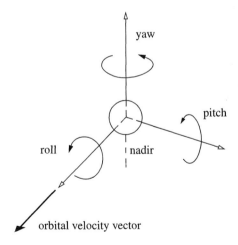

FIGURE 3-19. Conventional definitions for the three attitude axes of a sensor platform. Although the choice of axes direction is arbitrary (Puccinelli, 1976; Moik, 1980), a right-handed coordinate system should be used (the cross product of any two vectors points in the direction of the third).

Although the spacecraft's attitude does not behave in a predictable manner within the controlled limits of excursion, i.e. it is not systematic, it can usually be assumed to be a slowly changing function of time. Some success has been achieved by modeling the attitude variable, a (representing roll, pitch or yaw), with a power series polynomial, over time periods of several image frames for TM (Friedmann *et al.*, 1983), and for individual SPOT frames (Chen and Lee, 1993),

$$a = a_0 + a_1 t + \ldots \qquad\qquad (3-21)$$

The attitude of aircraft sensors is subject to large changes from wind and turbulence. If a gyro-stabilized platform is not used for the sensor, the resulting imagery can contain severe distortions (Fig. 3-20).

FIGURE 3-20. *ASAS imagery of Maricopa Farm near Phoenix, Arizona. The top image is taken at-nadir and the bottom image at +30° off-nadir. Note how the distortion pattern changes within and between the two images, indicating a continuous change in the aircraft platform attitude over time. (Imagery courtesy of Dr. Alfredo Huete, University of Arizona.)*

3.9.3 Scanner Models

Scanner-induced distortions are one of the easier factors to account for because they can usually be described by a fixed function of time. For example, the MSS has a non-linear scan mirror velocity that is well documented (Anuta, 1973; Steiner and Kirby, 1976). The effect of this is to cause a sinusoidal-like displacement of pixels across each scan, with a maximum error of about ± 400 m near the midpoint of the scan on either side of nadir. As long as such distortions are consistent throughout an image and from orbit-to-orbit, they can be easily calibrated and corrected.

Whiskbroom scanners have more inherent distortions than pushbroom scanners because they have more moving parts. Pixel positioning cross-track is determined by the scan mirror motion (MSS acquires data in one direction only, while TM acquires data in opposite directions on alternate scans), which couples with satellite motion to determine pixel positioning in-track. A pushbroom scanner, either of the linear or array type, on the other hand, has rigid cross-track geometry which is essentially decoupled from the in-track geometry. Some of the more important sources of scanner distortion are summarized in Table 3-2.

Scanner-induced distortions can be measured pre-flight and an appropriate functional model determined for application to flight imagery, assuming there is no change in system performance during flight. If the source of the distortion can be physically modeled (for example, a motion analysis of the MSS scan mirror), that can be expected to result in the best correction possible.

The earliest Landsat MSS data was provided in an uncorrected format,[10] so there were numerous efforts to model the distortions arising from the sensor and platform motion relative to the earth (Anuta, 1973; Steiner and Kirby, 1976). Because these distortions are consistent over time, they can be corrected in a deterministic manner using orbital models and scanner calibration data.

3.9.4 Earth Model

Although the earth's geometric properties are independent of the sensor's, they interact intimately via the orbital motion of the satellite. There are two factors to consider here. One is that the earth is not an exact sphere; it is somewhat oblate, with the equatorial diameter larger than the polar diameter. In many satellite imaging models, the intersection of the sensor view vector with the earth's surface is calculated (Puccinelli, 1976; Forrest, 1981; Sawada *et al.*, 1981). Therefore, the exact shape of the earth is important.

The earth ellipsoid is described by the equation,

---

[10] The infamous *X-format.*

TABLE 3-2. Examples of sensor-specific internal distortions. The reader is cautioned that some measurements of distortion were made from ground-processed imagery, and that a careful reading of the indicated reference is required before assuming the errors apply to all of the imagery from a given sensor. For example, the inter-focal plane misregistration in TM is given for early data; later data were found to be registered to within 0.5 pixel because of improved ground processing (Wrigley et al., 1985).

sensor	source	effect on imagery	max. error	reference(s)
MSS	non-unity aspect ratio sampling	cross-track versus in-track scale differential	1.41:1	(USGS/NOAA, 1984)
	nonlinear scan mirror velocity	nonlinear cross-track distortion	± 6 pixels	(Anuta, 1973) (Steiner and Kirby, 1976)
	detector offset	band-to-band misregistration	2 pixels between bands	(Tilton *et al.*, 1985)
TM	focal plane offset	misregistration between visible (bands 1 - 4) and IR (bands 5 - 7)	−1.25 pixel	(Bernstein *et al.*, 1984) (Desachy *et al.*, 1985) (Walker *et al.*, 1984)
SPOT	detector element misalignment	in-track and cross-track pixel-to-pixel positional error	± 0.2 pixel	(Westin, 1992)

$$\frac{p_x^2 + p_y^2}{r_{eq}^2} + \frac{p_z^2}{r_p^2} = 1 \qquad (3-22)$$

where (p_x, p_y, p_z) are the *geocentric* coordinates of any point P on the surface (Fig. 3-21), r_{eq} is the equatorial radius and r_p is the polar radius. The *geodetic latitude* φ and *longitude* λ, as given on maps, are related to the components of p by (Sawada *et al.*, 1981),

$$\varphi = \text{asin}(p_z/r) \qquad (3-23)$$

and

$$\lambda = \text{atan}(p_y/p_x), \qquad (3-24)$$

where r is the local radius of the earth at the point P. The *eccentricity* of the earth ellipsoid is a useful quantitity in map projections (Chapter 7),

$$\varepsilon = \frac{r_{eq}^2 - r_p^2}{r_{eq}^2} . \qquad (3-25)$$

The eccentricity of a sphere is zero.

Some basic properties of the earth are listed in Table 3-3 for reference. Parameters such as the radii are updated periodically as more precise spacecraft measurements become available. Furthermore, different countries use different values (Sawada *et al.*, 1981) to produce their maps.

TABLE 3-3. *Useful parameters for the "Figure of the Earth" and its rotational velocity. The dimensional values are from the current standard for map-making in North America, the Geodetic Reference System (GRS80) (Maling, 1992).*

parameter	value
equatorial radius	6,378.137 km
polar radius	6,356.752 km
equatorial circumference	40,075.02 km
polar circumference	39,940.65 km
eccentricity	0.00669
angular velocity	7.2722052 x 10^{-5} rad/sec

The second factor is that the earth rotates at a constant angular velocity, ω_e. While the satellite is moving along its orbit and scanning orthogonal to it, the earth is moving underneath from west to east. The velocity at the surface is,

$$v_0 = \omega_e r_e \cos \varphi \qquad (3-26)$$

where r_e is the earth's radius and φ is the geodetic latitude. Since satellites such as Landsat and SPOT have an orbit inclination, i, of about 9.1° to the poles (in order to set up the desired revisit period and sun-synchronism), the earth's rotation is not quite parallel to the cross-track scans. The projected earth rotational velocity in the scan direction is thus reduced,

$$v_e = v_0 \cos(i) = 0.98769 v_0 . \qquad (3-27)$$

The most important geometric parameters for earth-orbit modeling are shown in Fig. 3-21. The three vectors, *s*, *g,* and *p* form the "fundamental observation triangle" (Salamonowicz, 1986), and obey the "look vector equation" (Seto, 1991),

$$p = s + g \qquad\qquad (3-28)$$

A particularly simple algorithm to find *P* is given by (Puccinelli, 1976). The actual modeling can become quite involved mathematically, but all approaches use the above framework as a starting point.

There is a long and rich history of modeling satellite orbits and image distortion (Table 3-4). There is also a parallel body of literature on *measuring* the distortion in production imagery that have been geometrically corrected (Table 3-5). Note the lack of references for AVHRR production imagery geometric quality; this is because there is no central site for production and dissemination of AVHRR imagery. It can be received with a relatively low cost station, and therefore, users of the data have tended to create their own geometric correction software, which is evidenced by the large number of AVHRR modeling references.

There has been considerable interest in models for "navigation" within AVHRR imagery because of its large FOV and relatively low spatial resolution. A relatively simple sensor and geoid (earth shape) model incorporating ephemeris data has yielded positional accuracies of 1 to 1.5 pixels (Emery *et al.*, 1989), and with a more complex model involving TBUS ephemeris messages receivable at AVHRR ground stations, sub-pixel accuracies are claimed (Moreno and Meliá, 1993). A summary of AVHRR rectification work is given in Table 3-6. In all cases, the additional use of GCPs has been recommended to obtain the highest possible accuracy. In the case of TM, rectification accuracies of less than 0.5 pixel have been reported using a bi-directional scan model (Seto, 1991). For both TM and AVHRR, these accuracies apply only in areas of low topographic relief, i.e. they assume a spherical or ellipsoidal earth surface.

3.9.5 Line and Whiskbroom Scan Geometry

The cross-track pixel sampling is in fixed time increments for line or whiskbroom scanners, which, given a constant scan velocity, results in fixed angular increments, $\Delta\theta$, where θ is the scan angle from nadir. For line and whiskbroom scanners, the cross-track *GSI* therefore varies across the scan, increasing with increasing scan angle according to (Richards, 1993),

$$\textit{flat earth:} \quad GSI_f(\theta)/GSI(0) = [1/\cos(\theta)]^2 \ , \qquad (3-29)$$

again assuming a flat earth. This approximation is accurate for a rather large scan angle, even at the altitude of the AVHRR (Fig. 3-22). However, at larger angles,

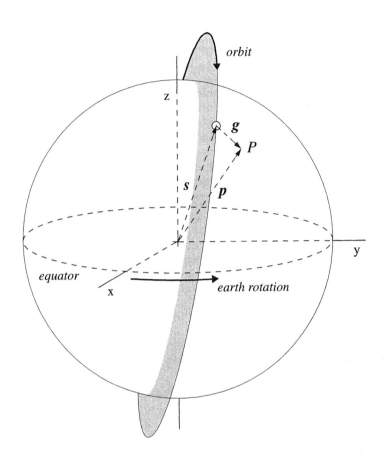

FIGURE 3-21. *The key parameters for modeling the imaging geometry for an earth-orbiting satellite in a near-polar orbit. The inclination is the angle between the orbital plane and the poles and is about 9° for sun-synchronous remote sensing satellites. A fixed (non-rotating) geocentric coordinate system is defined by (x,y,z). The three vectors in that system define the satellite location (**s**), the view direction to a particular point P on the earth's surface (**g**), and the location of P (**p**). Although we are ignoring topography here, it can be included in the vectors **g** and **p**.*

TABLE 3-4. Some of the published work on scanner and satellite orbit
modeling. The MSS geometric scanner model is essentially
unchanged for all five Landsats, and the MSS orbital model for
Landsat-4 and -5 is the same as that for TM.

sensor	reference
AVHRR	(Emery and Ikeda, 1984) (Brush, 1985) (Ho and Asem, 1986) (Brush, 1988) (Emery et al., 1989) (Bachmann and Bendix, 1992) (Moreno et al., 1992) (Moreno and Meliá, 1993) (Krasnopolsky and Breaker, 1994)
MSS (Landsat-1, -2, -3)	(Anuta, 1973) (Puccinelli, 1976) (Forrest, 1981) (Sawada et al., 1981) (Friedmann et al., 1983) (Salamonowicz, 1986)
TM (Landsat-4, -5)	(Seto, 1991)
SPOT	(Kratky, 1989) (Westin, 1990)

the curvature of the earth must be accounted for, and the equation for the cross-track *GSI* becomes (Richards, 1993),

$$\textit{spherical earth:} \quad GSI_e(\theta)/GSI(0) = \frac{[H + r_e(1 - \cos\phi)]}{H\cos(\theta)\cos(\theta + \phi)} \qquad (3-30)$$

where ϕ is the geocentric angle corresponding to the surface point at the scan angle θ and is given in (Brush, 1985) as,

$$\phi = \operatorname{asin}\{[(r_e + H)/r_e]\sin(\theta)\} - \theta \quad . \qquad (3-31)$$

The cross-track *GIFOV* varies similarly, yielding a ground-projected "bow-tie" scan pattern (Fig. 3-23). Since the surface distance from nadir is proportional to ϕ and the pixel index from the center of a scan is proportional to θ, Eq. (3 – 31) also describes the ground distance from nadir at each scanline pixel.

TABLE 3-5. *Some of the published work on satellite production image geometric quality. Unless otherwise noted, data are for system-corrected products, after least-squares polynomial adjustment to maps using Ground Control Points (GCPs).*

sensor	typical error (pixels)	reference
MSS (Landsat-1, -2, -3)	±2 – ±4 (bulk film product) ±1/2 – ±1 (after polynomial adjustment)	(Wong, 1975)
MSS (Landsat-4, -5)	±1.5	(Welch and Usery, 1984)
TM (Landsat-4, -5)	±1.5 (L-4), ±0.4 (L-5) ±1.1 (L-4) ±1 (L-4) ±1 (L-5) ±0.7 ±0.6 (L-4), ±0.3 (L-5) ±1.5 (Earth-orbit model)	(Borgeson *et al.*, 1985) (Walker *et al.*, 1984) (Welch and Usery, 1984) (Bryant *et al.*, 1985) (Fusco *et al.*, 1985) (Welch *et al.*, 1985) (Wrigley *et al.*, 1985)
SPOT	±0.3 (P data, Earth-orbit model)	(Westin, 1990)

TABLE 3-6. *AVHRR rectification experiments using sensor and geoid models.*

model and data	positional accuracy (pixels)	reference
circular orbit and earth	2 – 3	(Legeckis and Pritchard, 1976)
elliptical orbit and earth + ephemeris	1 – 2	(Emery and Ikeda, 1984)
elliptical orbit and earth + TBUS	<1	(Moreno and Meliá, 1993)

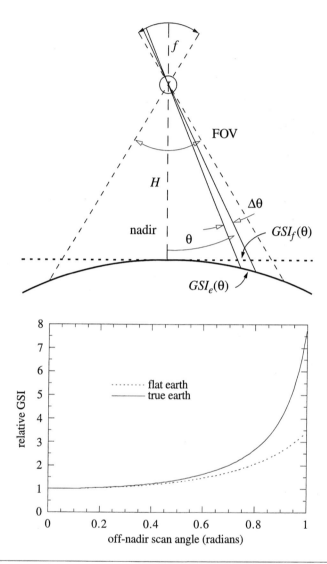

FIGURE 3-22. *Line and whiskbroom scanner geometry in the cross-track direction,*
used to derive Eq. (3 – 29) and Eq. (3 – 30). The data along the scan is sampled at a
fixed time interval to create pixels. Assuming the scan rotational velocity is constant,
that interval corresponds to a fixed angular interval, Δθ. The cross-track GSI
therefore increases with increasing θ, as shown below for an altitude of 850 km. The
flat earth approximation is good within 4% out to a scan angle of about 0.4 radians,
or 23°.

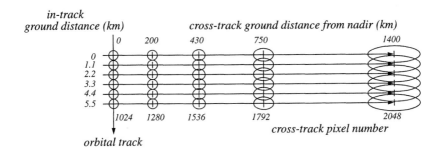

in-track
ground distance (km)   cross-track ground distance from nadir (km)

orbital track   cross-track pixel number

FIGURE 3-23. The "bow-tie" effect in AVHRR data, characteristic of line and
whiskbroom scanners. It is similar to distortion arising in wide-field panoramic
cameras. The horizontal and vertical scales above are linear in ground distance. The
ellipses represent the area of integration of the system spatial response function and
the crosses represent pixel samples. The GIFOV and GSI are greatly exaggerated in
size relative to the scan length, and only five cross-track samples are shown to
emphasize the effect in this drawing. (Adapted from (Moreno et al., 1992).)

3.9.6 Pushbroom Scan Geometry

The cross-track *GSI* for pushbroom scanners does not vary in the same way as for
whiskbroom scanners, assuming the imaging system has constant magnification
across the linear detector array (Fig. 3-24). In a pushbroom system (linear or area
arrays; see Chapter 1), each cross-track line of the image is formed optically as if
in a conventional frame camera. The detector elements are equally spaced at a
distance w (equal to the detector element width) across the array, and therefore
the cross-track IFOV changes across the array, i.e. as a function of the cross-track
view angle.[11] If the earth were flat, however, and the optical system exhibits no
distortion, the equal spacing of w in the focal plane corresponds to a uniform
cross-track *GSI* on the ground,

$$\text{flat earth:} \quad GSI_f = w \times \frac{H}{f}, \tag{3-32}$$

but a changing cross-track IFOV,

$$IFOV(\theta)/IFOV(0) = [\cos(\theta)]^2 . \tag{3-33}$$

---

[11] Notice the use of "view" angle for pushbroom scanners, in contrast to "scan" angle for
line and whiskbroom scanners.

With Eq. (3 – 33) and Eq. (3 – 30), it can be shown that for the true spherical earth, the cross-track *GSI* varies with view angle according to,

$$\text{\textit{spherical earth:}} \quad GSI_e(\theta)/GSI(0) \; = \; \frac{[H + r_e(1 - \cos\phi)]\cos(\theta)}{H\cos(\theta + \phi)} \quad\quad (3-34)$$

where ϕ is given by Eq. (3 – 31). This function is plotted in Fig. 3-24 with the flat earth approximation for the 832 km altitude of the SPOT satellite. Now, since the off-nadir FOV of the SPOT HRV sensor is only $\pm \text{atan}(30/832)$, or ± 0.036 radians, the flat earth approximation is valid within 0.03%. The HRV has the capability, however, to be pointed off-nadir by up to $\pm 26°$; the geometry in that configuration is considerably more complicated than the nadir-pointing condition assumed here.

3.9.7 Topographic Distortion

While the previous discussion accounts for the most significant geometric factors influencing satellite imaging, topography must be included for precise calculations. The height of any point on the ground, from a reference plane, or *datum*, affects its image location. A simple example is shown in Fig. 3-25. Intuitively, one can see that the displacement of an image point from its orthographic location depends on the off-nadir view angle and the height of the object from the reference plane; the greater the off-nadir angle and object height, the greater the image displacement.

Given only one image, it is impossible to infer pixel-to-pixel elevation differences. However, with two images, it is possible to determine elevation differences between two points from their *parallax* (Fig. 3-26). It is easy to show that the image parallaxes of ground points A and C are,

$$p_a \; = \; a_1 - a_2 \; = \; \frac{fB}{H - Z_A} \quad\quad (3-35)$$

$$p_c \; = \; c_1 - c_2 \; = \; \frac{fB}{H - Z_C}, \quad\quad (3-36)$$

where H and Z are measured relative to the datum plane and the image coordinates, a_1, a_2, c_1 and c_2 are all measured relative to the optical center (*principal point*) of the respective image (Wolf, 1983). Combining Eq. (3 – 35) and Eq. (3 – 36), we have,

$$\Delta Z \; = \; Z_A - Z_C \; = \; \frac{\Delta p(H - Z_C)}{p_a}, \qu\quad (3-37)$$

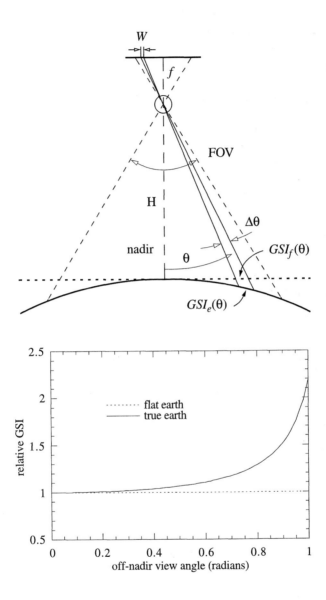

FIGURE 3-24. *Cross-track pushbroom scanner geometry for Eq. (3 – 32) and   Eq. (3 – 34). The cross-track GSI is constant under the flat earth approximation, but increases with increasing view angle θ for the true spherical earth. The altitude used is 832km, corresponding to the SPOT satellites.*

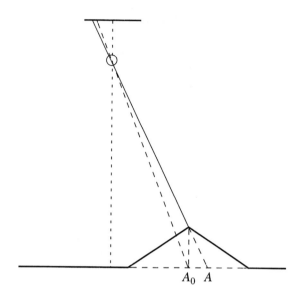

FIGURE 3-25. *Geometry for a pushbroom scanner imaging a topographic feature off-nadir in the cross-track direction. The in-track vector is pointing out of the page. The image of the peak appears to come from A because of optical parallax. If the image were orthographic, the image of the peak would appear to come from A_0.*

where Δp is the parallax difference, $p_a - p_c$. This equation can be simplified for high sensor altitude and moderate relief ($H \gg Z_C$) to,

$$\Delta Z \cong \Delta p \frac{H^2}{fB} \;=\; \Delta p \times \frac{H}{f} \times \frac{H}{B} \;=\; \frac{\Delta p/m}{B/H}, \qquad (3-38)$$

where B is the *base*, the ground distance between the centers of the two images (Ehlers and Welch, 1987). The quantity B/H is called the *base-to-height ratio*. For a given sensor focal length and altitude, the base determines the sensitivity of elevation measurements made from image parallax. The term $\Delta p/m$ is the image parallax difference scaled by the sensor magnification to the ground. From an image analysis viewpoint, the primary significance of Eq. (3 – 38) is that *the elevation difference between two ground points is proportional to the parallax difference between their image points*. This simple relationship is used to derive elevation maps from stereo imagery (Chapter 8).

    We assumed a rather simple situation here, namely a pushbroom scanner parallel to the datum. In reality, they may not be parallel, causing a *tilted* imaging plane. Such a situation occurs in SPOT off-nadir viewing, for example. The

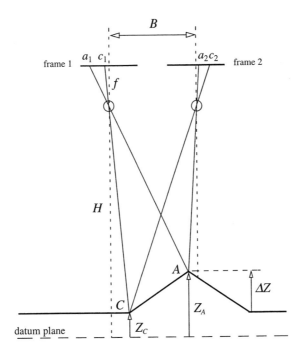

FIGURE 3-26. *Geometry of stereo imaging. The distance between the image points a_1 and c_1 in frame 1 is not equal to the distance between the image points a_2 and c_2 in frame 2 because of the elevation difference between ground points A and C and the different view points of the two frames. The distance between the two view points (called "camera stations" in aerial photography) is called the "base" B of the stereopair.*

resulting geometry is considerably more complex in detail but follows the basic principles outlined here.

3.10   Summary

In this chapter, we have seen how the sensor modifies the signal of interest in remote sensing. The sensor affects the spatial and radiometric quality of the signal. The important aspects of the sensor influence are:

- Scene spectral radiances received at the sensor are weighted by the sensor spectral response in each band.

- Scene spatial features are weighted by the sensor spatial response, which includes blurring by the optics, image motion, detector, and sensor electronics.

- The overall imaging process, under reasonable assumptions, is a linear transform of the at-sensor radiance.

- Geometric distortion arises from internal sensor and external platform and topographic factors.

In the next chapter, we discuss data models, which provide the link between the physical remote sensing models of Chapter 2, the sensor models of this chapter and image processing algorithms.

3.11   Exercises

3-1.   Verify the pixel *DN* values in Fig. 3-4.

3-2.   Calculate the convolution of a square sensor *GIFOV* and a square wave radiance target, either by integration using Eq. (3 − 2), or by a graphical approach. Allow the *GIFOV* to be 1/2, 1 and 2 times the width of a single bar in the pattern and discuss the results in terms of sample-scene phase.

3-3.   What is the east-west displacement, in pixels, due to earth rotation from the top to the bottom of a Landsat TM scene, centered on Tucson, Arizona? What is it for a scene centered on Anchorage, Alaska?

3-4.   Derive Eq. (3 − 34), starting with Eq. (3 − 30).

3-5.   Plot Eq. (3 − 30) and Eq. (3 − 34) as a function of H (over a range of 10 to 1000 km) for a fixed scan/view angle of 20°. Is the flat earth approximation better or worse for aerial scanners compared to satellite scanners?

3-6.   Derive Eq. (3 − 37) and Eq. (3 − 38).

CHAPTER 4

Data Models

4.1 Introduction

The remote-sensing analyst may often view the image only as "data," disconnected from the underlying physical processes that created the "data." This is often a mistake, because it can lead to less than optimal processing algorithms. The various radiation and sensor models discussed in Chapters 2 and 3 are often implicit in *data models*. Since data models are used for data processing algorithm design, they can be viewed as providing the link between the physics of remote sensing and image processing algorithms (Fig. 4-1). In this chapter, we discuss the common data models used in remote sensing and relate them to the underlying physical process models.

radiation models

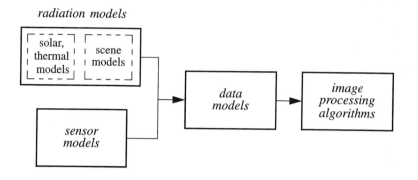

FIGURE 4-1. *Data models form the link between physical radiation and sensor models and image processing algorithms.*

4.2   A Word on Notation

The assortment of notations used to denote pixels and their locations in 2–D images is confusing, at best. Engineers use different mathematical symbology than do earth scientists, who in turn use different symbology than statisticians. Part of the problem is the inappropriate use of continuous notation for discrete functions. This is understandable at times, given the complexity of discrete notation, but is to be avoided wherever possible.

The real world of the scene is continuous in space, wavelength and time (Chapter 2), and all functions expressed in that context should use continuous notation. Beyond the point where the sensor converts the signal to sampled, quantized *DN* values in the A/D converter (Chapter 3), mathematical descriptions should become discrete. All digital image processing equations, for example, should use discrete notation.

How do we designate a particular pixel in a digital image? We could think of the image as an array of numbers, with indexes i and j. The array values are the pixel *DN*s. A pixel value at row i and column j would be denoted DN_{ij} (or simply DN_p as we have done in previous chapters). Rows and columns (or lines and samples) are conventionally numbered from $(1,1)$ at the upper left to (N,M) at the lower right of the image array, as viewed (Fig. 4-2).[1] This notation is natural for computer programs because of sequential data storage formats. One must be

---

[1]   The reader should be aware that a coordinate system beginning at $(0,0)$ and extending to $(N\text{-}1, M\text{-}1)$ is also used. If one is switching between two computer programs that use opposite conventions, mistakes can arise!

careful, however, in associating the row index with *y* and the column index with *x*, because that results in a left-handed coordinate system. If rows are associated with *x* and columns with *y*, as if the array were rotated –90°, the only difference between them (other than discrete versus continuous) is an offset in the origin from (0,0) to (1,1). This notation is used, for example, in (Gonzalez and Woods, 1992); it leads to a right-handed (*x,y*) system, but *y* is horizontal and *x* is vertical, contrary to a "natural" system. In this book, we will use both the (row, column) and the "natural" right-handed (*x, y*)–coordinate systems, as appropriate in a particular context.

If the spatial order of the pixels is unimportant in a particular calculation, it is convenient to replace the double subscript by a single subscript *p* and any accompanying two-dimensional spatial operators by a single operator, indexed over *p*. We will use this briefer notation as often as possible.

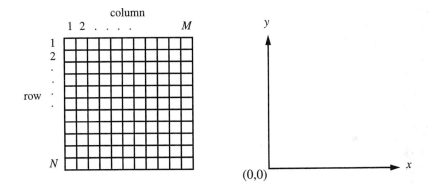

FIGURE 4-2. *A discrete (row, column) notation compared to a continuous, "natural" (x,y) notation.*

4.3   Univariate Image Statistics

We start with definitions for basic *univariate* image statistics, which apply to single band images, and then extend the discussion to *multivariate* statistics to include multispectral and multisensor images.

4.3.1 Histogram

The image histogram describes the statistical distribution of image pixels in terms of the number of pixels at each *DN*. It is calculated simply by counting the number of pixels in each *DN* "bin," and dividing by the total number of pixels in the image, *N*,

$$hist_{DN} = \text{count}(DN)/N . \qquad (4-1)$$

The histogram is often associated with the *Probability Density Function (PDF)* of statistics,

$$hist_{DN} \approx PDF(DN), \qquad (4-2)$$

but this association is mathematically problematic because (1) the *PDF* is defined for continuous variables, and (2) it is only properly used for statistical distributions from a random process. Aside from the issue of discrete versus continuous variables, images are seldom treated as instances of a random processes, but rather as individual data arrays.[2] The histogram is, therefore, a more appropriate description for images.

The histograms of images are typically unimodal (i.e., they have a single "peak"), with an extended tail toward higher *DN*s, i.e. higher scene radiances (Fig. 4-3). It is important to remember that an image histogram only specifies the number of pixels at each *DN*; it contains no information about the *spatial distribution* of those pixels. Sometimes, however, spatial information can be inferred from the histogram. For example, a strongly bimodal histogram usually indicates two dominant materials in the scene, such as land and water. What cannot be inferred is the extent to which pixels in each category are spatially connected. For example, NIR images of a scene with many small lakes and a scene of an ocean coastal area could have similar, bimodal histograms.

The image histogram is a useful tool for *contrast enhancement*. One contrast enhancement algorithm "stretches" the range of *DN*s and "clips" or thresholds it at one or both ends, resulting in a certain percentage of saturated pixels. The appropriate *DN* thresholds can be obtained from the histogram as percentages of the total number of pixels in the image.

Normal distribution

In remote sensing, as in most science and engineering fields, it is often mathematically convenient to assume a *normal* (Gaussian) distribution for independent, identically-distributed samples from a random process. In one dimension, this continuous distribution has the form,

---

[2]   A notable exception may be the treatment of certain noise processes.

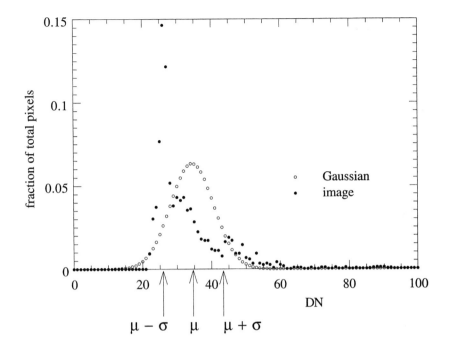

FIGURE 4-3. An example image histogram compared to a Gaussian distribution with the same mean and variance. Note the asymmetry in the histogram, which is not atypical, and that it appears to be multimodal.

$$N(f; \mu, \sigma) = \frac{1}{\sigma \sqrt{2\pi}} e^{-\left[\frac{(f-\mu)^2}{2\sigma^2}\right]} \qquad (4-3)$$

which has the familiar shape shown in Fig. 4-3. The centroid of the distribution is given by μ and its width is proportional to σ. While the normal distribution is not appropriate as a global model for most images, it is widely used to model the distribution of pixels having similar characteristics within an image (Chapter 9).

4.3.2 Cumulative Histogram

Some image processing algorithms, notably histogram equalization, histogram matching and destriping (Richards, 1993), require a function, the *cumulative histogram (chist)*, derived from the histogram as follows,

$$chist_{DN} = \sum_{DN = DN_{min}}^{DN} hist_{DN} .$$ (4 – 4)

The cumulative histogram is the fraction of pixels in the image with a *DN* less than or equal to the specified *DN*. It is a monotonic function of *DN*, since it can only increase as each histogram value is accumulated from the minimum *DN*. Because the histogram as defined in Eq. (4 – 1) has unit area, the asymptotic maximum for the cumulative histogram is one (Fig. 4-4). In this normalized form, the cumulative histogram is also called the *Cumulative Distribution Function (CDF)* (Castleman, 1996), although that association has the same theoretical problems as Eq. (4 – 2).

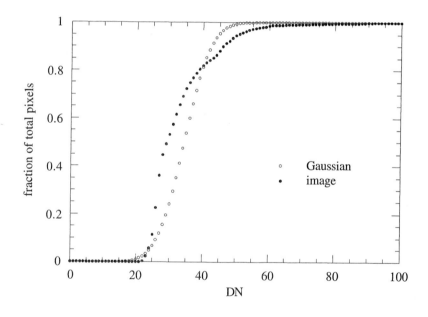

FIGURE 4-4. Comparison of an example image cumulative histogram to a Gaussian cumulative histogram using the same data as in Fig. 4-3.

4.3.3 Statistical Parameters

The *DN mean* can be calculated in two ways,

$$\mu = \frac{1}{N} \sum_{p=1}^{N} DN_p$$

$$= \sum_{DN=DN_{min}}^{DN_{max}} DN \times hist_{DN} \ . \tag{4-5}$$

The first approach adds the *DN*s of all the pixels in the image and divides by the total number of pixels to yield the average *DN*. The second approach weights each *DN* by the corresponding histogram value (the fraction of the image that has that *DN*) and sums the weighted *DN*s. If the histogram is already available, the latter form, known as the *first moment* of the histogram, is a more efficient calculation. The *DN* variance is given by,

$$\sigma^2 = \frac{1}{N-1} \sum_{p=1}^{N} (DN_p - \mu)^2$$

$$= \frac{N}{N-1} \sum_{DN=DN_{min}}^{DN_{max}} (DN - \mu)^2 \times hist_{DN} \ , \tag{4-6}$$

where either calculation may be used, as in Eq. (4 – 5). The second form is the *second moment* of the histogram. In the sum over pixels, N–1 is used rather than N when the mean must be estimated from the data, rather than being known *a priori* from a distribution (Press *et al.*, 1986). The *DN standard deviation*, σ, is the square root of the variance.

The mean and variance are sufficient to specify a normal, or Gaussian, distribution (Eq. (4 – 3)). If the histogram is unimodal and symmetric, a Gaussian distribution may not be a bad model for the actual data. However, as noted earlier, global image histograms tend to be asymmetric, and sometimes multimodal. Whether the distribution is normal or not, the *DN* mean and variance are useful. As discussed below, the image standard deviation can be used as a measure of image contrast, since it is a measure of the histogram width, i.e. the spread in *DN*s.

Other statistical parameters are sometimes useful, including the *mode* (*DN* at which the histogram is maximum), *median* (*DN* which divides the histogram area in half, with 50% of the pixels below the median and 50% above), and the higher order measures of *skewness* (asymmetry),

$$skewness = \frac{1}{N} \sum_{p=1}^{N} \left(\frac{DN_p - \mu}{\sigma} \right)^3$$

$$= \sum_{DN = DN_{min}}^{DN_{max}} \left(\frac{DN - \mu}{\sigma} \right)^3 \times hist_{DN}$$

$$(4-7)$$

and *kurtosis* (sharpness of the peak relative to a normal distribution) (Press *et al.*, 1986; Pratt, 1991),

$$kurtosis = \left[\frac{1}{N} \sum_{p=1}^{N} \left(\frac{DN_p - \mu}{\sigma} \right)^4 \right] - 3$$

$$= \left[\sum_{DN = DN_{min}}^{DN_{max}} \left(\frac{DN - \mu}{\sigma} \right)^4 \times hist_{DN} \right] - 3 .$$

$$(4-8)$$

Skewness is zero for any symmetric histogram. A histogram with a long tail towards larger *DN*s has a positive skewness, and this is typical of remote-sensing images. Kurtosis is zero for a normal distribution. If a histogram has a positive kurtosis, then the peak is sharper than that of a Gaussian; a negative kurtosis means the peak is less sharp than that of a Gaussian. Note that both skewness and kurtosis are unitless, unlike the mean and standard deviation. For the image data represented in Fig. 4-3, we have

$$\mu = 34.1$$
$$\sigma = 9.12$$
$$skewness = 1.78$$
$$kurtosis = 4.32$$

indicating a degree of asymmetry and sharpness in the peak of the histogram. Skewness and kurtosis are quite sensitive to *outliers*, pixels with *DN*s far removed from the majority distribution, because of their high order.

4.4   Multivariate Image Statistics

The extension of the image statistical measures of the previous section to K dimensions is straightforward. The data measurement variable, *DN*, becomes a measurement vector, **DN**, having K components (Fig. 4-5),

$$DN_{ij}^{[K]} = \begin{bmatrix} DN_{ij1} & DN_{ij2} & \cdots & DN_{ijK} \end{bmatrix}^T = \begin{bmatrix} DN_{ij1} \\ DN_{ij2} \\ \vdots \\ DN_{ijK} \end{bmatrix} \qquad (4-9)$$

As we suggested before, a simpler notation can be used, if we understand that the vector is K-dimensional and that the subscript p indicates a particular pixel,

$$DN_p = \begin{bmatrix} DN_{p1} & DN_{p2} & \cdots & DN_{pK} \end{bmatrix}^T = \begin{bmatrix} DN_{p1} \\ DN_{p2} \\ \vdots \\ DN_{pK} \end{bmatrix}. \qquad (4-10)$$

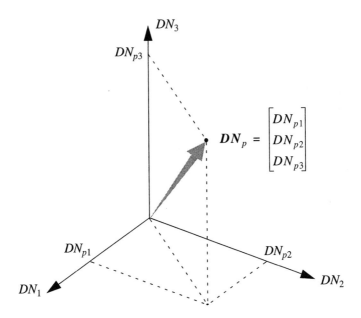

FIGURE 4-5. *Visualization of a three-band multispectral image pixel* DN_p *as a vector in three-dimensional space.*

One way to visualize two- or three-dimensional data is the *scatterplot*. An example for bands 2, 3 and 4 of a Landsat TM image is shown in Fig. 4-6. This is a binary plot which shows a dot if a particular multispectral vector has a histogram count of one or more. Therefore, the number of pixels having a particular multispectral vector is not shown. Since the histogram is three-dimensional, different viewing angles reveal different features in the data; some software programs allow interactive rotation to assist interpretation. The 3-D plot can be reduced to three 2-D plots by projecting every point onto one of the bounding planes (Fig. 4-7). While these 2-D scatterplots are a convenient way to probe the higher dimensions in the data, they do not contain all the information of the original K dimensions. The operation of projection removes some of that information because all points along a given projection line are represented by only a single point in 2-D.

Even with the large amount of data reduction achieved with scatterplots, visualization of multispectral data can be challenging. There are 21 possible unique scatterplots of band pairs for a seven-band Landsat TM image. An example is shown in Fig. 4-8.

Some additional information about the *density* of the distribution along a projected line can be preserved by representing the integrated value as a grey level and displaying the scatterplot as a greyscale image (we call this a *scattergram*). Some examples from various sensor band combinations are shown in Fig. 4-9. It is possible to see the distribution of pixel counts over the multispectral space in these representations. The 2-D scattergrams can be reduced to 1-D histograms in each band by projection onto one of the axes.

In K dimensions, we write the histogram as,

$$hist_{DN} = \text{count}(DN)/N \ . \tag{4-11}$$

Note that this is a *scalar* function of a K-D *vector*. In one dimension, the normal distribution is given by Eq. (4 − 3) and requires only two parameters, the data mean and variance, to specify the function completely. Similarly, the only parameters of a K-dimensional normal distribution are the *mean vector*, the expected value of DN,[3]

$$\mu = \begin{bmatrix} \mu_1 \ ... \ \mu_K \end{bmatrix}^T = \langle DN^T \rangle \tag{4-12}$$

where the mean in band k is,

---

[3]   The expected value of a random variable q is sometimes denoted $E\{q\}$. For simplicity, we are assuming the sample mean and covariance of a finite number of samples from a random process equals the true mean and covariance of an infinite number of samples. In practice, only the limited sample statistics are available. See (Press *et al.*, 1986) and (Fukunaga, 1990) for further discussion.

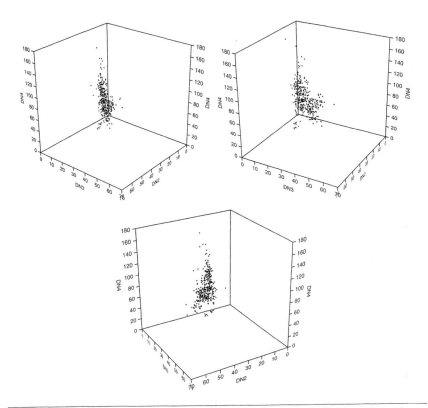

FIGURE 4-6. *Three-band scatterplots of bands 2, 3 and 4 of a TM image, viewed from three different directions. Only every 20th sample and line of the image are used to calculate these scatterplots, so that they are not too dense. Every dot in the scatterplot represents one or more pixels with a particular spectral vector. Note that the image data occupies a small fraction of the total DN volume.*

$$\mu_k = \langle DN_k \rangle = \frac{1}{N} \sum_{p=1}^{N} DN_{pk} \qquad (4-13)$$

and the *covariance matrix,*

$$\boldsymbol{C} = \begin{bmatrix} c_{11} & \cdots & c_{1K} \\ \vdots & & \vdots \\ c_{K1} & \cdots & c_{KK} \end{bmatrix} = \langle (\boldsymbol{DN} - \mu)(\boldsymbol{DN} - \mu)^T \rangle \qquad (4-14)$$

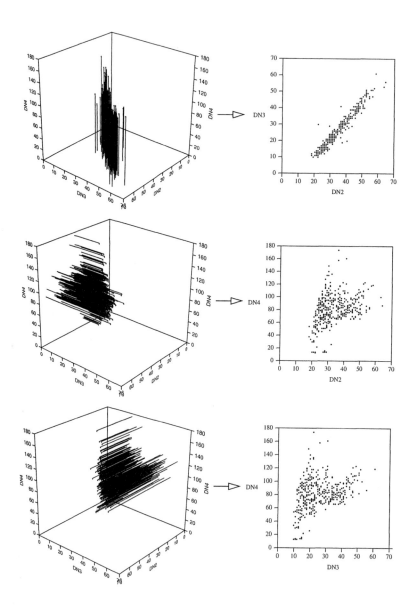

FIGURE 4-7. Reduction of 3-D scatterplots to 2-D scatterplots by projections onto the
three bounding planes. The 2-D scatterplots provide multiple views of the data, but do
not contain all the information that exists in 3-D.

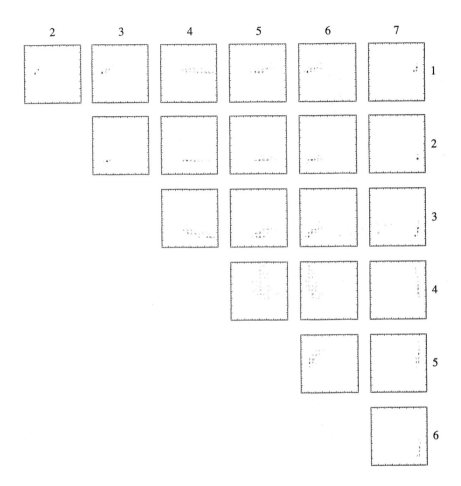

FIGURE 4-8. *A full set of band-pair scatterplots for a seven-band TM image. The row of numbers across the top indicates the TM band whose DN is plotted along the abscissa of each scatterplot, and the column of numbers on the right indicates the TM band whose DN is plotted along the ordinate.*

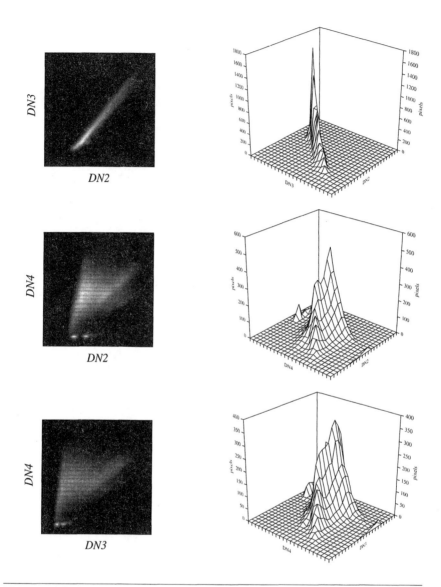

FIGURE 4-9. Two-dimensional scattergrams with density coded as grey levels and displayed as surfaces. The data are from the TM image of Fig. 2-13. Note the two bright "clusters" of pixels at the bottom of the DN4 versus DN2 and versus DN3 greyscale representations. These are from the Briones (left) and San Pablo (right) Reservoirs. The units of the surface plots are number of pixels per DN.

where the *covariance* between bands m and n is,

$$c_{mn} = \frac{1}{N-1}\sum_{p=1}^{N}(DN_{pm}-\mu_m)(DN_{pn}-\mu_n) . \qquad (4-15)$$

The covariance matrix is *symmetric*, i.e. c_{mn} equals c_{nm}. Also note that because the diagonal elements, c_{kk}, are the variances of the distribution along each dimension (refer to Eq. (4 – 6)), they are always positive; however, *there is no such positive constraint on the off-diagonal elements*. In other words, two bands of a multispectral image may have a negative covariance. That means that pixels with relatively low *DN*s in one band will have relatively high *DN*s in the other, and vice versa.

The significance of the off-diagonal terms of the covariance matrix may be better appreciated by defining the *correlation matrix*,

$$\boldsymbol{R} = \begin{bmatrix} 1 & \cdots & \rho_{1K} \\ \vdots & & \vdots \\ \rho_{K1} & \cdots & 1 \end{bmatrix}, \quad -1 \le \rho_{mn} \le 1 \text{ or } |\rho_{mn}| \le 1 \qquad (4-16)$$

where the *correlation coefficient* between two bands m and n is defined as,

$$\rho_{mn} = c_{mn}/(c_{mm}\cdot c_{nn})^{1/2} , \qquad (4-17)$$

i.e. the covariance between two bands, divided by the product of their standard deviations. The value of ρ_{mn} must be between minus one and plus one; the diagonal terms, for which m equals n, are each normalized to one. Examples of the shape of a two-dimensional normal distribution for different correlation coefficients are shown in Fig. 4-10. Note that values of ρ_{mn} close to plus or minus one imply a strong linear dependence between the data in the two dimensions, whereas if ρ_{mn} is near zero there is little dependence between the two dimensions. We will see later (Chapter 5) how the off-diagonal elements in the covariance and correlation matrices may be changed to zero by appropriate transformation of the K-dimensional image. The K features of the transformed image are then *uncorrelated*, a useful property for data analysis.

A K-dimensional, discrete, normal (Gaussian) distribution may be defined similarly to a continuous, normal distribution as,

$$N(\boldsymbol{DN};\mu, \boldsymbol{C}) = \frac{1}{|\boldsymbol{C}|^{1/2}(2\pi)^{K/2}}e^{-(\boldsymbol{DN}-\mu)^T\boldsymbol{C}^{-1}(\boldsymbol{DN}-\mu)/2} \qquad (4-18)$$

which is a scalar function of the *DN* vector. The normal distribution is displayed in 2–D as a greyscale image and with overlaying equal probability contours in

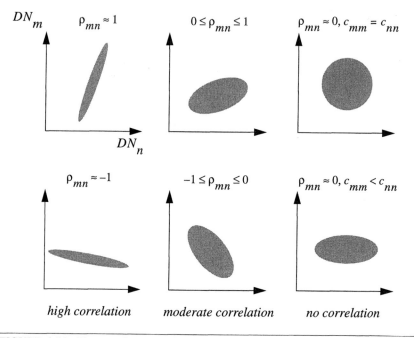

FIGURE 4-10. The correlation coefficient indicates the shape of the scatterplot (or scattergram) of a multispectral image.

Fig. 4-11. Note that the contours are elliptical (see Ex. 4-2). One of the normal distribution's important characteristics is that the probability never goes identically to zero, no matter how far away the **DN** vector is from the mean. A consequence of this is that a **DN** vector always has a nonzero probability under the normal distribution assumption. This fact impacts certain classification algorithms (Chapter 9).

4.4.1 Reduction to Univariate Statistics

In general, the full vector-matrix form of multivariate statistics is necessary. The mathematics reduces to a 1-D description *only if,*

$$PDF(\mathbf{DN}) = PDF(DN_1) \cdot PDF(DN_2) \cdot \dots \cdot PDF(DN_K) \quad , \quad (4-19)$$

i.e. the *probability density functions are independent of each other* in all dimensions. This is equivalent to the definition of a *separable* function, used in signal processing (Chapter 3 and Appendix A). The 1-D distributions of each image

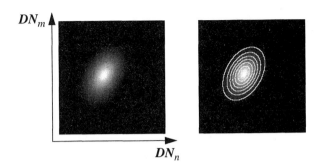

FIGURE 4-11. The normal probability density function in 2–D, displayed as a greyscale image and with overlaying contours of equal probability.

band are called the *marginal distributions.* An important consequence of independence between bands is that the covariance matrix is then *diagonal,* with zeros in all off-diagonal elements. The correlation coefficient is also zero in this case. In practice, Eq. (4 – 19) can be approximated with Eq. (4 – 1), Eq. (4 – 2) and Eq. (4 – 11).

4.5   Noise Models

Noise is introduced into the data by the sensor. It is a variation in the sensor output that interferes with our ability to extract scene information from an image. Image noise occurs in a wide variety of forms (Fig. 4-12 and Fig. 4-13) and is difficult to model; for these reasons, many noise reduction techniques are *ad hoc.* It is beneficial to categorize noise types and generalize their descriptive models, which we pursue in the following.

The simplest model for noise is an *additive, signal-independent* component at each pixel, *p,*

$$DN_p = int[a_p + n_p] \qquad (4-20)$$

where all three quantities are vectors in multispectral space. The noise can originate at several places along the imaging chain described in Chapter 3 and should be modeled accordingly when a specific source is known. We are treating it as being added to the amplified detector signal, a_p (see Eq. (3-16)).

The function n_p in Eq. (4 – 20) is unspecified at this point and can be tailored to describe several common types of noise. In most cases, it is reasonable to

2X

FIGURE 4-12. An example of photographic granularity. The image is an aerial photograph of Dulles Airport near Washington D.C., digitized with a GSI of about one meter. Note the randomness of the noise, but that there also seems to be some spatial "clumping" of neighboring pixels, corresponding to film grains.

aerial video aperiodic scanline noise

Landsat–4 MSS coherent noise

Landsat–1 MSS bad scanline noise

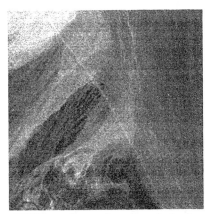

AVIRIS scanline and random pixel noise

Landsat–4 TM banding noise

FIGURE 4-13. A gallery of scanner noise. The image contrast has been increased in all cases to emphasize the noise pattern.

assume the noise has zero mean over a large area of the image, and is therefore manifested as positive or negative fluctuations about the noiseless *signal*, a_p.

A *signal-dependent*, but still additive, noise model is useful in some situations,

$$DN_p = int[a_p + n_p(a_p)] \ . \tag{4-21}$$

For example, the noise in photographic film (*granularity*), caused by randomly-located silver halide grains, is signal-dependent because the probability of a grain developing at any given point in the image depends on the exposure at that point. To illustrate, we take some sample areas, with no apparent scene detail, from the Dulles Airport photograph (Fig. 4-12) and calculate the *DN* mean and standard deviation, which are plotted against each other in Fig. 4-14. The noise, given by σ_{DN}, is clearly dependent on the signal, μ_{DN}, in this case.

FIGURE 4-14. *Plot of DN standard deviation versus DN mean for five featureless areas in Fig. 4-12. The exact form of this relationship is the combined result of granularity and film scanner characteristics, which are not known for this image.*

Global, random noise like that in Fig. 4-12 is, fortunately, not prominant in electro-optical scanner systems, because the quantization interval (Fig. 3-18) is typically twice as large as the detector noise standard deviation. *Periodic* noise related to scanning is more likely. Whiskbroom scanners with multiple detectors per scan, like MSS and TM, are subject to *striping*, caused by differences in calibration and response of each of the detectors. The SPOT pushbroom sensor can show striping along-track in some images. TM also exhibits *banding* noise, in which *all* of the detectors can show a response change from scan to scan. Since the TM scans in both cross-track directions, the banding often appears in alternate scans and is usually associated with high contrast radiance transitions (Fig. 4-13).

Detector scan noise is modeled easily with either Eq. (4 – 20) or Eq. (4 – 21). As an example, we will model different (linear) gains and offsets for each of the 16 detectors in each band of a TM image. This scan noise is represented by modifying Eq. (3-17) for each band as follows,

$$DN_{ij}^{det} = int[gain^{det} \times e_{ij}^{det} + offset^{det}] \qquad (4-22)$$

where the superscript, *det*, refers to the detector index 1, 2, 3, . . ., 16. The vector symbol for *DN* is dropped here, since Eq. (4 – 22) is band specific, and the row-column index, *ij*, is made explicit, since we need to model the noise along the in-track direction only. The row (image line) index for a given detector is given by,

$$i = (scan - 1) \times 16 + det, \quad scan = 1, 2, 3, \ldots \qquad (4-23)$$

as a function of the scan index (16 lines per scan) and detector index, *det*. Noise correction techniques based on this model are described in Chapter 7.

4.5.1 Statistical Measures of Image Quality

The quality of a digital or displayed image is influenced by many factors and is impossible to express in a single numerical *metric*. In this section, we present some of the common metrics for image quality.

Contrast

Numerical contrast may be defined in several ways, e.g.,

$$C_{ratio} = \frac{DN_{max}}{DN_{min}}, \qquad (4-24)$$

$$C_{range} = DN_{max} - DN_{min} \qquad (4-25)$$

or,

$$C_{std} = \sigma_{DN} \qquad (4-26)$$

where DN_{max} and DN_{min} are the maximum and minimum *DN*s in the image, and σ_{DN} is the *DN* standard deviation. One definition or another may be appropriate in particular applications. For example, C_{ratio} and C_{range} may be inappropriate for some noisy images because one or two bad pixels could result in deceptively high contrast values; C_{std} would be much less affected. Note that, of these three definitions for contrast, only C_{ratio} does not depend on the units of the data.

The *visual* contrast of a displayed image is another indicator of the quality of the image. In this context, the quantitites in Eq. (4 – 24) to Eq. (4 – 26) should not be the *DN*s of the digital image, but rather the displayed *GL*s (Chapter 1) or preferably, direct measurements of the radiance from the display. However, the visual contrast depends not only on the digital image's *DN* range, but also on psychophysical factors such as the spatial structure within the image (see (Cornsweet, 1970) for examples), the ambient viewing light level, and the display monitor characteristics. Furthermore, visual and numerical contrast are both spatially-varying quantities, in that an image which has high global contrast may have local regions of low, intermediate or high contrast.

Modulation

Another easily measured image property is modulation, M, defined as,

$$M = \frac{DN_{max} - DN_{min}}{DN_{max} + DN_{min}} \qquad (4-27)$$

Because *DN*s are always positive, this definition insures that modulation is always between zero and one, and unitless.

Modulation is most appropriate for periodic (spatially repetitive) signals. An example, a periodic *sinusoidal* function (Fig. 4-15), is described as,

$$f(x) = mean + amplitude \times \sin\left(2\pi\frac{x}{period} - phase\right) . \qquad (4-28)$$

The *spatial frequency* of such a function is related to its period by,

$$frequency = \frac{1}{period} \qquad (4-29)$$

and has units of reciprocal distance.[4] If, for the periodic signal in Fig. 4-15, we divide the numerator and denominator in Eq. (4 – 27) by two, we obtain,

$$M = \frac{(DN_{max} - DN_{min})/2}{(DN_{max} + DN_{min})/2} = \frac{amplitude_{DN}}{mean_{DN}} \qquad (4-30)$$

---

[4]   This type of function plays an essential role in Fourier transforms, as described in Chapter 6.

which expresses modulation as one-half the ratio of the data range to the data mean. For aperiodic signals, such as most images, Eq. (4 – 30) is still meaningful, but must be interpreted more carefully.

Note that modulation is related to one measure of contrast, C_{ratio}, as follows,

$$M = \frac{C_{ratio} - 1}{C_{ratio} + 1} \, .$$

(4 – 31)

In general, modulation suffers from the same sensitivity to outliers as C_{ratio} or C_{range}.

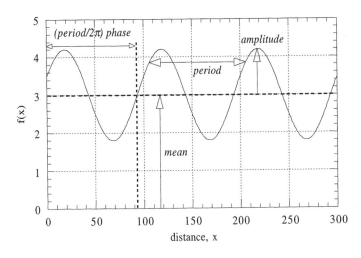

FIGURE 4-15. A 1-D periodic function of spatial coordinates, with a period of 100 spatial units, a mean of 3 units, an amplitude of 1.2 units, and a phase of 5.8 radians, corresponding to an offset of 92.3 spatial units. The modulation of this function is 0.4.

Signal-to-Noise Ratio (SNR)

A *signal* is the noiseless part of a measurement, i.e., the component that carries information of interest. The specific definition of a signal therefore changes with the application. In optical remote sensing, we might call reflectance the signal if our goal is to measure reflectance. Or, we might define the signal as the at-sensor radiance, if our purpose is to analyze the sensor's influence on the radiation it receives. Since an image consists of spatial variations, we need to think of the signal as spatially variant, as well.

Numerous types of *noise* can be introduced into the data by the sensor. Noise corrupts the signal and makes information extraction more difficult. Some measure of the relative amounts of signal and noise is necessary for engineering design, data quality assessment, noise reduction algorithms and certain information extraction algorithms. The *Signal-to-Noise Ratio (SNR)* is such a measure; it is unitless and therefore independent of the data units. The difficult problem is how to meaningfully define signal and noise. For an image contaminated by random noise at every pixel, an "amplitude" *SNR* can be defined as the ratio of the noise-free image contrast to the noise contrast,

$$SNR_{amplitude} = \frac{C_{signal}}{C_{noise}} \qquad (4-32)$$

where C is any of the definitions in Eq. $(4-24)$ to Eq. $(4-26)$. Because of the problem of outliers, the measure C_{std} is generally the most reliable. We then have *SNR* defined in terms of the ratio of the signal standard deviation to the noise standard deviation,

$$SNR_{std} = \frac{\sigma_{signal}}{\sigma_{noise}}. \qquad (4-33)$$

A "power" *SNR* is given by,

$$SNR_{power} = (SNR_{amplitude})^2 = \left(\frac{C_{signal}}{C_{noise}}\right)^2 \qquad (4-34)$$

and, if we again choose to use C_{std} as the measure of contrast, this yields

$$SNR_{var} = \frac{\sigma_{signal}^2}{\sigma_{noise}^2} \qquad (4-35)$$

which is probably the most commonly-used definition of *SNR*. The use of variance in the *SNR* measure is compatible with other statistical descriptors of images, such as the *semivariogram* described later in this chapter, and with statistically-based image transforms, such as the *Maximum Noise Fraction* transform described in Chapter 5.

The *SNR* expressed in decibels (dB) is given by,

$$SNR_{dB} = 10\log(SNR) \qquad (4-36)$$

where *SNR* can be any of the previous definitions. This is not another definition of *SNR*, but simply a nonlinear transform of a given *SNR* measure; it is particularly useful for compressing the large dynamic range of SNR_{power} and also has some computational benefits in systems analysis.

By now it should be clear that there is no single definition of *SNR*. The definition should be chosen to be meaningful to the problem at hand. The reader should likewise always question the definition of *SNR* used in any publication; unfortunately it is not uncommon to simply specify *SNR* values without being specific about the definition. In some cases, it may be possible to infer the definition being used from the context of the discussion.

The overriding practical problem in trying to estimate *SNR* from data is, of course, that one does not usually have access to the pure signal. If the noise level is not high, it may suffice to use the noisy image as an approximation of the signal. The noise level (for uniform, random noise) can be estimated from uniform areas of the image which are assumed to have no signal content (see the discussion about Fig. 4-14). Simulations of an image with different amounts of global, additive random noise at each pixel are shown in Fig. 4-16.

Any type of periodic noise is more visible than random noise (Fig. 4-17), but is generally easier to correct (Chapter 7). Robust measures of *SNR* for striping or isolated (local) random noise have not been developed, however (see Ex. 4-5.).

4.5.2 Noise Equivalent Signal

Sometimes one is interested in the input signal level to an optical sensor that results in an output signal equal to the noise level. This is the *noise equivalent signal*; it can also be interpreted as the input signal level that results in an output *SNR* of one. Examples in remote sensing usage are *Noise Equivalent Radiance* (*NER*), the input radiance to an optical sensor that produces an output *SNR* of one, and *Noise Equivalent Reflectance Difference* (*NE$\Delta\rho$*), a differential measure that indicates the ground reflectance difference that produces an image contrast (C_{range} in Eq. (4 – 25)) of one (Schott, 1997). It is therefore implied that the value $\Delta\rho$ is just detectable, but that is an arguable point.

4.6 Spatial Statistics

The statistical properties of spatial data have been studied and modeled for years in applications such as mining and nonrenewable resource (oil, natural gas) exploration (Journel and Huijbregts, 1978). The techniques have been generalized to other spatial data in recent years (Isaaks and Srivastava, 1989; Carr, 1995), including remote-sensing data (Curran, 1988). The field of spatial statistics is called *geostatistics* by its practitioners.

A key concept in geostatistics is the notion of *spatial continuity*, which expresses the likelihood of a particular data value at a particular location (a pixel in an image in our case) given neighboring or regional data values. Techniques have been devised to use second-order statistics to measure data continuity

$$SNR_{std} = 1$$
$$SNR_{var} = 1$$
$$SNR_{dB} = 0$$

$$SNR_{std} = 2$$
$$SNR_{var} = 4$$
$$SNR_{dB} = 6$$

$$SNR_{std} = 5$$
$$SNR_{var} = 25$$
$$SNR_{dB} = 14$$

$$SNR_{std} = 10$$
$$SNR_{var} = 100$$
$$SNR_{dB} = 20$$

FIGURE 4-16. The effect of random noise on image quality is simulated in this aerial photograph of a residential neighborhood in Portland, Oregon, by adding a normally-distributed random number to each pixel. Even for a SNR of unity, there is discernable image structure because of the spatial correlation between neighboring pixels. This correlation means that the mean DN is fairly constant over local neighborhoods, but changes globally across the image. Therefore, one is still able to detect the roofs, trees and streets even at very low SNR, but smaller objects, such as cars, are obscured. (The original image is courtesy of the U.S. Geological Survey.)

noiseless

global random noise   *detector striping noise*

FIGURE 4-17. Comparison of the visual effect of equal amounts of random and striping noise. The aerial photograph is used in a simulation of global, additive random noise and unequal detector offsets. The random noise is normally distributed, and the striping noise has a period of 16 lines (as the TM). Both noise distributions have a mean of zero and a standard deviation of eight; the standard deviation of the original image is 18.

(*covariance* and *semivariogram*) and to optimally estimate missing values by interpolation (*kriging*).

In the world of image processing, many spatial descriptors have been proposed to address the recognition of different spatial *textures*. If one thinks of an image as a tangible, physical surface defined by $DN(x,y)$, then texture describes the visual "roughness" of that surface, analogous to the sensation of moving a finger across a real surface. Texture is manifested in an image as local, quasi-periodic variations in DN and is caused by the spatial distribution of scene reflectance, and shading and shadows from topography. One texture descriptor is based on properties of the *spatial co-occurrence matrix* (Haralick *et al.*, 1973).

Others are based on *fractal geometry* (Mandelbrot, 1983; Pentland, 1984; Feder, 1988). We will look at these as well as the geostatistical measures in the following sections.

4.6.1 Visualization of Spatial Covariance

The joint probability of a pair of *DN* values at two pixels, separated by an amount *h* in a given direction, may be approximated by the joint histogram,

$$hist_h = count(DN_1, DN_2)/N_h \qquad (4-37)$$

which is a special case of Eq. $(4-11)$; N_h is the number of pixel pairs used in the count operation for the given *h*. We can visualize the histogram as a 2-D scattergram with one axis corresponding to the *DN* of a pixel and the other corresponding to the *DN* of another pixel which is *h* pixels away in the given direction. An example is shown in Fig. 4-18. At short separations (small *h*), the scattergram shows high correlation, i.e. close neighbors are similar. As the separation increases (larger *h*), the scattergram becomes less correlated and eventually nearly isotropic, indicating the dissimilarity of pairs of pixels. We will see later that such scattergrams are identical to the spatial co-occurrence matrix.

4.6.2 Covariance and Semivariogram

In geostatistics, the *covariance function* of a 1-D data set is defined as (Isaaks and Srivastava, 1989),

$$C(h) = \frac{1}{N(h)} \sum_{(i,j)|h_{ij}=h} DN_i DN_j - \mu_{-h}\mu_{+h} \; . \qquad (4-38)$$

The parameter, h_{ij}, is called the *lag* and is the distance between the pairs of data points located at *i* and *j*. The quantities, μ_{-h} and μ_{+h} are the mean *DN*s of all data points at a distance of $-h$ or $+h$ from some other data point; these two values are seldom equal in practice. $N(h)$ is the number of data point pairs separated by *h*.

The *semivariogram* is given by (Isaaks and Srivastava, 1989),

$$\gamma(h) = \frac{1}{2N(h)} \sum_{(i,j)|h_{ij}=h} (DN_i - DN_j)^2 \qquad (4-39)$$

or, as expressed in the equivalent notation of (Woodcock *et al.*, 1988a),

$$\gamma(h) = \frac{1}{2N(h)} \sum_{i=1}^{N} (DN_i - DN_{i+h})^2 \; . \qquad (4-40)$$

Equation (4 – 40) measures the moment of inertia about the 45° line in the spatial scattergrams of Fig. 4-18. It therefore is a measure of the spread in the scattergram and the degree of spatial correlation between different pixels.

The covariance function and semivariogram are related by,

$$\gamma(h) = C(0) - C(h) \qquad (4-41)$$

if a sufficiently large number of data samples are used so that the $C(0)$ is the same over any set of samples (this is the *stationarity* assumption from statistics). Thus, the covariance function and the semivariogram actually contain the same information about the image.

The semivariogram usually exhibits a characteristic shape, increasing from small lags to larger lags. The plateau where γ becomes more or less constant is the *sill*. The distance from zero lag to the the onset of the sill is the *range*. The semivariogram is, by definition, zero for zero lag. However, for real data, noise may cause nonzero semivariogram values for lags smaller than the sample interval. This residual value at small lags is known as the *nugget* value. The nugget can be used to estimate spatially-uncorrelated image noise (Curran and Dungan, 1989). The sill, on the other hand, is the other extreme, i.e., the value of the semivariogram for very large lags. The sill is usually reached more or less asymptotically at the range, which is one measure of *correlation length* in the data.

The spatial characteristics of the data can be emphasized by normalizing the covariance functions by their value at the origin, namely the variance of the data. This normalization removes differences due to contrast among data sets, and more clearly shows the spatial correlation relative to that at zero lag. Examples are shown later.

Several continuous models for the 1-D covariance function and semivariogram are used and are summarized in Table 4-1. The exponential Markov covariance model, widely used in image processing (Jain, 1989; Pratt, 1991), is equivalent (with a scale change on h) to the exponential model for semivariograms, in view of Eq. (4 – 41). The various models of Table 4-1 are plotted in Fig. 4-19.

Geophysical data are often available only on an irregular sample grid, with large gaps between measured data points. Semivariograms are used in that context to estimate the missing values using a procedure known as *kriging* (Isaaks and Srivastava, 1989). Continuous models are needed to interpolate values of the semivariogram at locations other than the discrete values used in Eq. (4 – 40). Some interesting applications of the semivariogram in remote-sensing data analysis are summarized in Table 4-2.

As an example of semivariograms derived from an image, the aerial photograph of a desert area in Fig. 4-20 will be used. This image contains several distinct regions of different vegetation densities. Three transects were extracted in each of three characteristically different regions, the semivariogram was

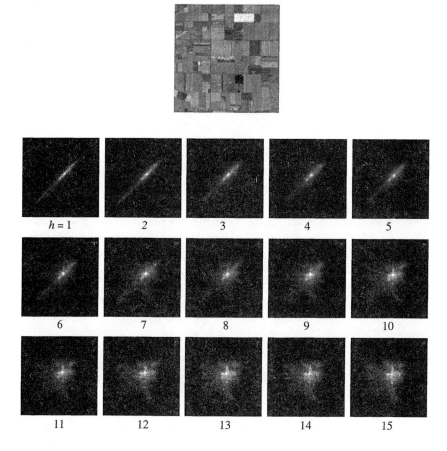

FIGURE 4-18. A TM band 4 image of agricultural fields (top) and a series of scattergrams between pairs of horizontally-separated pixels. The scattergrams estimate the joint probability of a pair of DN values in separated pixels within a row of data. For separations greater than about five pixels, the data appears to lose spatial correlation rapidly. Each scattergram is contrast stretched independently to emphasize the shape of the distributions; in reality, as the distributions spread outward, the peak value decreases since the number of pixel pairs available to calculate the scattergram decreases. These scattergrams are the density analogues to the "h-scatterplots" described in (Isaaks and Srivastava, 1989). (TM image courtesy of Larry Biehl, Purdue University, and Space Remote Sensing Center, Institute of Technology Development, NASA/Stennis Space Center)

TABLE 4-1. Some 1-D continuous models for the discrete spatial covariance and semivariogram (Isaaks and Srivastava, 1989; Pratt, 1991; Carr, 1995). Note each of the semivariogram models is normalized to a sill value of one.

spatial statistic	model	equation
covariance	exponential (Markov model)	$C(h) = C(0)e^{-\alpha h}$
semivariogram	Gaussian	$\gamma(h) = 1 - e^{-3(h/\alpha)^2}$ where $\alpha = \gamma(0.95 sill)$
	exponential	$\gamma(h) = 1 - e^{-3(h/\alpha)}$ where $\alpha = \gamma(0.95 sill)$
	spherical	$\gamma(h) = \begin{cases} 1.5h/\alpha - 0.5(h/\alpha)^3, & h \le \alpha \\ 1, & h > \alpha \end{cases}$ where $\alpha = \gamma(sill)$

calculated for each transect and averaged for each region. The result clearly shows distinguishing features among the three vegetation density categories (Fig. 4-21). For example, the low density vegetation area shows the largest sill since it has the greatest contrast in the image. It also has a greater range than that for the high density vegetation area, indicating that the average plant size is larger in the low density area. These empirical data are consistent with the disks-on-a-uniform-background model described and analyzed in (Woodcock *et al.*, 1988a).

To isolate size variations from contrast, the covariances are normalized at zero lag in Fig. 4-22. From these spatial correlation curves, we can estimate the correlation length, or range, for each of the three vegetation density categories as the lag distance over which the data are correlated. We will use the exponential model of Table 4-1 and fit it to the covariance functions of Fig. 4-22. Only the covariances for smaller values of lag will be used; the exponential model clearly does not describe the behavior at large lags. The fitted model curves are shown in Fig. 4-23 and are reasonably good approximations to the data in each case. The correlation length may be defined as the reciprocal of the parameter α in the model, corresponding to the lag at which the covariance function falls to $1/e$ of its value for zero lag (Table 4-3).

These values can be used to quantify and characterize the spatial patterns of the different areas in the image. The slope of the covariance function or

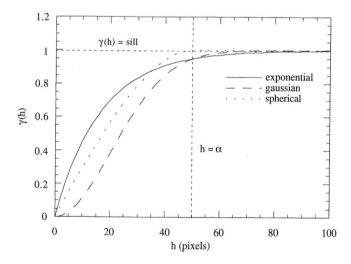

FIGURE 4-19. Graphs of the exponential covariance model (top) and three normalized
semivariogram models (bottom).

TABLE 4-2. Example applications of the semivariogram in remote sensing.

application	data source	reference
agricultural fields forestry	TM aerial photography TM airborne simulator	(Woodcock et al., 1988b)
landcover categorization	airborne MSS	(Curran, 1988)
estimate SNR	AVIRIS	(Curran and Dungan, 1989)
sea surface temperature estimate random pixel noise	AVHRR CZCS, AVHRR	(Wald, 1989)
measure misregistration between images	simulated MODIS	(Townshend et al., 1992)
sea surface temperature	AVHRR	(Gohin and Langlois, 1993)
landcover categorization	field spectroradiometer HCMM thermal SPOT TM	(Lacaze et al., 1994)
estimate landcover classes within shadows	TM	(Rossi et al., 1994)
kriging of stereo disparity maps	SPOT	(Djamdji and Bijaoui, 1995)

semivariogram near zero lag indicates the average plant size, because it measures how rapidly the pixels lose correlation as we move away from a given pixel. The slope for low density vegetation is less than that for high density vegetation because the plants are larger. The slope for the nonvegetated area is much lower than that for either of the areas with vegetation due to the lack of spatial structure.

Finally, note that the curves in Fig. 4-21 do not quite satisfy Eq. (4 – 41), because a finite (and fairly small) sample set was used. The relationship is satisfied, however, for all practical purposes.

Separability and anisotropy

Extension of the 1-D spatial statistics models of this section to 2-D images can be done in two ways. One is to simply replace the 1-D lag, h, by a 2-D radial distance, r. For example, the exponential covariance model (Table 4-1) becomes an isotropic function,

FIGURE 4-20. Scanned aerial image used to illustrate spatial statistics. Three groups of three transects each are extracted from regions of distinctly different vegetation density and plant size: no vegetation (top), low density (middle) and high density (bottom). Each transect is 100 pixels long.

$$C(r) = C(0)e^{-\alpha r} \qquad\qquad (4-42)$$

where

$$r = (h_x^2 + h_y^2)^{1/2} . \qquad\qquad (4-43)$$

An isotropic function is the same in all directions from a given origin. Generally, however, remote-sensing images are not isotropic, and therefore, neither are their spatial statistics. Alternatively, we can create a *separable* model by multiplication of two 1-D functions, in orthogonal directions,

$$C(x, y) = C(0)e^{-\alpha|x|}e^{-\beta|y|} = C(0)e^{-(\alpha|x| + \beta|y|)} . \qquad\qquad (4-44)$$

FIGURE 4-21. *Covariance functions and semivariograms for the transects of Fig. 4-20. The average function for the three transects in each category is plotted.*

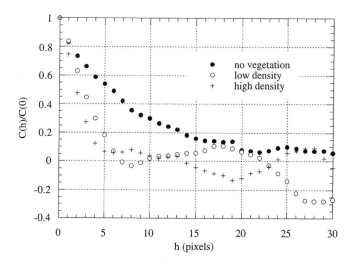

FIGURE 4-22. Normalized covariance functions. In effect, the data variance is set to one for each class.

A more realistic *nonseparable, anisotropic* model has been suggested in (Jain, 1989) for images and in (Isaaks and Srivastava, 1989) for geostatistical quantities,

$$C(h_x, h_y) = C(0)e^{-\sqrt{\alpha h_x^2 + \beta h_y^2}} \, . \qquad (4-45)$$

The isotropic, separable and anisotropic 2-D forms of the exponential covariance function model are shown in Fig. 4-24. Which is more realistic? The answer, of course, depends on the image. To model images *in general*, the isotropic form should be used, since there is no *a priori* reason to believe the image statistics are different in different directions.[5] The anisotropic exponential model is probably a good choice if one is modeling a *particular* image. The axes directions can be rotated if needed to fit the actual covariance. Combinations of these models to fit more complex data are discussed in (Isaaks and Srivastava, 1989).

---

[5] If the sensor has an asymmetry between the in-track and cross-track directions due to an asymmetric PSF (Chapter 3), a consistent asymmetry can be expected also in the image spatial statistics. A broader PSF profile in one direction will cause a broader image covariance function. Any asymmetry in *GSI* (for example in the MSS; see Fig. 1-10) will lead to a scaling of the covariance or semivariogram abscissa if plotted in pixel units. Such "asymmetry" should disappear, however, if the functions are plotted in absolute distance units.

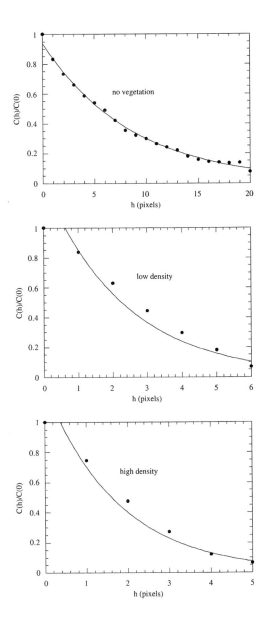

FIGURE 4-23. Exponential model fitted to covariance functions from Fig. 4-22. Note that the horizontal axis is scaled differently in each graph.

TABLE 4-3. Correlation lengths obtained from the exponential model fits in Fig. 4-23.

category	α (pixels^{-1})	correlation length (pixels)
no vegetation	0.114	8.78
low density	0.422	2.37
high density	0.560	1.79

4.6.3 Power Spectral Density

The Fourier transform of the covariance function is called the *Power Spectral Density (PSD)*, or *power spectrum*.[6] It characterizes the spatial frequency content in the covariance function. The width of the power spectrum is proportional to the inverse of the spatial correlation length. If the correlation length of the data is small, the covariance function is narrow, and the power spectrum will be wide. Conversely, data that is correlated over a substantial distance will have a narrow power spectrum, with little power at high spatial frequencies.

Using the data of the previous section, we can calculate the 1-D spatial power spectrum for each of the cover types in Fig. 4-20 as the Fourier transform of the respective covariance function in Fig. 4-22. Figure 4-25 demonstrates that the widths of the power spectra are inversely proportional to the widths of the covariance functions.

Some example image blocks and their power spectra are shown in Fig. 4-26. The image was partitioned into adjacent 128-by-128 pixel blocks, and the power spectrum calculated for each block as the square of the Fourier transform amplitude of the image block.[7] The power spectrum has directional information about linear features (edges, lines) in the image. The highest power components are orthogonal to the direction of such objects in the image.

The exponential Markov covariance model (Table 4-1) leads to a corresponding model for the power spectrum,

$$PSD(u) = \frac{2\alpha}{\alpha^2 + (2\pi u)^2} \qquad (4-46)$$

---

[6] If the reader is unfamiliar with Fourier transforms, it may be useful to first read the appropriate sections in Chapter 6 before reading this section.

[7] This alternate use of the term *power spectrum* is quite common in image processing. It is not a proper usage in a statistical context, but is so common, that we defer to the majority! The two quantities, the Fourier transform of the covariance function and the squared amplitude of the Fourier spectrum of a data block, are related and convey approximately the same information.

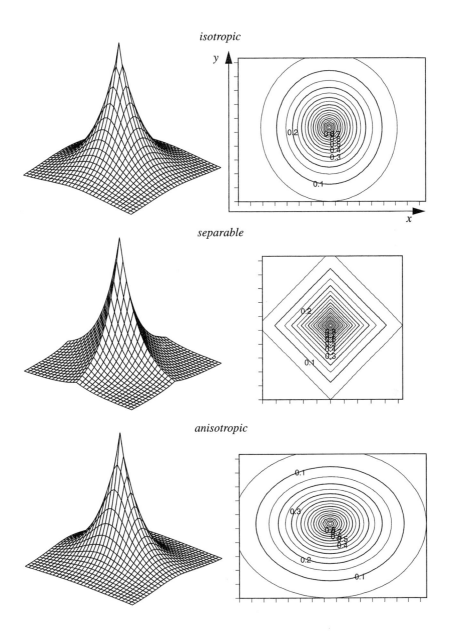

FIGURE 4-24. Three possible forms for 2-D exponential covariance models, in perspective and as contour maps. For the anisotropic model, $\alpha = \beta/4$.

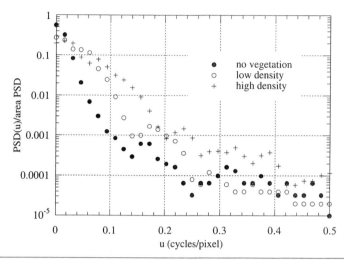

FIGURE 4-25. *Power spectra calculated from the data in Fig. 4-22. The normalization of the covariance functions corresponds to normalization of the areas of the power spectra. The latter therefore more clearly show the relative distribution of variance at each spatial frequency. Note the inverse ordering of the three curves compared to Fig. 4-22.*

where u is spatial frequency. This model is graphed in Fig. 4-27 for the three vegetation density categories. Just as the exponential model was reasonable for the covariance function, the model of Eq. (4 – 46) fits the PSD reasonably well.

Although we have gone into some detail to illustrate the modeling of image covariance and power spectral density functions, the reader is advised that most of the time *any given image* will not necessarily be modeled well by the functions we have used (note Fig. 4-26). They are best thought of as *average models*, that can be used as simple functional representations for analytic work.

4.6.4 Co-occurrence Matrix

The 2-D matrix of joint probabilities $p(DN_1, DN_2)$ between pairs of pixels, separated by a distance, h, in a given direction, is called the *Co-occurrence Matrix* (CM).[8] The scattergrams of Fig. 4-18 are examples. The CM contains a large amount of information about the local spatial information in an image; the major problem is how to extract that information. A set of texture features derived from

---

[8] The *CM* goes by several names in the literature, including *Grey-Tone Spatial-Dependence Matrix* (Haralick *et al.*, 1973), *Grey-Level Co-occurrence Matrix* (Gonzalez and Woods, 1992), *Grey Level Dependency Matrix* (Pratt, 1991), and *Concurrence Matrix* (Jain, 1989).

FIGURE 4-26. *Landsat TM band 4 image of sand dunes and irrigated agriculture near Yuma, Arizona, partitioned into 16 128-by-128 pixel blocks and the corresponding power spectra of each block. See Chapter 6 for a description of the spatial frequency coordinate system used for power spectra.*

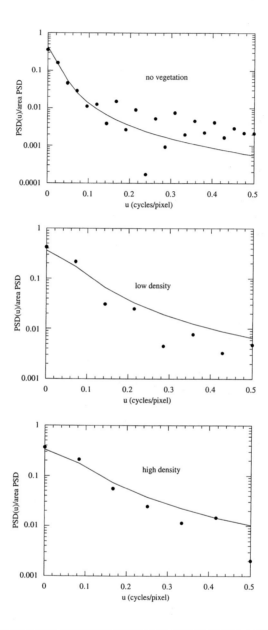

FIGURE 4-27. *Power spectra calculated by the Fourier transform of the covariance functions in Fig. 4-23, compared to the model of Eq. (4 – 46). The value of α used in each case is from Table 4-3.*

the *CM* was suggested in (Haralick *et al.*, 1973) (Table 4-4). These features measure various properties of the *CM*, but all are not independent of each other.

The *CM*, for a given lag h and direction, is 2^Q-by-2^Q, where Q is the number of bits/pixel. With a Q of 8 bits/pixel, the *CM* is therefore 256-by-256. For a complete description of the image, the *CM* should be calculated for all lags and in all directions (at least horizontal, vertical and diagonally). To reduce the computation burden, it is common practice to reduce the *DN* quantization (by averaging over adjacent *DNs*) to some smaller value, say 16 levels, and to average the *CMs* for different spatial directions.

To illustrate the calculation of texture features from the *CM*, three images with widely different spatial characteristics will be used (Fig. 4-28); one is the

TABLE 4-4. *Some of the spatial texture features derivable from the CM, as originally proposed in (Haralick et al., 1973). The quantities p_x and p_y are the marginal distributions of the CM, defined as $p_x = \sum_j p_{ij}$ and $p_y = \sum_i p_{ij}$.*

name	equation	related image property
Angular Second Moment (ASM)	$f_1 = \sum_i \sum_j p_{ij}^2$	homogeneity
Contrast	$f_2 = \sum_i \sum_j (i-j)^2 p_{ij}$	semivariogram
Correlation	$f_3 = \dfrac{\sum_i \sum_j ij p_{ij} - \mu_1 \mu_2}{\sigma_1 \sigma_2}$	covariance
Sum of Squares	$f_4 = \sum_i \sum_j (i-\mu)^2 p_{ij}$	variance
Inverse Difference Moment	$f_5 = \sum_i \sum_j \dfrac{p_{ij}}{1+(i-j)^2}$	—
Sum Average	$f_6 = \sum_{i=2}^{2^{Q+1}} i p_{x+y}$	—
Entropy	$f_9 = -\sum_i \sum_j p_{ij} \log(p_{ij})$	—

agriculture image used for Fig. 4-18. The entropy and contrast features were calculated for the original *DN* images, with requantization of the *DN* scale to 32 levels between the minimum and maximum *DN* of each image. The spatial parameter was varied in the horizontal direction from a shift of one pixel to 15 pixels.

In Fig. 4-29, the behavior of the *CM* contrast and entropy features with lag, *h*, are shown. There appears to be distinct differences among the three images, but the fact that their *DN* histogram characteristics are quite different confuses the comparison. To normalize this difference, a histogram equalization transform has been recommended (Haralick *et al.*, 1973); the resulting images are shown in Fig. 4-28 and the new contrast and entropy features in Fig. 4-30. Their shapes have not changed much, but they are now on a similar scale, at large *h*, for each image.

FIGURE 4-28. Three sample TM images used for co-occurrence matrix analysis. The histogram equalization process produces an image with an approximately equal pixel count-per-DN density in all parts of the DN range. It is described in Chapter 6.

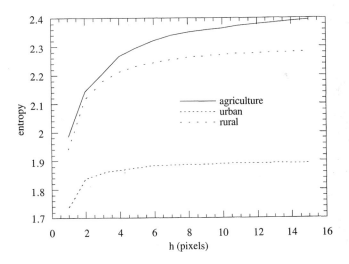

FIGURE 4-29. The contrast and entropy spatial features from the sample images of Fig. 4-28. Note the similarities to the semivariogram functions in Fig. 4-21 (which were calculated from an aerial image), namely the relatively rapid change as h increases from zero and the asymptotic behavior for large h.

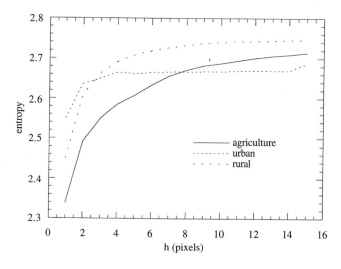

FIGURE 4-30. Contrast and entropy CM features after equalization of the image histograms. The shapes of the features as a function of h have not changed much from that in Fig. 4-29, but both features are now normalized to approximately the same value at large h for the three images.

4.6.5 Fractal Geometry

Fractal geometry is a way to describe the "texture" of a surface.[9] There are four topological dimensions in traditional Euclidean geometry: 0-D for points, 1-D for straight lines, 2-D for planes, and 3-D for volumetric objects like cubes and spheres. An object that is "fractal" has an intermediate dimensionality, such as 1.6 for an irregular line or 2.4 for an image "surface". Generally, the higher the fractal dimension, the finer and "rougher" the texture.

Objects that are "fractal" also obey the notion that they are "self-similar" in their statistics. This means that the object is statistically the same, no matter what scale it is viewed at. A classic example is measurement of the length of the perimeter of the coastline of Britain (Mandelbrot, 1967). If the perimeter is measured with rulers of different lengths (scales), different values are obtained. Shorter rulers lead to longer values, and vice versa. The logarithm of the measured perimeter is found to obey a linear relationship to the logarithm of the ruler length (Feder, 1988); the slope of that log-log relation is related to the fractal dimension. Most natural objects and surfaces are found to be approximately fractal over a limited range of scales.

Several tools have been developed for estimating the fractal dimension, D, from an image. One approach uses 1-D data transects. The fractal dimension is related to either the semivariogram or power spectrum of the transect (Carr, 1995). For either the semivariogram or the power spectrum, a log-log plot is calculated, and a linear function is fit to the log-transformed data (equivalent to fitting a power function to the data before taking the logarithms). The fractal dimension, D, is then related to the slope of the fitted line, as given in Table 4-5. If one wishes to use data in multiple directions, but anisotropy is *not* of interest, then the 2-D power spectrum can be calculated over the image area of interest, averaged azimuthally to yield an average radial power spectrum, and the fractal dimension calculated in the 1-D manner. This technique was used in (Pentland, 1984) to produce fractal maps from images. The fractal dimension was measured over 8-by-8 pixel blocks across the image, and the result used to segment the image into regions of different texture.

An interesting technique for estimating the fractal dimension of imagery was tested by (Pelig *et al.*, 1984) and described in detail in (Peli, 1990). A "covering blanket" is calculated above and below the image surface, for different resolution scales. One visualizes the blanket as being draped over (and under) the image, so that it bounds the *DN* surface. As the resolution is reduced, the two "blankets" become smoother. The change in volume between the blankets at two successive

---

[9] We do not have space here to discuss the rich mathematical background of fractal geometry. Suffice it to say that there are many and varied examples in nature; the reader is recommended to (Mandelbrot, 1983) for the classic description and to (Feder, 1988) for a particularly readable discussion of applications.

scales is plotted versus the scale index in a log-log plot, and the slope used to estimate the fractal dimension as previously described. Although the processing is 2-D, the end result is an isotropic measure, since only the volume information is used.

TABLE 4-5. *Some ways to estimate fractal dimensions from images. The slope of the log-log straight line fit is denoted s.*

measurement	fractal dimension (D)	reference
semivariogram	$2 - s/2$	(Carr, 1995)
power spectrum	$2.5 - s/2$	(Carr, 1995), (Pentland, 1984)
covering blanket volume	$2 + s$	(Pelig *et al.*, 1984)

An example of fractal dimension calculation from imagery is shown in Fig. 4-31. The TM image has two characteristic texture regions, the desert sand dunes in the upper portion and the agricultural area in the lower portion. The technique of Pentland was used, with 32-by-32 pixel blocks and the DC (zero) frequency component excluded from the power spectral analysis. The resulting fractal dimension map appears to correlate with the image texture regions. Now, if we look at the histogram of the original image, we do not see an obvious point at which a *DN* threshold might separate the two regions (Fig. 4-32). If we look at the fractal dimension histogram, however, a valley appears to signify two classes. In fact, if we threshold the fractal dimension image at the valley, the resulting map appears to reasonably map the two regions (Fig. 4-33).

4.7   Topographic and Sensor Effects

The most appropriate data models and associated parameters for given imagery are, in part, determined by terrain and sensor effects. We have seen in Chapter 3 how topography and sensor spectral and spatial resolution affect the reproduction of scene information in the recorded imagery. In this section, we will visualize some of these effects, with the aid of simulation and the spatial and spectral models discussed in this chapter.

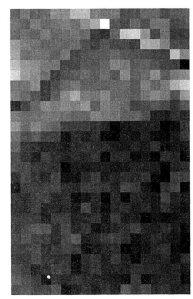

TM band 4   *fractal dimension map*

FIGURE 4-31. A TM band 4 image (the same area as Fig. 4-26) and the 32-by-32 pixel block fractal dimension map. (Fractal map calculation courtesy of Per Lysne, University of Arizona.)

4.7.1 Topography and Spectral Scattergrams

Assuming the surface has a Lambertian reflectance characteristic, its radiance is proportional to the cosine of the angle between the solar vector and the surface normal vector (Eq. (2 – 6)). Suppose we have such a scene consisting of two materials, soil and vegetation, each having a certain within-class variance but no band-to-band correlation. We can generate reflectance images in the red and NIR spectral regions from random Gaussian distributions for each class in each band. A real topographic DEM will be used (Fig. 4-34), and we want to distribute the soil and vegetation pixels equally across the scene, assuming 50% coverage by each class.

To make the spatial characteristics more realistic, a soil mask and a complementary vegetation mask (Fig. 4-35) are created.[10] Each mask is filled with

---

[10] The simulated spatial distribution of soil and vegetation approximates the actual landcover distribution for this DEM, which is an area of the Oakland, California, hills. They are uniformly covered by soil and grass, with patches of woodlands.

FIGURE 4-32. The original image and fractal dimension map histograms. The arrow in the lower graph is the location of the threshold for Fig. 4-33.

pixels from the respective class, and the result multiplied by the topographic shaded relief image (Fig. 4-36), as indicated mathematically by Eq. (2 − 7). Finally, the two sets of data are composited into a two band, simulated scene (Fig. 4-37).

Looking at the NIR-red scattergram of the simulated images (Fig. 4-38), we see that the originally uncorrelated distributions are now strongly correlated thresholded image   thresholded fractal map

FIGURE 4-33. Threshold maps of the original image (DN = 170), and the fractal dimension map (D = 2.2). Note, the two regions of the original image appear inseparable with simply a DN threshold, while they are separable in the fractal dimension map. The average fractal dimension of the sand area is 2.45 and that of the agriculture area is 1.88, which, being less than two, actually implies a nonfractal structure (Pentland, 1984). Pentland found that contrast boundaries generally were nonfractal; the irrigation canal in the upper half of the image and the dark-light field boundaries in the lower half are in that category.

along lines passing through the scattergram origin, or zero reflectance. This happens because the incident radiation is reduced by a cosine factor, scaled in this case to the range [0,1] by saturating negative values at zero. This *topographic modulation* has long been recognized as an influence on spectral signatures in remote sensing data, and detailed modeling of its characteristics has been done (Proy *et al.*, 1989). Although most work has been directed at reducing topographic effects (Table 4-6), topography can be extracted from multispectral imagery using models (Eliason *et al.*, 1981), and physical models for topographic modulation can be used to decompose imaging spectrometer data into its spectral constituents (Gaddis *et al.*, 1996).

FIGURE 4-34. The shaded relief image generated from a DEM, part of the same area shown in Fig. 2-10. The algorithm to generate the relief image is simply a cosine calculation as in Eq. (2 − 6).

soil   vegetation

FIGURE 4-35. The complementary soil and vegetation spatial masks. The process used to generate these patterns consisted of creating a random, Gaussian noise pattern with zero mean, smoothing that by convolution with a Gaussian low-pass filter (see Chapter 6), and thresholding at a DN of zero. The convolution step introduces spatial correlation in the noise, leading to the patterns above after thresholding.

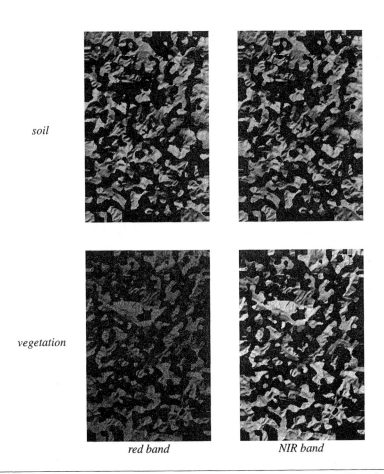

soil

vegetation

red band NIR band

FIGURE 4-36. The simulated red and NIR images for soil and vegetation. The relative
brightness of each class and band is preserved in this figure.

4.7.2 Sensor Characteristics and Spatial Statistics

The sensor introduces many, but two particularly notable, effects into the imagery: noise and blurring. In this section, we want to see how these two factors affect the spatial statistics derived from the imagery. We will first look at image noise. The aerial photograph of Fig. 4-20 was modified by the addition of normally-distributed, spatially-independent noise at each pixel, with a *DN* standard deviation of 20. The noise level is set rather high to illustrate its effect.

red band   *NIR band*

FIGURE 4-37. The final simulated images composited from the images in Fig. 4-36.

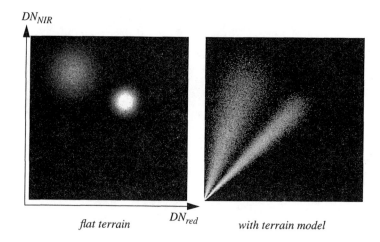

flat terrain DN_{red} *with terrain model*

FIGURE 4-38. NIR versus red scattergrams for the original two class soil and vegetation data, before and after merger with the terrain model. The signature of each class is now correlated between the two bands and their distributions extend through the zero reflectance origin of the scattergram.

TABLE 4-6. Examples of research on measurement and correction of topographic effects in analysis of remote sensing imagery.

data	reference
field measurements	(Holben and Justice, 1980)
MSS	(Holben and Justice, 1981) (Eliason *et al.*, 1981) (Justice *et al.*, 1981) (Teillet *et al.*, 1982) (Kawata *et al.*, 1988)
TM	(Civco, 1989) (Conese *et al.*, 1993) (Itten and Meyer, 1993)

noisy sensor

larger GIFOV

FIGURE 4-39. The two simulated images used to investigate the influence of the sensor on spatial statistics measured from its imagery. On the left, normally-distributed, uncorrelated random noise with a DN standard deviation of 20 was added to the original image of Fig. 4-20. This corresponds to the noise model of Eq. (4 – 20) and a C_{std} of about 2 by Eq. (4 – 26). On the right, the original noiseless image was convolved with a 5-by-5 uniform spatial response function, simulating a GIFOV of that size. The same transects of Fig. 4-20 were extracted for the analysis.

The spatial covariances for the transects in Fig. 4-20 are shown in Fig. 4-40. The uncorrelated noise is seen in the spike at zero lag, which in each case is about 400 units higher than the original covariance, i.e., an amount equal to the added noise variance. At nonzero lags, the image noise introduces some covariance noise but no overall shift because the added noise is uncorrelated from pixel-to-pixel. The semivariograms shown in Fig. 4-41 contain the same information in a different form; the sills are about 400 *DN* units higher at all lags, but still zero at zero lag, by definition. The nugget value, obtained by extrapolating the lowest, nonzero values, to a lag of zero, now equals about 400. The fact that the semivariogram nugget value represents uncorrelated noise can be used to estimate image noise (Curran and Dungan, 1989).

To see the effect of the sensor *GIFOV* on spatial statistics, we return to the original aerial photograph, without added noise, and introduce a uniform, square *GIFOV* of 5-by-5 pixels by convolution (Fig. 4-39). The scene details are blurred, and we see a large reduction in covariance at all lags (Fig. 4-42). The shapes of the covariance functions, as seen earlier in (Fig. 4-22), are more easily compared by normalization to one for zero lag (Fig. 4-43). Now, the *GIFOV* is seen to make the covariance function broader because the spatial averaging within the *GIFOV* adds correlation to neighboring points. Also, the *GIFOV* has changed the shape of the covariance for small lags; it now is not as sharp and approaches zero slope at zero lag. This implies that composite data models may be a better fit to image data than any one of the three discussed earlier. The semivariograms exhibit a decrease in the sills and an increase in range, corresponding to the increase in width of the covariance functions (Fig. 4-44).

An experimental investigation of semivariogram dependence on *GIFOV*, in which an airborne sensor was flown at two different altitudes over the same test site, is reported in (Atkinson, 1993). The semivariogram data presented there exhibit the same behavior as in our simulation, even to the point of suggesting a composite model for the semivariogram of the lower resolution imagery.

4.7.3 Sensor Characteristics and Spectral Scattergrams

Having seen how the sensor affects spatial statistics, we now turn to its effect on spectral statistics. In this case, we create a two-band, three-class scene with a computer paint program. The scene may be thought of as showing asphalt roads traversing a landscape of soil and vegetation in the form of grass, shrubs and trees. The radiance values in the simulated scene approximate the relative values that might be recorded in red and near IR spectral bands (top row, Fig. 4-45).

Now, we want to add some spatial "texture" to each class. First, spatially- and spectrally-uncorrelated random variation is added, with a standard deviation of five in each band for the asphalt and soil classes, and ten for the vegetation class (middle row, Fig. 4-45). Then, spatial correlation over a distance of four scene points is introduced within each class (bottom row, Fig. 4-45).[11] The scatter-

---

[11] The correlation is achieved by convolving the added noise with an exponential function, $e^{-r/2}$. This is not quite the same as the exponential autocovariance model discussed earlier, but is satisfactory for the present demonstration.

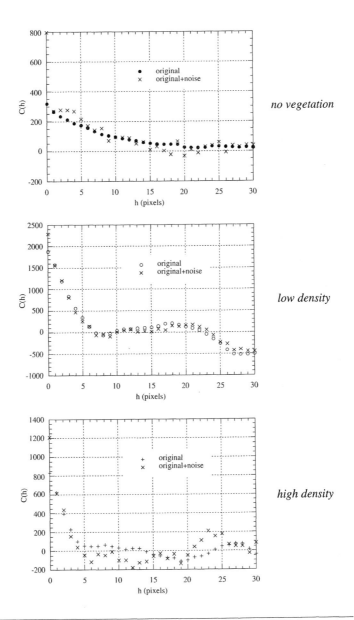

FIGURE 4-40. *Covariance functions for the original image and with added spatially-uncorrelated noise. Note that the vertical scale is different for each category (compare to Fig. 4-21).*

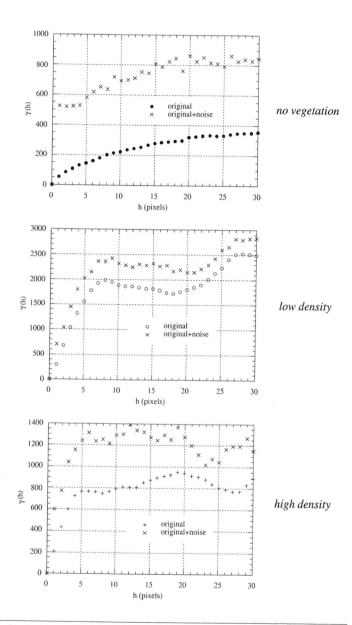

FIGURE 4-41. Semivariogram functions for the original image with added spatially-uncorrelated noise. Note that the vertical scale is different for each category (compare to Fig. 4-21).

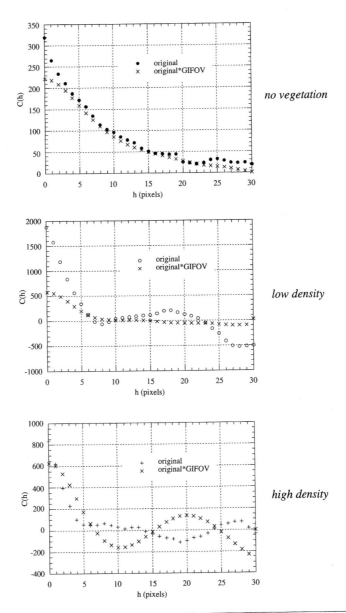

FIGURE 4-42. *Covariance functions for the original image with a simulated GIFOV of 5-by-5 pixels. Note that the vertical scale is different for each category.*

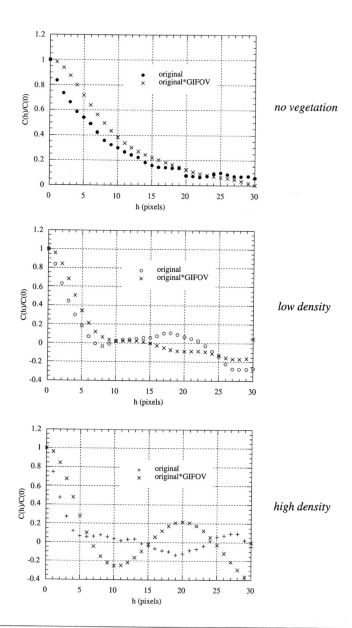

no vegetation

low density

high density

FIGURE 4-43. Normalized covariance functions for the original image with a simulated GIFOV of 5-by-5 pixels. Note that the vertical scale is different for each category.

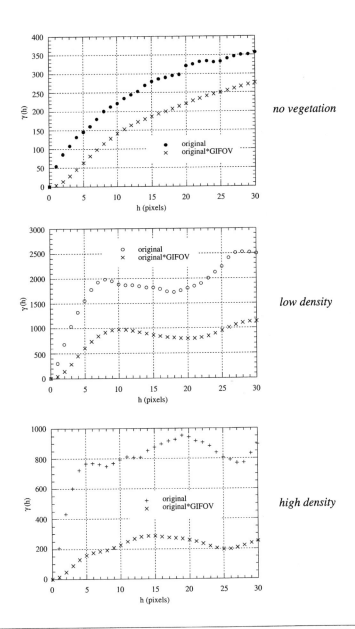

FIGURE 4-44. Semivarigram functions for the original image with a simulated GIFOV of 5-by-5 pixels. Note that the vertical scale is different for each category (compare to Fig. 4-21).

grams for these cases are shown in Fig. 4-46. Note that the introduction of spatial correlation, while clearly changing the visual appearance of the scene, is not particularly evident in the spectral domain.

Finally, spatial averaging by the sensor (Chapter 3) is simulated by convolution with a Gaussian system spatial response having a diameter of five scene elements, (Fig. 4-47). The effect on the data in the spectral domain, shown in Fig. 4-48, is interesting and illustrates two significant points:

- The sensor's spatial response reduces the texture variance in the spatially-uncorrelated case by averaging neighboring scene points which are independent of each other. In the spatially-correlated case, the reduction of within-class variance is much less because the scene data are already correlated over several points before the sensor integration occurs.

- For both types of scene texture, the sensor's spatial response *mixes* the spectral signatures wherever two classes are spatially adjacent. Thus, the scattergram, which originally consisted of only three points, begins to fill in with mixed signatures. We will discuss this phenomenon in terms of image classification in Chapter 9.

In this simulation, we have not included the 3-D nature of the scene, specifically shading and shadowing. If topographic variation were also simulated, the net scattergrams would exhibit characteristics of both Fig. 4-38 and Fig. 4-48. Even without topography, the final images, particularly the one with spatially-correlated scene texture, are visually quite believable. That is not to say, however, that they necessarily represent accurate models of the real world! More detailed and scene-specific modeling is presented in the literature (Woodcock *et al.*, 1988a; Woodcock *et al.*, 1988b; Jupp *et al.*, 1989a; Jupp *et al.*, 1989b). A detailed analysis of information about semivegetated landscapes contained in the red-NIR reflectance scattergram is given in (Jasinski and Eagleson, 1989; Jasinski and Eagleson, 1990).

red band NIR band

class maps, no texture

uncorrelated texture

spatially-correlated texture

FIGURE 4-45. *Creation of a simulated two-band scene with three classes. Spatial correlation was introduced in the texture patterns by convolution with a LPF, separately for each class. The results were then combined so there was no blurring introduced across class boundaries.*

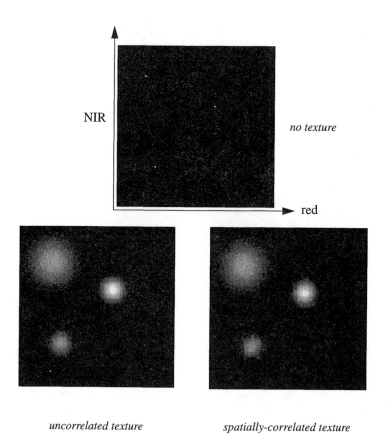

FIGURE 4-46. Near IR versus red band scattergrams for the original scene, the scene with spatially-uncorrelated texture and the scene with spatially-correlated texture. The class variances were renormalized to five and ten after the calculation of the correlated texture case. In each scattergram, the vegetation signature is in the upper left, the soil signature on the right, and the asphalt signature at the bottom left. In these, and the later scattergram figures, the contrast has been greatly increased to make relatively less-populated scattergram bins visible.

red band   *NIR band*

uncorrelated texture

spatially-correlated texture

FIGURE 4-47. The simulated images resulting from spatial averaging of the two textured-scene cases in Fig. 4-45. The spatial averaging used a 5-by-5 pixel Gaussian weighting function to simulate a sensor PSF.

4.8 Summary

Some of the common models used in image data analysis have been described and illustrated with examples. The connection between radiation and sensor models and data models was also discussed and explored by simulation. The important points of this chapter are:

- The spectral statistics of remote-sensing image data are influenced by the topography in the scene. The topographic effect tends to correlate the data between spectral bands along a straight line through the origin of the reflectance scattergram.

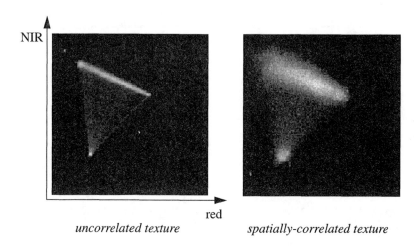

NIR

red

uncorrelated texture   *spatially-correlated texture*

FIGURE 4-48. The effect of the sensor's spatial integration on the scattergrams of Fig. 4-46. On the left, the sensor has reduced the spectral variance of each class for spatially-uncorrelated scene texture. On the right, the spectral variance is much less reduced for spatially-correlated scene texture. In both cases, new spectral vectors have been created by the mixing of the three classes within the spatial response function of the sensor. The new vectors are interior to the triangle defined by the original three signatures (Adams et al., 1993); these mixed pixels are significant in thematic classification, as discussed in Chapter 9.

- The spectral statistics of remote-sensing image data are also influenced by the sensor's spectral passband locations and widths, and noise characteristics. Random, spatially-uncorrelated sensor noise increases the within-class variance of all surface materials equally.

- The spatial *and* spectral statistics of remote-sensing image data are influenced by the sensor's spatial response function, which increases the spatial correlation length, reduces within-class variance, and creates mixed spectral vectors.

With the background provided by Chapters 2, 3 and 4, we are now ready to look at image processing techniques and algorithms in the physical context of remote sensing. Chapters 5 and 6 present tools for spectral and spatial transforms that are useful in the correction and calibration of images for atmospheric and sensor effects, described in Chapter 7, and in the fusion of images, described in Chapter 8. Finally, many of the data models discussed in this chapter will prove useful in our discussion of thematic classification in Chapter 9.

4.9   Exercises

4-1. If a multispectral image dataset is uncorrelated among spectral bands, are the data necessarily statistically-independent? What if the data are normally distributed in each band?

4-2. Show mathematically that the equal probability contours of a 2-D Gaussian distribution are ellipses.

4-3. Verify the modulation and phase values for the function in Fig. 4-15.

4-4. For 100 data points in an image transect, how many pairs of points can be used to calculate the semivariogram for h equal to 2? How many pairs for h equal to 50?

4-5. *SNR*, as defined by Eq. (4 – 32), Eq. (4 – 34) or Eq. (4 – 36), is a useful metric for evaluating image quality in the presence of noise that is distributed more or less uniformly across an image. These definitions are not well-suited to isolated noise, such as dropped pixels or lines, and detector striping, however. Suggest *SNR* measures that would be meaningful in these three cases and describe how they could be measured.

4-6. Cover the TM image in Fig. 4-26 with a sheet of paper and examine the power spectra to identify the image blocks that have

   • few or no linear features
   • a relatively small correlation length

   Confirm your selections for both cases by comparison to the image.

4-7. What value of horizontal spatial lag, h, provides the best discrimination among the three types of images, based on the contrast and entropy *CM* features of Fig. 4-30?

4-8. Sketch the scattergrams of Sect. 4.7.1 if the incident cosine factor is in the range [0.2,0.5].

CHAPTER 5

Spectral Transforms

5.1   Introduction

Image processing and classification algorithms may be categorized according to the *space* in which they operate. The multispectral *image space* is $DN(x,y)$, where the spatial information is implicit. In Chapter 4, we developed the concept of a multidimensional *spectral space*, defined by the multispectral vectors, where spatial information is ignored. Spectral transformations, discussed in this chapter, alter the spectral space; and spatial transformations, discussed in the next chapter, alter the image space. Many of these transformed spaces are useful for thematic classification (Chapter 9), and are collectively called *feature spaces* in that context. In this chapter, we discuss various feature spaces that can be derived from the spectral space. These derived spaces do not represent the addition of any new information to the image, but rather a redistribution of the original information into a more useful form. Here we will discuss various linear and nonlinear transformations of the *DN* vector, motivated by the possibility of finding a new coordinate system that may have advantages over the original spectral bands.

5.2   Feature Space

A general linear matrix transformation of the spectral vector,

$$DN' = W \cdot DN$$

$$= \begin{bmatrix} w_{11} & \cdots & w_{1K} \\ \cdots & \cdots & \cdots \\ w_{K1} & \cdots & w_{KK} \end{bmatrix} \begin{bmatrix} DN_1 \\ \cdots \\ DN_K \end{bmatrix} \qquad (5-1)$$

where W is a transformation matrix, consisting of weights on each of the original spectral bands, that corresponds to a coordinate rotation and scaling of the original DN spectral space.[1] Examples of W are discussed in the following sections.

A general, nonlinear transformation of the data vector is denoted,

$$DN' = f(DN) . \qquad (5-2)$$

Useful transformations can be interpreted in terms of the physical aspects of remote sensing imaging, which we will attempt to do.

5.3   Multispectral Ratios

One of the earliest feature extraction techniques applied to remote sensing images was calculation of the *ratio* of DNs in spectral band m to those in band n, pixel-by-pixel,

$$R_{mn}(x, y) = \frac{DN_m(x, y)}{DN_n(x, y)} . \qquad (5-3)$$

This is a nonlinear transformation of the type in Eq. $(5-2)$. Now, based on the analyses of Chapters 2 and 3 (in particular, Eq. $(2-10)$ and Eq. $(3-20)$), the DN in band b is approximately a linear function of earth surface reflectance,[2]

$$DN_b(x, y) \approx a_b \rho_b(x, y) \cos[\theta(x, y)] + b_b . \qquad (5-4)$$

The ratio of Eq. $(5-3)$ is then,

---

[1]   In some literature, for example (Fukunaga, 1990), the linear transformation is defined $DN' = U^T \cdot DN$, where $U = W^T$. This is completely equivalent to our definition, since $(W^T)^T = W$.

[2]   Only the solar reflective radiation component is considered here.

$$R_{mn}(x, y) \approx \frac{a_m \rho_m(x, y) \cos[\theta(x, y)] + b_m}{a_n \rho_n(x, y) \cos[\theta(x, y)] + b_n} \, . \qquad (5-5)$$

If the *DN* bias, b_b, can be estimated and subtracted from the data in each band, for example with an atmospheric haze correction technique (Chapter 7), then,

$$bias\text{-}corrected: \quad R_{mn}(x, y) \approx \frac{a_m \rho_m(x, y)}{a_n \rho_n(x, y)} = k_{mn} \frac{\rho_m(x, y)}{\rho_n(x, y)} \, . \qquad (5-6)$$

The incident topographic irradiance factor does not appear in Eq. (5 – 6); in fact, this is one of the primary benefits of band ratios, namely that they can suppress topographic shading if a bias correction is done first. The ratio of Eq. (5 – 6) is directly proportional to the ratio of reflectances in the two bands, since k_{mn} is a constant for any two bands, and is more representative of surface properties than Eq. (5 – 3). The dynamic range of the ratio image is normally much less than that of the original image because the radiance extremes caused by topography have been removed. Thus, the *reflectance contrast* between surface materials can be enhanced by color composites of different band ratios (Chavez *et al.*, 1982).

Alternatively, the coefficients a_b and b_b can be fully calibrated, as described in Chapter 7, for solar irradiance, atmospheric transmittance and scatter, and sensor gain and offset. We then obtain the ratio directly in terms of reflectances,

$$fully\text{-}calibrated: \quad R_{mn}(x, y) = \frac{[DN_m(x, y) - b_m]/a_m}{[DN_n(x, y) - b_n]/a_n} = \frac{\rho_m(x, y)}{\rho_n(x, y)} \, . \qquad (5-7)$$

Correction for spatial irradiance variation due to topography is not necessary because the cosine term cancels. The various vegetation indexes used to estimate biomass and vegetative land cover are defined in terms of surface reflectances, although they are also widely used on uncalibrated *DN* values.

Multispectral ratios of NIR to visible bands can enhance radiance differences between soils and vegetation. Soil and geology will exhibit similar ratio values near one, while vegetation will show a relatively larger ratio of two or more. A TM image of an agricultural area (Plate 9-1) provides an illustration of this mapping in Fig. 5-1. A ratio value of about one corresponds to the concentration in the data arising from the large amount of bare soil in the image. In fact, a *soil line* has been defined in the NIR-red spectral space (Richardson and Wiegand, 1977). Spectral vectors along this line represent the continuum of dark to light soils in this image. As the ratio increases, the isolines rotate counterclockwise towards the vegetation region of the scattergram. Ratio values around three in this image represent the active crop fields, and values between one and three represent areas of partial vegetative cover, such as emergent crops and the desert area in the upper right of the image.

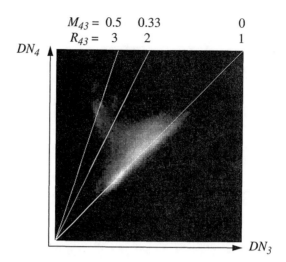

FIGURE 5-1. The DN scattergram for band 4 versus band 3 of a TM image of an agricultural scene, with superimposed ratio isolines. Values of the modulation and simple ratios (based on DNs) are indicated along the top. The data are uncalibrated.

The *modulation ratio* is a useful nonlinear variant of the simple ratio,

$$M_{mn} = \frac{\rho_m - \rho_n}{\rho_m + \rho_n} = \frac{R_{mn} - 1}{R_{mn} + 1}. \tag{5-8}$$

The relation between the two is shown in Fig. 5-1 and Fig. 5-2. Any gain factor that is common to both bands is removed by the modulation ratio, as it is by the simple ratio.[3]

5.3.1 Vegetation Indexes

As shown in the previous section, the ratio of a NIR band to a red band can indicate vegetation. We'll call that specific ratio, the *Ratio Vegetation Index (RVI)*,

$$RVI = \frac{\rho_{NIR}}{\rho_{red}}. \tag{5-9}$$

---

[3] We're assuming that the *DN* data is fully calibrated to surface reflectance, and that *R* is given by Eq. (5 – 7).

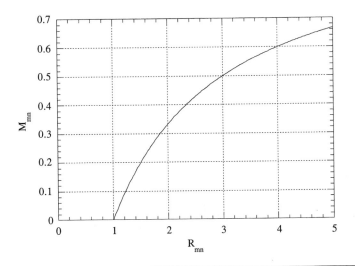

FIGURE 5-2. *Plot of the modulation ratio as a function of the simple ratio. If we think of this as a contrast enhancement transform of a simple ratio image, the modulation ratio provides increased contrast of lower ratio values relative to higher values.*

Similarly, the modulation ratio for NIR and red bands is called the *Normalized Difference Vegetation Index (NDVI)* ,[4]

$$NDVI = \frac{\rho_{NIR} - \rho_{red}}{\rho_{NIR} + \rho_{red}} . \qquad (5-10)$$

The *NDVI* is just a transformed *RVI*, as was shown previously. It is used extensively to monitor vegetation on continental and global scales using AVHRR data (Townshend and Justice, 1986; Tucker and Sellers, 1986; Justice *et al.*, 1991), but appears to be a poor indicator of vegetation biomass if the ground cover is low, as in arid and semi-arid regions (Huete and Jackson, 1987).

The *Soil-Adjusted Vegetation Index (SAVI)* is a superior vegetation index for low cover environments (Huete, 1988),

$$SAVI = \left(\frac{\rho_{NIR} - \rho_{red}}{\rho_{NIR} + \rho_{red} + L} \right)(1 + L) \qquad (5-11)$$

---

[4] Note the bias correction discussed earlier is less important here, since the atmospheric path radiance is low in the red and NIR bands. It should not be ignored in precise work, however.

where L is a constant that is empirically determined to minimize the vegetation index sensitivity to soil background reflectance variation. If L is zero, $SAVI$ is the same as $NDVI$. For intermediate vegetation cover ranges, L is typically around 0.5. The factor $(1 + L)$ insures the range of $SAVI$ is the same as $NDVI$, namely $[-1,+1]$.

A *transformed vegetation index (TVI)* has been applied to biomass estimation for rangelands (Richardson and Wiegand, 1977). Two indices derived from Landsat MSS data are given by,

$$MSS: \quad \begin{aligned} TVI_1 &= \left(\frac{DN_4 - DN_2}{DN_4 + DN_2} + 0.5 \right)^{1/2} = \left(\frac{R_{42} - 1}{R_{42} + 1} + 0.5 \right)^{1/2} \\ TVI_2 &= \left(\frac{DN_3 - DN_2}{DN_3 + DN_2} + 0.5 \right)^{1/2} = \left(\frac{R_{32} - 1}{R_{32} + 1} + 0.5 \right)^{1/2} \end{aligned} \quad (5-12)$$

where the 0.5 bias term automatically prevents negative values under the square root for most images. As seen from Eq. $(5 - 12)$, the TVI is a function of RVI, and consequently contains no additional information. The TVI shows an advantage over the $NDVI$ in some situations, in that it tends to be more linearly related to biomass, thus simplifying regression calculations.

A *Perpendicular Vegetation Index (PVI)*, orthogonal to the soil line in two dimensions, was also defined in (Richardson and Wiegand, 1977),

$$PVI = \left[(\rho_{red}^{soil} - \rho_{red}^{veg})^2 + (\rho_{NIR}^{soil} - \rho_{NIR}^{veg})^2 \right]^{1/2} . \quad (5-13)$$

The PVI is interpreted geometrically as the perpendicular distance of a pixel vector from the soil line in a NIR versus red reflectance space (Fig. 5-3). Since the PVI is defined in terms of reflectance, without scaling constants, it can be applied to calibrated data from any sensor.

A general approach to defining vegetation indexes in K dimensions has been described by (Jackson, 1983). The methodology provides a unified description of the various 2-D and K-D indexes, such as the *Tasseled Cap Transform* described later.

5.3.2 Image Examples

One of the benefits of the spectral ratio transform is suppression of the topographic shading factor. The sensor data must first be calibrated, however, for atmospheric bias (haze). The importance of this is illustrated in Fig. 5-4. The original data were calibrated to at-sensor radiance using the sensor gain and offset coefficients, and then to surface radiance by subtracting a dark object value from each band; the process is described in Chapter 7. Notice how the

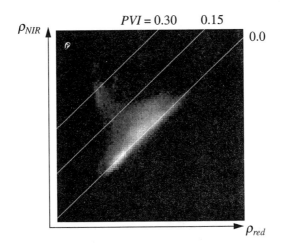

FIGURE 5-3. *Isolines for the PVI. The data are the same as that in Fig. 5-1 and are uncalibrated DNs. Because the PVI is defined in terms of a physical scene property, namely the soil line, it is only strictly valid in reflectance space. A reflectance range from zero to 0.5 is arbitrarily assumed here for illustration. The line where the PVI is zero corresponds to the soil line for this image.*

topographic shading remains in the *DN* ratio of band 2/1 and band 3/2, while it is not evident in the surface radiance ratios.

The Cuprite area is highly mineralized, with little vegetative cover. The enhancement of different mineral signatures is particularly apparent in the 5/4 and 7/5 ratios. The dark areas in the latter ratio image indicate clay-rich minerals in general, which absorb strongly in band 7 and have high reflectance in band 5 (Avery and Berlin, 1992); in this case they correspond to the clay mineral, kaolinite, as indicated by analysis of imaging spectrometer data of Cuprite (Kruse *et al.*, 1990). Another property of band ratios can be seen in these data. When the reflectances in the two bands are similar (as for the TM 2/1 and 3/2 ratios of geologic materials), and particularly for low reflectances, any image noise is amplified by the ratio calculation.

The *RVI* and *NDVI* vegetation indexes are demonstrated with an agriculture scene (Fig. 5-5). Both indexes clearly indicate the presence of vegetation and appear correlated with the amount of vegetation in particular fields (compare to Plate 5-1). The degree and nature of those correlations cannot be determined without careful ground measurements, however.

DN band ratios

2/1

3/2

surface radiance band ratios

2/1

3/2

4/3

5/4

7/5

FIGURE 5-4. *Spectral band ratios for a Landsat TM image of Cuprite, Nevada, acquired on October 4, 1984.*

RVI   *NDVI*

FIGURE 5-5. *The RVI and NDVI index images for the TM agriculture image of Plate 5-1. Bands 4 and 3 were used and the indexes are calculated from uncalibrated DNs.*

5.4 Principal Components

Multispectral image bands are often highly correlated, i.e. they are visually and numerically similar (Fig. 5-6). The correlation between spectral bands arises from a combination of factors:

- *Material spectral correlation.* This correlated component is caused by, for example, the relatively low reflectance of vegetation across the visible spectrum, yielding a similar signature in all visible bands. The wavelength range of correlation is determined by the material spectral reflectance.

- *Topography.* For all practical purposes, topographic shading is the same in all solar reflectance bands and can even be the dominant image component in mountainous areas and at low sun angles. It therefore leads to a band-to-band correlation in the solar reflective region, which is independent of surface material type. The effect is different in the thermal region (see Fig. 2-22).

- *Sensor band overlap.* Ideally, this factor is minimized in the sensor design stage, but can seldom be avoided completely. The amount of overlap is typically small (Fig. 3-8), but is nevertheless important for precise calibrations.

Analysis of all of the original spectral bands is inefficient because of this redundancy. The *Principal Component Transformation (PCT)*[5] is a feature space

---

[5]   Also known as the *Karhunen-Loeve (KL)* or *Hotelling* transformation.

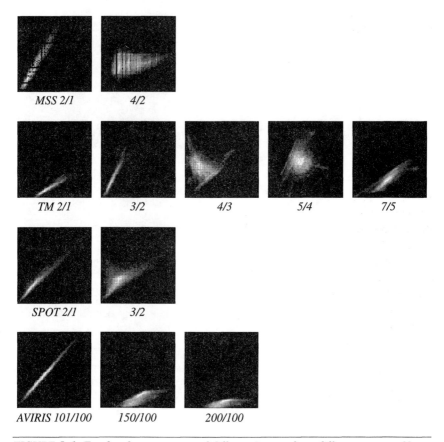

FIGURE 5-6. Two-band scattergrams of different images from different sensors. Note the similarities between the NIR/visible scattergrams for different sensors (implying vegetation in the respective images) and the relatively high degree of correlation between visible bands for all sensors.

transformation designed to remove this spectral redundancy (Ready and Wintz, 1973). It is a linear transform of the type in Eq. (5 – 1), with an image-specific matrix W_{PC},

$$PC = W_{PC} \cdot DN. \qquad (5-14)$$

This transformation alters the covariance matrix as follows,

$$C_{PC} = W_{PC}CW_{PC}^{T} \qquad (5-15)$$

and the PCT is optimum in the sense that, of all possible transformations, we choose the particular W_{PC} that diagonalizes the covariance matrix of the original multispectral image, so that,

$$C_{PC} = \begin{bmatrix} \lambda_1 & \cdots & 0 \\ \vdots & & \vdots \\ 0 & \cdots & \lambda_K \end{bmatrix}. \tag{5-16}$$

The K *eigenvalues* λ_k are found as the K roots of the *characteristic equation*,

$$|C - \lambda I| = 0, \tag{5-17}$$

where C is the original data covariance matrix and I is the (diagonal) identity matrix. Each eigenvalue is equal to the variance of the respective *PC* image along the new coordinate axes, and the sum of all the eigenvalues must equal the sum of all the band variances of the original image, thus preserving the total variance in the data. Since C_{PC} is diagonal, the principal component images are uncorrelated and, by convention, are ordered by decreasing variance, such that PC_1 has the largest variance and PC_K has the lowest. The result is removal of any correlation present in the original K-dimensional data, with a simultaneous compression of most of the total image variance into fewer dimensions.

The *PC* coordinate axes are defined by the K *eigenvectors*, e_k, obtained from the vector-matrix equation for each eigenvalue λ_k,

$$(C - \lambda_k I)e_k = 0, \tag{5-18}$$

and form the rows of the transformation matrix W_{PC},

$$W_{PC} = \begin{bmatrix} e_1^t \\ \vdots \\ e_K^t \end{bmatrix} = \begin{bmatrix} e_{11} & \cdots & e_{1K} \\ \vdots & & \vdots \\ e_{K1} & \cdots & e_{KK} \end{bmatrix}, \tag{5-19}$$

where e_{ij} is the jth element of the ith eigenvector. The eigenvector components are the direction cosines of the new axes relative to the original axes. This completes the eigenanalysis, and the transformation of Eq. (5 – 14) is applied; the eigenvector component, e_{ij}, becomes the weight on band j of the original multispectral image in the calculation for the principal component i.[6] For a particularly clear description of how to find the eigenvalues and eigenvectors of a given data covariance matrix, see (Richards, 1993).

The PCT has several interesting properties:

---

[6] The eigenvector weighting is called a *loading* or *factor loading* in statistics circles.

- *It is a rigid rotation in K-D of the original coordinate axes to coincide with the major axes of the data* (Fig. 5-7). The data in the PC images result from projecting the original data onto the new axes. In this example, the PC_2 component will have negative DNs. That is not a problem because the origin of the PC space can be arbitrarily shifted to make all PC values positive (routinely done for display), with no other change in the PCT properties (Richards, 1993).

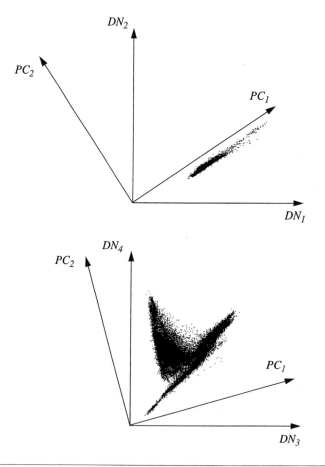

FIGURE 5-7. 2-D PCTs of two highly-correlated ($\rho = 0.97$) and two nearly-uncorrelated ($\rho = 0.13$) TM bands from a nonvegetated desert scene and a partially-vegetated agriculture scene, respectively. In the latter case, there are approximately equal amounts of vegetative cover and bare soil, which explains the orientation of the PC axes.

- *Although the PC axes are orthogonal to each other in K-D, they are generally not orthogonal when projected to the original multispectral space.* The depiction in Fig. 5-7 is for a 2-D PCT; these orthogonal PC axes do not align with the projection of a higher dimensional transformation as shown in Fig. 5-8. This makes it difficult to interpret the eigenvectors of a high dimensional PCT in terms of 2-D scattergrams.

- *It optimally redistributes the total image variance in the transformed data.* The first PC image contains the maximum possible variance for any linear combination of the original bands, the second PC image contains the maximum possible variance for any axis orthogonal to the first PC, and so forth. The total image variance is preserved by the PCT.[7] This property, illustrated in Fig. 5-9, is why principal components are important as a data compression tool. If the low variance (low contrast) information in the higher order components can be ignored, significant savings in data storage, transmission and processing time can result. Also, any uncorrelated noise in the original image will usually appear only in the higher order components (see Fig. 7-12), and can therefore be removed by setting those PC

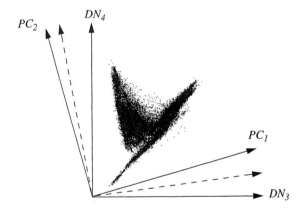

FIGURE 5-8. *Comparison of the 2-D PC axes from Fig. 5-7 (dashed arrows) to the 6-D PC_1 and PC_2 axes (solid arrows) projected to the band 4 - band 3 data plane. They are not the same, and the axes projected from 6-D are not orthogonal.*

---

[7] The *trace*, or sum of diagonal elements, of the original covariance matrix and the new covariance matrix are equal.

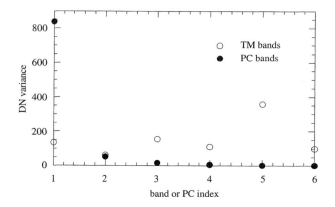

FIGURE 5-9. Distribution of total image variance across the original spectral
bands and across the principal components. The image is the nonvegetated scene
of Fig. 5-13. The TIR band 6 is excluded; band index 6 above is actually TM band
7.

images to a constant value. Caution is advised, however, because
small, but significant band differences may also appear only in the
higher-order components.

- *The transform, W_{PC}, is data-dependent.* The eigenvalues of a PC calculation for parts of two TM scenes, one largely vegetated (north of
San Francisco, California) and the other almost without vegetation (a
desert area near Cuprite, Nevada) are compared in Fig. 5-10. The
vegetated scene has a relatively higher contribution in the second
eigenvalue, reflecting the higher dimensionality of that data. If the
TIR band is included in the analysis of the vegetated scene, we see
that it contributes some small amount of additional variance, which
appears primarily in PC_4 for this particular image (Fig. 5-11). The
small contribution of the TIR band is expected because it usually has
much lower variance (contrast) than the non-TIR bands.

Other multispectral sensors exhibit similar behavior in the PC eigenvalues. In
Fig. 5-12, a comparison is made between the TM and the MSS eigenvalues for
the nonvegetated scene of Fig. 5-13. The TM appears to have somewhat higher
dimensionality than MSS, probably because of spectral contrast between different minerals that TM senses in the SWIR (see Fig. 5-15 and related discussion).

An example of principal components transformation of Landsat TM images is
shown in Fig. 5-13 and Fig. 5-14. These are the six non-TIR bands of the nonvegetated scene used in the preceding discussion. The positive-negative

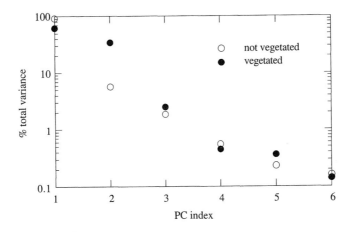

FIGURE 5-10. *Scene dependence in the percentage of total TM image variance captured by each eigenvalue. The TIR band 6 is excluded.*

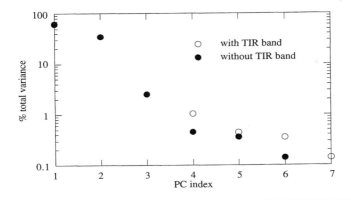

FIGURE 5-11. *Contribution of the TM TIR band to the PC eigenvalues for a vegetated scene.*

relationship of the first two components, a common characteristic of this transformation, expresses the uncorrelated nature of the new coordinates. The higher order components typically contain less image structure and more noise, indicating the data compression that has occurred. The contrast of each of the images in Fig. 5-13 and Fig. 5-14 has been stretched for display, but their variances reveal the redistribution of image contrast achieved by the PCT (Fig. 5-9). The eigenvectors for the various cases discussed are plotted in Fig. 5-15.

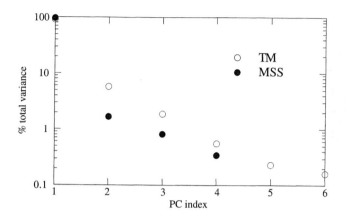

FIGURE 5-12. TM and MSS compared in terms of percent of total image variance captured in each eigenvalue for the same nonvegetated area.

The PCT can also be applied to multitemporal datasets (Table 5-1). If the multispectral images from multiple dates are simply "stacked," as in a single, multispectral image, the PCT performs a *spectral-temporal* transformation and can be difficult to interpret because of the coupling of the two dimensions. If it can be assumed that the percentage area of change is small relative to areas of no change, then the latter will dominate the lower-order PCs, and the areas of change will be emphasized in the higher-order PCs (Richards, 1984; Ingebritsen and Lyon, 1985). The analysis of temporal changes may be facilitated if a feature such as the difference between the two multispectral images is first calculated, followed by a PCT on the multispectral difference image.

TABLE 5-1. Example applications of the PCT to multitemporal imagery.

data	description	PCT	reference
MSS	landcover on two dates	8-D	(Byrne *et al.*, 1980)
MSS	fire burn, two dates	8-D	(Richards, 1984)
MSS	surface mining, two dates	8-D	(Ingebritsen and Lyon, 1985)
MSS	landcover, two dates	8-D	(Fung and LeDrew, 1987)
AVHRR-derived *NDVI*	36 monthly *NDVI* datsets	36-D	(Eastman and Fulk, 1993)

FIGURE 5-13. PC transformation of a nonvegetated TM scene.

FIGURE 5-14. PC transformation of a vegetated TM scene.

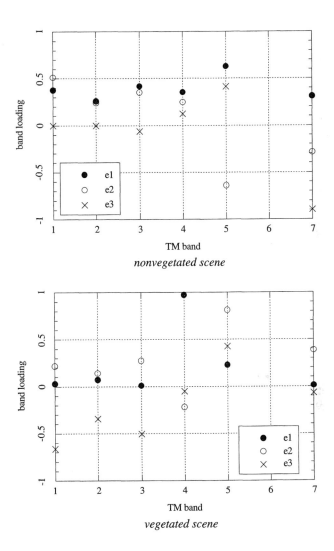

FIGURE 5-15. *The first three eigenvectors for the nonvegetated and vegetated scenes. These are for a six band PCT, excluding the thermal band 6. In both cases, the first eigenvector has all positive band components. For the nonvegetated scene the major spectral contrast appears to be in bands 5 and 7 (where mineral signatures are found; see Fig. 5-4), while for the vegetated scene, bands 3, 4 and 5 contain most of the spectral contrast because of the vegetation reflectance "edge" at 700nm.*

5.4.1 Standardized Principal Components (SPC)

Some researchers suggest that a PCT based on the *correlation*, rather than covariance, matrix has advantages for remote sensing analysis (Singh and Harrison, 1985). In using the correlation matrix, we are effectively normalizing the original bands to equal and unit variance (the covariance matrix is then identically the correlation matrix). This may have advantages when data with different dynamic ranges are combined in a PCT. The SPCT does not exhibit the optimum compression characteristic of the PCT, of course.

5.4.2 Maximum Noise Fraction (MNF)

Also known as the *Noise-Adjusted Principal Components (NAPC)* transform (Lee *et al.*, 1990), this technique was introduced in (Green *et al.*, 1988). The MNF is a modification of the conventional PCT to improve the isolation of image noise that may occur in one or only a few of the original bands. Unfortunately, the noise covariance matrix must be available (or estimated) for the MNF transform.

5.5   Tasseled-Cap Components

The PCT is data dependent; the coefficients of the transformation are a function of the data spectral covariance matrix. While this feature allows the PCT to adapt to a given dataset and produce the best transformation from a compression standpoint, it makes the comparison of PCTs from different images difficult. The PCs can usually be interpreted in terms of physical characteristics of a given scene, but the interpretation must change from scene to scene. Early in the Landsat era, it was recognized that a *fixed* transformation, based on physical features, would be useful for data analysis.

A fixed-feature space, designed specifically for agricultural monitoring, was first proposed for Landsat MSS data by Kauth and Thomas (Kauth and Thomas, 1976). They noted that the *DN* scattergrams of Landsat MSS agricultural scenes exhibit certain consistent properties, for example, a triangular-shaped distribution between band 4 and band 2 (Fig. 5-6). Visualization of such distributions in K dimensions as crops grow yields a shape described as a "tasseled cap," with a base called the "plane of soils." Crop pixels move up the tasseled cap as the crops grow and back down to the plane of soils as the crop progresses to a senescent stage. Kauth and Thomas first derived a linear transformation of the four MSS bands that would yield a new axis called "soil brightness," defined by the signature of nonvegetated areas. A second coordinate axis, orthogonal to the first and

nonvegetated PC_1 PC_2 PC_3

PC_4 PC_5 PC_6

vegetated PC_1 PC_2 PC_3

PC_4 PC_5 PC_6

FIGURE 5-16. Standardized principal components; compare to the conventional PCs for the two scenes in Fig. 5-13 and Fig. 5-14.

called "greenness," was then derived in the direction of vegetation signatures. The third and fourth transformed axes, called "yellow stuff" and "non-such," respectively, were derived to be orthogonal to the first two axes. The *Tasseled - Cap Transformation (TCT)* is a special case of Eq. $(5 - 1)$, with a specific transformation matrix W_{TC},

$$TC = W_{TC} \cdot DN. \tag{5-20}$$

Unlike the PCT, the weights in the TCT matrix are *fixed and independent of the scene*. Like the PCT, however, the new axes produced by the TCT are orthogonal in the K-dimensional space, but are *not* orthogonal when projected onto a lower-dimension data space (Fig. 5-17). The matrix transformation coefficients, row-by-row, are given for the Landsat-1 and -2 MSS and the Landsat-4 and -5 TM in Table 5-2. The Landsat-5 TM coefficients were derived by reference to the Landsat-4 coefficients; they include extra additive terms that must be used to compare the TC features from the two sensors (Crist *et al.*, 1986).

The motivation behind the development of the TCT was to find a fixed set of axes in feature space that represented the physical properties of growing crops in the American midwest. The soil axes represents the pre-emergence time regime and as crops grow, their feature space signature moves away from this axis in the direction of greenness. Then, as crops mature and become senescent, their signature increases in the yellow stuff dimension, and decreases in greeness. If this transformation, derived for midwestern agriculture, is applied to other types of agriculture and vegetation in different climate zones, the transformation coefficients can still be interpreted in a similar fashion, but some new scene components may not be represented, such as non-green vegetation and different soil types (Crist, 1996). The yellow stuff and non-such dimensions for MSS data have been shown to indicate changes in atmospheric haze conditions and therefore are useful for relative atmospheric calibration between images (Lavreau, 1991). In the TM TCT space, the fourth component is called "haze," and has a similar interpretation.

In Fig. 5-18, we provide a comparison of the PCT and TCT for a TM agricultural scene (see the CIR composite in Plate 5-1 for a visual reference). The most highly-correlated components of the two transformations are PC_1 and TC_1 (correlation of 0.971) or TC6 (0.868), and PC_4 and TC_5 (0.801). It is not possible to draw general conclusions from these observations, however, because the PCT will change from scene to scene, while the TCT will not. In general, it appears that the TCT leads to superior separation of scene components, particularly soil and vegetation. A drawback is that it is sensor-dependent (primarily on the spectral bands), and the coefficients must be rederived for every sensor.

TABLE 5-2. *Tasseled-cap coefficients for Landsat-1 MSS (Kauth and Thomas, 1976), Landsat-2 MSS (Thompson and Whemanen, 1980), Landsat-4 TM (Crist and Cicone, 1984) and Landsat-5 TM (Crist et al., 1986). The MSS coefficients are for a 0-63 DN scale in band 4 and a 0-127 DN scale in the other bands, i.e. the same as data supplied to users.*

sensor	axis name	W_{TC}					
		MSS band 1	2	3	4		
L-1 MSS	soil brightness	+0.433	+0.632	+0.586	+0.264		
	greenness	−0.290	−0.562	+0.600	+0.491		
	yellow stuff	−0.829	+0.522	−0.039	+0.194		
	non-such	+0.223	+0.120	−0.543	+0.810		
L-2 MSS	soil brightness	+0.332	+0.603	+0.676	+0.263		
	greenness	+0.283	−0.660	+0.577	+0.388		
	yellow stuff	+0.900	+0.428	+0.0759	−0.041		
	non-such	+0.016	+0.428	−0.452	+0.882		
		TM band 1	2	3	4	5	7
L-4 TM	soil brightness	+0.3037	+0.2793	+0.4743	+0.5585	+0.5082	+0.1863
	greenness	−0.2848	−0.2435	−0.5436	+0.7243	+0.0840	−0.1800
	wetness	+0.1509	+0.1973	+0.3279	+0.3406	−0.7112	−0.4572
	haze	−0.8242	+0.0849	+0.4392	−0.0580	+0.2012	−0.2768
	TC5	−0.3280	+0.0549	+0.1075	+0.1855	−0.4357	+0.8085
	TC6	+0.1084	−0.9022	+0.4120	+0.0573	−0.0251	+0.0238
L-5 TM	soil brightness	+0.2909	+0.2493	+0.4806	+0.5568	+0.4438	+0.1706
	greenness	−0.2728	−0.2174	−0.5508	+0.7221	+0.0733	−0.1648
	wetness	+0.1446	+0.1761	+0.3322	+0.3396	−0.6210	−0.4186
	haze	+0.8461	+0.0731	+0.4640	−0.0032	−0.0492	+0.0119
	TC5	+0.0549	−0.0232	+0.0339	−0.1937	+0.4162	−0.7823
	TC6	+0.1186	−0.8069	+0.4094	+0.0571	−0.0228	+0.0220
	soil brightness			additive terms:	+10.3695		
	greenness				−0.7310		
	wetness				−3.3828		
	haze				+0.7879		
	TC5				−2.4750		
	TC6				−0.0336		

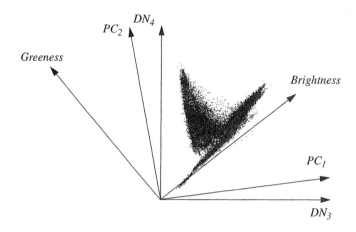

FIGURE 5-17. Projection of two of the component axes from 6-D PC and TC transformations onto the TM band 4 versus band 3 data plane. The Brightness-Greenness projection is fixed and independent of the data. Notice how the Brightness-Greenness space "captures" the physical components of soil and vegetation better than does the PC_1-PC_2 space.

5.6   Contrast Enhancement

The spectral transforms discussed so far are motivated by quantitative analysis of multispectral imagery. Contrast enhancement, on the other hand, is only intended to improve the visual quality of a displayed image. The transformations used for enhancing contrast are generally not desirable for quantitative analysis.

The need for contrast enhancement derives from the requirements on sensor design described in Chapter 3. A remote sensing system, particularly a satellite sensor with global coverage, must image a wide range of scenes, from very low radiance (oceans, low solar elevation angles, high latitudes) to very high radiance (snow, sand, high solar elevation angles, low latitudes). Therefore, the sensor's dynamic range must be set *at the design stage* to accomodate a large range of scene radiances, and it is desirable to have as many bits/pixel as possible over this range for precise measurements. However, any particular scene will generally have a radiance range much less than the full range. When such a scene is imaged and converted to *DN*s, it therefore uses less than the full quantization range, which is typically eight bits/pixel or more. Since most display systems use eight bits/pixel in each color, with the extremes of that range adjusted to yield black

PC_1 PC_2 PC_3

PC_4 PC_5 PC_6

TC_1 *(Brightness)* TC_2 *(Greeness)* TC_3 *(Wetness)*

TC_4 *(Haze)* TC_5 TC_6

FIGURE 5-18. The PC and TC components for the Yuma agricultural scene. The PCT achieves more data compression (PC_5 and PC_6 contain almost no scene content), but the TCT better represents soil and vegetation in TC_1 and TC_2. Note PC_2 is inverted relative to TC_2.

and white on the monitor, the displayed image will have low contrast because it is not using the full range available in the display.

5.6.1 Global Transforms

Contrast enhancement is a mapping from the original *DN* data space to a *GL* display space. Each *DN* in the input image is transformed through the mapping function to a *GL* in the output (display) image. In the simplest case, the mapping function is determined by the global (full image) statistical properties, and is the same for every pixel. Since the primary goal is usually to expand the *DN* range to fill the available display *GL* range, the transform is called a *contrast stretch*; typical transformations are shown in Fig. 5-19. The slope of the transformation, $d(GL)/d(DN)$, can be considered the *gain* in contrast, much like the gain parameter in an electronic amplifier. For the linear stretch, the gain is the same at all *DN*s. In the piecewise-linear and histogram equalization stretches, the gain varies with *DN*, thus preferentially enhancing contrast in certain *DN* ranges (low water radiances, for example).

Linear stretch

Examples of linear contrast stretching and the effect on the image histogram are depicted in the top two rows of Fig. 5-20 and Fig. 5-21. The *min-max stretch* expands the image *DN* range to fill the dynamic range of the display device, e.g. [0,255]. Since it is based on the image minimum and maximum *DN*, it is sensitive to outliers, i.e. single pixels that may be atypical and outside the normal *DN* range. The resulting contrast will then be less than that possible if the outliers were excluded in determining the minimum and maximum *DN* in the image.

To achieve a greater contrast increase, a *saturation stretch* may be implemented with a linear stretch for all pixels within a *DN* range smaller than the min-max range. Pixels with a *DN* outside that range are then transformed to a *GL* of either zero or 255. This saturation ("clipping") at the extremes of the *DN* range is acceptable unless important image structure is lost in the saturated areas of the image. Typically, saturation of one or two percent of the image pixels is a safe level, but the allowable saturation depends of course on the image and intended application. If one is primarily interested in viewing cloud structure, for example, there should be little saturation at high *DN* values.

A linear transformation also can be used to *decrease* image contrast if the image *DN* range exceeds that of the display. This situation occurs for radar imagery, some multispectral sensors such as AVHRR (10 bits/pixel), and most hyperspectral sensors (12 bits/pixel). The displayed image will not contain all the radiometric information in the raw image, but on the other hand, the human eye can not usually distinguish more than about 50 *GL*s at once, anyway.

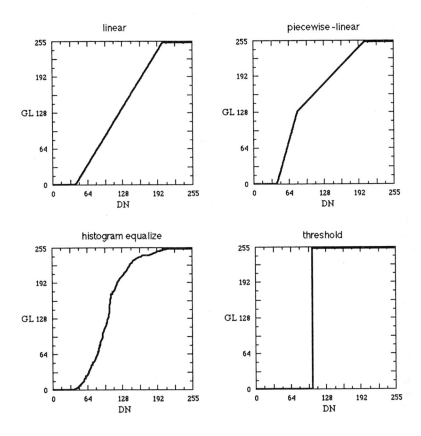

FIGURE 5-19. Some types of DN-to-GL transformations for contrast enhancement.

Nonlinear stretch

If the image histogram is asymmetric, as it often is, it is impossible to simulta-neously control the average display *GL* and the amount of saturation at the ends of the histogram with a simple linear transformation. With a *piecewise-linear transformation*, more control is gained over the image contrast, and the histo-gram asymmetry can be reduced, thus making better use of the available display range (lower left, Fig. 5-20). This example is a two segment stretch, with the left segment having a higher gain than the right segment. The transformation parame-ters are selected to move the input minimum and maximum *DN*s to the extremes of the display *GL* range, and to move the mode of the histogram to the center of the display range (128). More than two linear segments may be used in the transformation for better control over the image contrast.

Histogram equalization is a widely-used nonlinear transformation (lower right, Fig. 5-20). It is achieved by using the *Cumulative Distribution Function (CDF)* of the image as the transformation function, after appropriate scaling of the ordinate axis to correspond to output *GL*s. Equalization refers to the fact that the histogram of the processed image is approximately uniform in density (number of pixels/*GL*) (Gonzalez and Woods, 1992). Because of the near-Gaussian shape of most image histograms, equalization tends to automatically reduce the contrast in very light or dark areas and to expand the middle *DN*s toward the low and high ends of the *GL* scale. Where the *CDF* increases rapidly, the contrast gain also increases. The highest gain therefore occurs at *DN*s with the most pixels. This effect is seen in Fig. 5-21 as the variable spacing of *GL*s in the enhanced image histogram. The contrast of an equalized image is often rather harsh, so equalization is not recommended as a general purpose stretch. However, no parameters are required from the analyst to implement the transformation, making it easy to apply.

Normalization stretch

The normalization stretch is a robust contrast enhancement algorithm. The image is transformed with a linear stretch such that it has a specified *GL* mean and standard deviation, and then is clipped at the extremes of the display *GL* range. It is therefore a variation on the saturation stretch. Two parameters are required to achieve the desired second-order statistics, given the input *DN* mean and standard deviation (see Table 5-3). The clipping is necessary because the resulting *GL*s will usually extend beyond the limits of the display range. This two-parameter algorithm provides a convenient way to explicitly control the mean of the resulting image, and at the same time, vary the contrast by controlling the standard deviation. Examples are shown in Fig. 5-22.

Reference stretch

Differences between atmospheric conditions at different times, differences between sensors, and scene changes over time make it difficult to compare multisensor and multitemporal imagery. Even adjacent scenes from two orbits of the same sensor can vary significantly in their radiometric characteristics because they normally are acquired several days apart, and the sensor view angle is different through different atmospheric conditions. There is a need therefore to match the global radiometric characteristics of one image to another. It must of course be presumed and reasonable that the histograms of the two images *should in fact* be similar. That is, if the presumption of similarity is not justified, for example if

original

min-max

3% saturation

8% saturation

piecewise-linear

histogram equalization

FIGURE 5-20. Examples of contrast enhancement using point transformations and global statistics. The test image is from the GOES visible wavelength sensor and shows cloud patterns over North America.

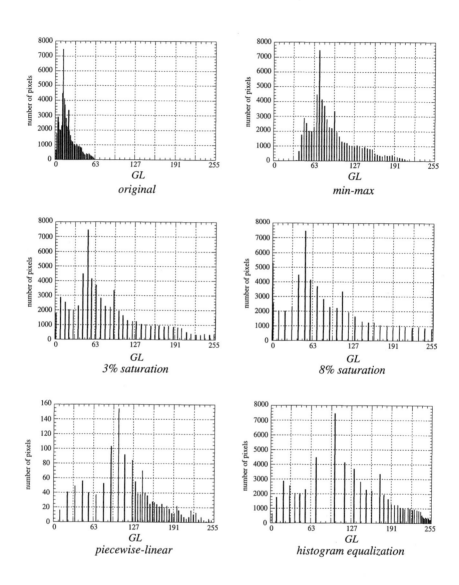

FIGURE 5-21. Histograms of the enhanced images in Fig. 5-20. As the gain in the stretch transformation increases, the occupied DN bins spread further apart. This is generally not noticeable in the image, unless only a few DNs are spread across the whole range. Also note the piecewise-linear transformed histogram resembles the equalized histogram in the sense of larger gaps between DNs for the lower DNs relative to the higher DNs.

original $\mu = 128,\ \sigma = 32$

$\mu = 128,\ \sigma = 48$ $\mu = 128,\ \sigma = 64$

FIGURE 5-22. *Examples of the normalization stretch. Note how the GL mean can be held constant, while the contrast is controlled by a single parameter, the GL standard deviation.*

there have been substantial surface changes between the two image dates, then reference matching is not justified.

One way to match the contrast and brightness characteristics of two images is with the linear normalization stretch of the previous section. However, that can only match the means and standard deviations of the two images. A natural generalization of the normalization stretch is to match histogram *shapes*. The *CDF reference stretch* does this by matching the *CDF* of an image to a reference *CDF* (Richards, 1993). This is achieved by forward mapping the image *DNs* through the image *CDF*, and then backward mapping through the reference CDF_{ref} to DN_{ref} (Fig. 5-23).[8] CDF_{ref} can be from another image, or a hypothetical image

---

[8]   This technique is sometimes called *histogram matching* or *histogram specification*.

with a specified *CDF*, for example one with a Gaussian histogram. The two-stage mapping creates a nonlinear look-up table for $DN_{ref}(DN)$. One complication is that the matching of the *CDF* levels does not necessarily lead to an integer DN_{ref}. The CDF_{ref} must therefore be *interpolated* between *DN*s to obtain the output DN_{ref} to be used. Linear interpolation is sufficient for this purpose since the *CDF* is normally a smooth function of *DN*.

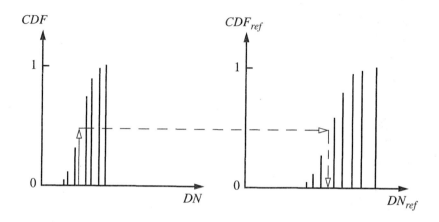

FIGURE 5-23. *The procedure to match the CDFs of two images. Note the calculated* DN_{ref} *must be estimated by interpolation of the* CDF_{ref} *between existing values.*

Note that the first transformation, through the *CDF* of the image to be stretched, is the same as a histogram equalization. The second transformation represents an inverse of that to an image having a *CDF* the same as that of the reference. The concept here is that the *CDF* of *any* histogram-equalized image is the same (i.e., a linear ramp), except for possibly a horizontal shift along the *DN* axis. An example of this contrast matching technique is shown in Fig. 5-24. The two images are eight months apart and show significant differences, in addition to that due to the change in solar irradiance. The *CDF* reference stretch result is compared to that from a linear stretch, with parameters determined from two targets in both images, one dark and one light. The latter is an empirical calibration transform for two images from different dates (see Chapter 7) and is prefered for radiometric comparison. The *CDF* stretch achieves a better visual match of the two images, but is a nonlinear transform that may not be appropriate because of the apparent changes between the two dates. For example, notice how the airport pavement is dark in December and bright in August. This indicates either a

TM band 3, December 31, 1982   *TM band 3, August 12, 1983*

dark-light target linear stretch   *CDF reference stretch*

FIGURE 5-24. *Contrast matching of TM images of San Jose, California. The December image is adjusted to match the August image using two techniques described in the text.*

non-Lambertian reflectance characteristic or perhaps an actual change in pavement material. The image histograms are compared in Fig. 5-25.

A second technique for reference stretching, *statistical pixel matching*, was originally developed for medical images (Dallas and Mauser, 1980). It requires that the two images are registered and cover the same area on the ground (not a requirement for the *CDF* reference stretch). Therefore, in remote sensing, it is only useful for either multitemporal or multisensor images of the same scene.

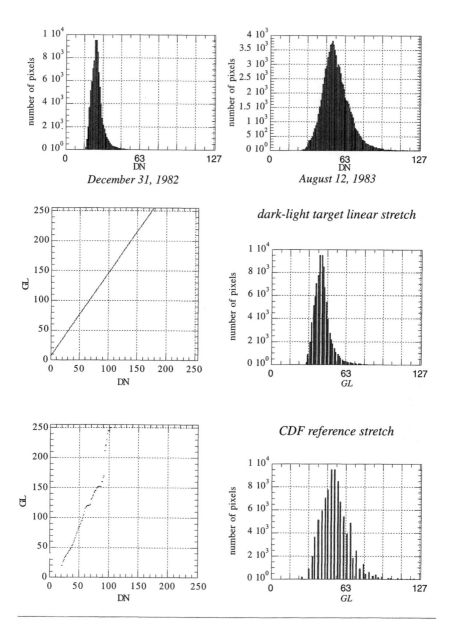

FIGURE 5-25. DN histograms of original multitemporal images and the contrast transformations and resulting histograms for matching the December image to the August image.

For mosaicing applications, the adjacent images would have to overlap, and only the region of overlap can be used in the algorithm. The procedure is,

- Create an accumulator table with an entry for each *DN* of the image to be stretched and initialize its elements to zero.

- For every pixel in the image to be stretched, accumulate the DN_{ref} of the corresponding pixel of the reference image, in the appropriate *DN* bin of the table.

- After all pixels have been counted, find the average DN_{ref} at each *DN* of the table. After a linear scaling of the DN_{ref} to *GL*, this becomes the desired *GL(DN)* transformation function.

- Do the transform.

This algorithm works best for images which are already quite similar everywhere except in small areas.

Thresholding

Thresholding is a type of contrast manipulation that *segments* an image into two categories defined by a single *DN* threshold. The use of a binary threshold on certain types of images, such as those with land and water or snow, results in sharply-defined spatial boundaries that may be used for masking portions of the image. Separate processing may then be applied to each portion and the results spatially recombined. Thresholding is really a simple classification algorithm, as we will see in Chapter 9. An example of thresholding of a GOES weather satellite image is shown in Fig. 5-26. The lowest threshold separates areas of water from clouds and land, the middle threshold isolates most clouds and the highest threshold selects only the brightest and most dense clouds.

   DN thresholding can be easily implemented in an interactive mode with a display and operator-controlled cursor, but selection of the "best" threshold level is a difficult task and requires prior knowledge about the scene to be meaningful. There are techniques for the quantitative determination of optimum threshold levels using only information from the image histogram (Pun, 1981).

   A summary of the point contrast transformations is given in Table 5-3. They are extendable to spatially-adaptive enhancement, since they can be used in place of the min-max transform in the LRM algorithm, described below.

5.6.2 Local Transforms

The contrast enhancement examples discussed thus far are point transformations using global image statistics; the same contrast stretch is applied to all image pixels. Obviously, contrast can vary locally within an image. A more optimal

$DN_T = 50$

$DN_T = 100$ $DN_T = 150$

FIGURE 5-26. Examples of thresholding on the GOES image of Fig. 5-20.

enhancement may therefore be achieved by using an *adaptive* algorithm whose parameters change from pixel-to-pixel according to the local image contrast. We will describe a robust algorithm, *Local Range Modification (LRM)*, which illustrates the nature of adaptive processing (Fahnestock and Schowengerdt, 1983).

The essential idea is to partition the image into adjoining blocks (designated A, B, etc.) and derive a contrast stretch, different at each pixel, which is dependent on the local contrast within the corresponding block and surrounding blocks (Fig. 5-27). The stretch must change smoothly from pixel-to-pixel; otherwise brightness discontinuities occur at the boundaries between blocks (Fig. 5-28). Also, one premise in LRM is that the final *GL* range of the enhanced image should be predictable and not exceed specified minimum and maximum *GL*s.

TABLE 5-3. Summary of contrast enhancement algorithms. A display GL range of [0,255] is assumed.

algorithm	equation	remarks
min-max	$GL = \dfrac{255}{DN_{max} - DN_{min}}(DN - DN_{min})$	sensitive to outliers
histogram equalization	$GL = 255\,CDF(DN)$	produces uniform histogram
normalization	1. $GL = \dfrac{\sigma_{ref}}{\sigma}(DN - \mu) + \mu_{ref}$ 2. $GL = 255, GL > 255$ $GL = 0, GL < 0$	matches means and variances
threshold	$GL = 255, DN \geq DN_T$ $GL = 0, DN < DN_T$	binary output
reference	$GL = CDF_{ref}^{-1}[CDF(DN)]$	matches histograms

The first step is to find the lowest and highest *DN*s, *L* and *H*, within each block. The overall minimum and maximum *DN*s of four adjoining blocks are then found by (using node 6 as the example),

$$MIN_6 = minimum(L_A, L_B, L_D, L_E)$$
$$MAX_6 = maximum(H_A, H_B, H_D, H_E)$$

$$(5-21)$$

and assigned to their shared node. This is equivalent to using overlapping blocks of twice the width and height in the intial image partitioning. At the corner of the image there is only one block, so in the upper left, for example,

$$MIN_1 = L_A$$
$$MAX_1 = H_A$$

$$(5-22)$$

and along the left image edge,

$$MIN_5 = minimum(L_A, L_D)$$
$$MAX_5 = maximum(H_A, H_D).$$

$$(5-23)$$

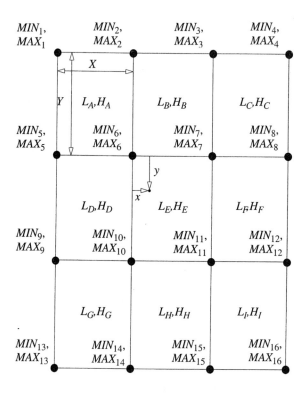

FIGURE 5-27. Blocking parameters for the LRM adaptive contrast enhancement.

The values *MIN* and *MAX* set the *GL* range of enhanced pixels within the corresponding set of blocks. They are then linearly interpolated in two dimensions at each pixel location (x,y) within a block,

$$GL_{min} = \left[\frac{x}{X}MIN_7 + \left(\frac{X-x}{X}\right)MIN_6\right]\left(\frac{Y-y}{Y}\right)$$

$$+ \left[\frac{x}{X}MIN_{11} + \left(\frac{X-x}{X}\right)MIN_{10}\right]\frac{y}{Y}$$

$$GL_{max} = \left[\frac{x}{X}MAX_7 + \left(\frac{X-x}{X}\right)MAX_6\right]\left(\frac{Y-y}{Y}\right) \quad (5-24)$$

$$+ \left[\frac{x}{X}MAX_{11} + \left(\frac{X-x}{X}\right)MAX_{10}\right]\frac{y}{Y}$$

to obtain *estimated* range extremes, GL_{min} and GL_{max}, at each pixel. Using linear interpolation insures that these values are within the range of the surrounding *MIN* and *MAX*, i.e.,

$$MIN \leq GL_{min} \leq MAX$$
$$MIN \leq GL_{max} \leq MAX$$

$$(5-25)$$

The estimated range values found from Eq. $(5-4)$ are then used to transform the pixel at (x,y) by a linear stretch,

$$GL' = \frac{255}{GL_{max} - GL_{min}}(DN - GL_{min}) \qquad (5-26)$$

Figure 5-28 is an example of adaptive contrast enhancement with the LRM algorithm. Note how the LRM algorithm provides a smooth contrast transition across blocks without obvious artifacts. Although the global, relative radiometry of the image is altered by the processing, it is much easier to distinguish previously low contrast features in the darker and brighter portions of the enhanced image, and there are no transition artifacts between blocks.

Other artifacts can appear in adaptive enhancements, however. If the area used in the calculation for each pixel is smaller than some uniform objects in the scene (such as a lake), the algorithm will stretch that very low contrast region excessively, resulting in high contrast noise at those pixels.

Any global stretch can be used within an adaptive framework, for example histogram equalization (Pizer *et al.*, 1987). For adaptive histogram equalization, the nonlinear *CDF* is calculated for each block and interpolated between blocks to yield a smooth transition across the image, as is done in LRM.

5.6.3 Color Images

The use of color in display and enhancement of remote-sensing images is an important aspect of image processing. Color may be used simply for display of multispectral images or may be manipulated directly by processing techniques to enhance visual information extraction from the images. At first glance, it would seem that one could simply extend the techniques used for black and white image contrast enhancement to color images. This is true to some extent, but it is often important to specifically consider the color aspects of the processing to achieve the best results. In this section we describe some relatively simple heuristic techniques for numerical color manipulation; theories that model the visual perception of color can be quite complex and will not be discussed here.

original   *blockwise min-max (16 blocks)*

LRM (16 blocks)   *LRM (64 blocks)*

FIGURE 5-28. *Adaptive contrast enhancement with blockwise stretching and the LRM algorithm, which does not introduce block discontinuities.*

Min-max stretch

The most obvious way to enhance a color image is to simply stretch the histogram of each band according to its minimum and maximum *DN*. This will indeed improve the contrast of the color composite, but the color balance is likely to change in an unpredictable way from that of the original image. The problem is that the min-max stretch is controlled by the image minimum and maximum *DN*, which can be sensitive to outliers, as noted earlier.

Normalization stretch

The normalization stretch discussed previously is a good, robust tool for consistent color contrast enhancement. It overcomes the shortcomings of the min-max stretch by incorporating the image mean and standard deviation. For color images, the normalization stretch is applied to each band independently, setting their means and standard deviations equal across the bands (Fig. 5-29). The *average* color of the resulting color composite is therefore grey, and other spectral information in the image appears as deviations about that mean color. For some images, dominated by a single cover type such as vegetation or water, this may not be good for interpretation because of the bias towards a grey color. For most images that contain a mixture of cover types, it does appear to be a broadly useful tool.

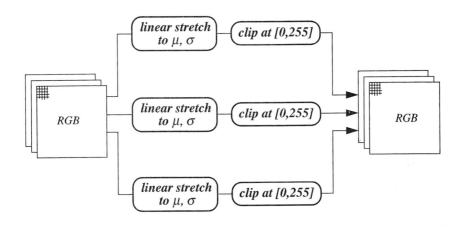

FIGURE 5-29. *The normalization algorithm for color images.*

Decorrelation stretch

One of the difficulties in color contrast enhancement is spectral band correlation. If three correlated bands are displayed, their data distribution lies nearly along a line in the color cube, from the darkest pixels to the brightest ones, and very little of the available color space is utilized. If we can decorrelate the bands, stretch the PCs to better fill the color space, and then inverse transform to the RGB color space, we will enhance whatever spectral information is present in the data. This idea is the basis of the PCT decorrelation stretch (Gillespie *et al.*, 1986; Durand

and Kerr, 1989; Rothery and Hunt, 1990). The process is indicated in Fig. 5-30. The contrast stretch applied to the PCs shifts the data to all positive values and equalizes the variances, thereby insuring that the data occupies a spherical volume in the PC space. The inverse PCT then calculates transformed bands that are decorrelated. Since the eigenvector transformation matrix is orthogonal, its inverse is equal to its transpose, and is therefore easily calculated by swapping rows and columns.

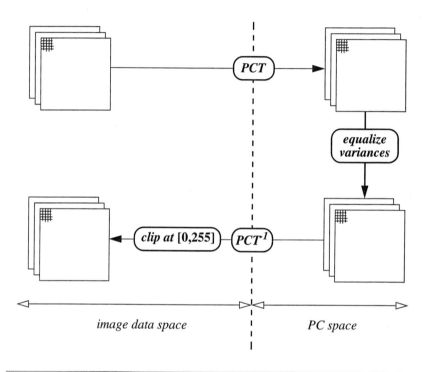

FIGURE 5-30. The PCT decorrelation contrast stretch algorithm. The image produced by the inverse PCT is not constrained to an 8-bit range and may need to be clipped for display purposes.

Color-space transforms

To describe the visually-perceived color properties of an image, we do not naturally use the proportions of red, green and blue components, but rather terms such as "intensity," "hue" and "saturation" for the subjective sensations of "brightness," "color" and "color purity," respectively. Similarly, it is often easier to anticipate the visual results of an intensity or hue manipulation on an image than it is for the results of red, green and blue manipulation. A transformation of the RGB components into hue, saturation and intensity (HSI) components before processing may therefore provide more control over color enhancement. Whatever algorithm is used, we will call the conversion of RGB to HSI a *Color-Space Transform (CST)*. In the perceptual color space of HSI, we can modify any or all of the components somewhat predictably. For example, the intensity component can be stretched as for any single band image, or the saturation might be increased at all pixels by an appropriate transformation. The processed images are then converted back to RGB for display using the inverse of the CST (Fig. 5-31).[9]

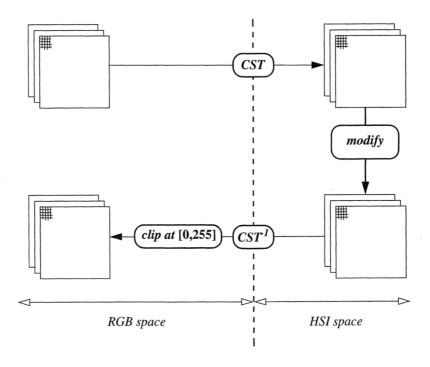

FIGURE 5-31. *The use of a Color-Space Transform (CST) to modify the perceptual color characteristics of an image.*

---

[9] The acronym HSI is sometimes permutated to IHS in the literature.

One particular RGB-to-HSI tranformation will be used to illustrate the concepts involved. This transformation, the hexcone model, is heuristic and not based on any particular color theory, but it is representative of most algorithms used in color image processing and color graphics (Smith, 1978; Schowengerdt, 1983).

The *hexcone CST* algorithm is based on the RGB color cube introduced in Chapter 1. We consider internal RGB subcubes defined by their vertex location along the greyline (Fig. 5-32). Imagine the projection of each subcube onto a plane perpendicular to the greyline at the subcube vertex. Moving the vertex from black to white, the projection onto a plane results in a series of hexagons of increasing size, as shown in Fig. 5-32. The hexagon at black degenerates to a point; the hexagon at white is the largest. This series of hexagons define a *hexcone*. The distance along the gray line defines the intensity of each hexagonal projection.[10] For a pixel with a given intensity, the color components, hue and saturation, are defined geometrically in the appropriate hexagon. The hue of a point is determined by the angle around the hexagon and the saturation is determined by the distance of the point from the center, i.e. the gray point. Points further from the center represent purer colors than those closer to the gray point. The use of simple linear distances for defining hue and saturation make the hexcone algorithm more efficient than similar transformations involving trigonometric functions.

Examples of processing in the HSI space and its effect on the RGB space are shown in Fig. 5-33 and Plate 5-1. By reducing the saturation component (second row), we obtain a "pastel" RGB image. By reducing the intensity component (third row), we obtain a darker RGB image. The last row shows a double ramp for hue that rotates the colors in RGB space in a cyclic fashion since the hue dimension is periodic.

Image examples

In Plate 5-1, two different three-band composites of a TM image of Cuprite, Nevada, are subjected to some of the enhancement algorithms described in this section. The visible-band composite shows very little color because of the high correlation among the three bands. Only the decorrelation stretch is able to enhance the subtle spectral information. In the NIR-SWIR composite, on the other hand, there is considerable spectral content due to a variety of minerals with different spectral reflectances. The color enhancement produced by the

---

[10] Smith used an alternate quantity, *value*, given by the maximum of R, G and B (Smith, 1978). Value is more closely related to artist's terminology for describing color. The distinction between value and intensity is important, but not crucial for our discussion.

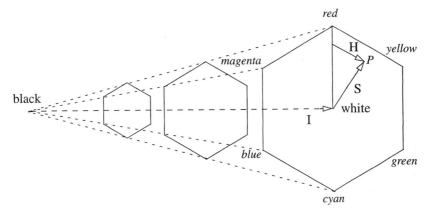

FIGURE 5-32. *Generation of the hexcone CST. Three possible RGB cubes are shown in the top figure and the resulting projections onto the plane perpendicular to the greyline are shown below. Such projections for all possible subcubes define the hexcone. The projection of a particular RGB point, P, into one of the hexagons is shown. At P, the intensity is the center grey value, the hue is the angle around the hexagon, and the saturation is the fractional distance from the center to the perimeter. The hue and saturation were defined as simplified linear relations to make the original algorithm efficient (Smith, 1978).*

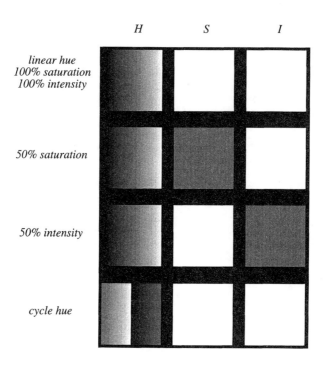

FIGURE 5-33. Four test images for color space processing examples. The resulting RGB images obtained from the inverse hexcone model are shown in Plate 5-1.

min-max and decorrelation stretches can be interpreted by an experienced geologist who is familiar with the area.

In Plate 5-2, a CIR TM image composite is subjected to different contrast transformations. The scatterplots of the data before and after the processing are shown in Fig. 5-34. The following comments can be made in the context of this particular image. The min-max stretch makes little change since the original image has a few outlier pixels at high *DN* levels (buildings with white roofs). The histogram equalization stretch produces higher contrast than is probably desired, and the PCT decorrelation stretch greatly exaggerates the color differences. Note how the scattergram for the latter image is expanded to a spherical (uncorrelated) distribution that fills all but the corners of the display color cube. The normalization and HSI stretches produce some enhancement without seriously distorting the original colors.

Color transformations are useful for displaying diverse, spatially-registered images. For example, a high-resolution visible band image may be assigned to

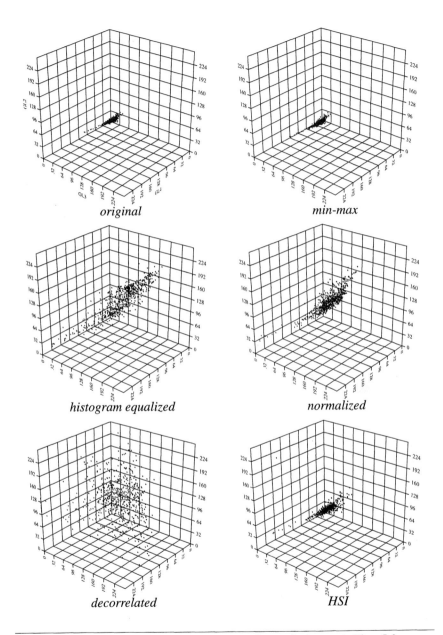

FIGURE 5-34. Three-band scatterplots corresponding to the images in Plate 5-2.

the intensity component, a lower-resolution thermal band image to the hue component, and combined with a uniform saturation component (Haydn *et al.*, 1982). After the inverse HSI transform, the resulting color image contains the detail structure of the scene expressed as intensity, with the thermal structure superimposed as color variations. This is a type of *image fusion,* discussed in detail in Chapter 8.

5.7   Summary

A variety of spectral transformations were examined, ranging from nonlinear spectral band ratios to linear transformations of various types. Some are designed to improve quantitative analysis of remote-sensing images, while others simply enhance subtle information so that it is visible. The following important points can be made:

- Spectral band ratios can help isolate spectral signatures from imagery by reducing topographic shading. Ratios involving NIR and red spectral bands are useful for vegetation measurements.

- The principal components transform is optimal for data compression, but because it is data dependent, the resulting features have different interpretations for different images.

- The tasseled-cap transform provides a fixed, but sensor-specific, transform based on soil and vegetation signatures.

- Color-image composites can be visually enhanced by a variety of spectral transforms.

In the next chapter, we will look at spatial transforms, which can be combined with spectral transforms for certain applications such as image fusion and feature extraction for classification.

5.8   Exercises

5–1.   Suppose a *GIFOV* contains soil and vegetation and that the net reflectance measured at the pixel is a weighted sum of the reflectances of each material, where the weights are simply the fractional coverages of each (i.e. the linear mixing model; see Chapter 9). For the two cases below, calculate and plot the *NDVI* and *SAVI*, with an *L* value of 0.5, versus the fraction of vegetation within the *GIFOV*. Which index has a more linear relation to vegetation coverage?

	Case I: dry soil		Case II: wet soil	
	ρ_{red}	ρ_{NIR}	ρ_{red}	ρ_{NIR}
soil	0.35	0.40	0.20	0.20
vegetation	0.10	0.50	0.10	0.50

5–2. Fill in the details on the facts and approximations that result in Eq. $(5-4)$. What underlying physical constants or quantities do the two parameters, a_b and b_b, contain?

5–3. Explain with a hypothetical 2-D scatterplot, why small areas of change between two images will be "captured" in the multitemporal PC_2 component.

5–4. Suppose you do a PC decorrelation stretch of a color image and compare the histograms of the result in HSI space to the histograms of the original image in HSI space. What differences do you expect and why?

5–5. Given the following histograms of the (actual) HSI components of a three-band image, specify three *DN* transformations that will make the saturation of every pixel equal to 200, linearly stretch the intensity to increase the contrast, and leave the hue unchanged.

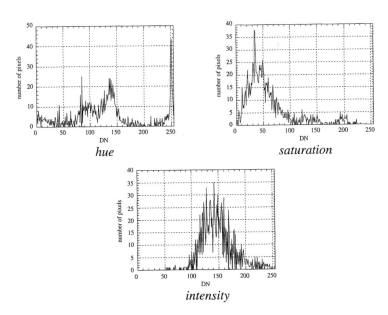

CHAPTER 6

Spatial Transforms

6.1 Introduction

Spatial transforms provide tools to extract or modify the spatial information in remote-sensing images. Some transforms, such as convolution, access only local image information, i.e. within relatively small neighborhoods of a given pixel. Others, for example the Fourier transform, can process the full image for global spatial structure. Between these two extremes, an increasingly important family of scale-space filters, including Gaussian and Laplacian pyramids and the wavelet transform, provide a data representation that allows access to spatial information over a wide range of scales, from local to global.

229

6.2   An Image Model for Spatial Filtering

A useful concept for understanding spatial filtering is that any image is made from spatial components at different scales. As an example, suppose we take an image and process it such that the value at each output pixel is the average of a small neighborhood of input pixels (say 3 x 3), as illustrated in Fig. 6-1. The result is a blurred version of the original image. Now we subtract this result from the original, yielding the image on the right in Fig. 6-1, which represents the difference between each original pixel and the average of its neighborhood. To look at spatial information at a different scale, we repeat the process with a larger neighborhood, such as 7 x 7 pixels. Anticipating our later discussion, we call the blurred image a *Low-Pass (LP)* version of the image and the difference between it and the original image a *High-Pass (HP)* version, and write the mathematical relation,

$$\text{image}(x, y) \ = \ LP(x, y) + HP(x, y) \qquad\qquad (6-1)$$

which is valid for any size (scale) neighborhood. As the neighborhood size is increased, the LP image isolates successively larger and larger structures, while the HP image picks up the smaller structures lost in the LP image, to maintain the relation in Eq. $(6-1)$.

This *decomposition* of an image into a sum of components at different scales is the basis of all spatial filtering. The inverse process, namely adding the components together to synthesize the image, is *superposition.*[1] In Eq. $(6-1)$, each of the two components actually contains a range of scales. In a similar fashion, scale-space filtering decomposes an image into several components, each containing a range of scales. We will see later that it is also possible to decompose an image into a large set of components, each representing only a single scale, using the Fourier transform.

6.3   Convolution Filters

The underlying operation in a convolution filter is the use of a moving window on the image. An operation is performed on the pixels within the window, a calculated value is put in the output image, usually at the same coordinate as the current center coordinate of the window, and the window is then moved one pixel along the same line to process the next neighborhood. When a line of pixels is finished, the window is then moved down one row and the process is repeated

---

[1]   These dual transformations are sometimes called *analysis* and *synthesis*, particularly in the wavelet literature.

$$image(x,y) \quad - \quad LP(x,y) \quad = \quad HP(x,y)$$

3 x 3 neighborhood

7 x 7 neighborhood

FIGURE 6-1. Examples of the global spatial frequency image model at two scales. The difference image has both positive and negative DNs and must be scaled to positive GLs for display.

(Fig. 6-2). Almost any function can be programmed within the moving window; some examples are listed in Table 6-1 and discussed below in the following sections.

6.3.1 Linear Filters

In Chapter 3 we introduced *convolution* as a fundamental physical process behind instrument measurements. Convolution in that context described the effect of the system response function (spectral or spatial) on the resolution of data provided by the instrument. The same type of operator is also quite useful in processing digital images. In this case, however, we *specify* the response function, and have nearly complete freedom to use *any* weighting function suited to the application at hand. That flexibility makes convolution one of the most useful tools in image processing. In effect, we are creating a *virtual instrument,*

FIGURE 6-2. A moving window for spatial filtering. The first output pixel is calculated (left), the next output pixel is calculated for the same row (center), and after the row is completed the process is repeated for the next row (right). The output pixel is located at the coordinate of the shaded pixel.

TABLE 6-1. Catalog of local filter types.

type	output	examples	applications
linear	weighted sum	LPF, HPF, HBF.BPF	enhancement sensor simulation
statistical	given statistic	minimum, maximum, median, standard deviation, mode	noise removal feature extraction *SNR* measurement
gradient	vector gradient	Sobel, Roberts	edge detection

with a given response, and applying it to an input image. The output image represents the output of the virtual instrument.

Convolution

A linear filter is calculated in the spatial domain as a weighted sum of pixels within the moving window. This *discrete convolution* between the input image f and the window response function w, both of size N_x x N_y, is written mathematically for the output pixel g_{ij},

$$g_{ij} = \sum_{m=0}^{N_x-1} \sum_{n=0}^{N_y-1} f_{mn} w_{i-m,j-n} \qquad (6-2)$$

and expressed symbolically in the convenient form,

$$g = f * w. \qquad (6-3)$$

Since the non-zero extent of the window is typically much smaller than the image, the sum in Eq. (6 – 2) does not have to be over every pixel. If the window is W_x x W_y pixels, we can write an alternate expression,[2]

$$g_{ij} = \sum_{m = i - \frac{W_y}{2}}^{i + \frac{W_y}{2}} \sum_{n = j - \frac{W_x}{2}}^{j + \frac{W_x}{2}} f_{mn} w_{i-m, j-n}, \qquad (6-4)$$

where w is centered at (0,0) and is nonzero over $\pm W_x/2$ and $\pm W_y/2$. In this form, we can clearly see that the output pixel is a weighted sum of pixels within a neighborhood of the input pixel.

The distinguishing characteristic of a linear filter is the *principle of superposition*, which states that the output for a sum of two or more inputs is equal to the sum of the individual outputs that would be produced by each input separately. This is achieved with a convolution because Eq. (6 – 2) is a linear weighted sum of the input pixels. Furthermore, the filter is *shift–invariant* if the weights do not change as the window moves across the image. The reader is advised to compare Eq. (6 – 3) and Eq. (6 – 4) with Eq. (3 – 1) and Eq. (3 – 2) to better appreciate the relationship between the continuous convolution of physical functions and the discrete convolution of data arrays.

The implementation of Eq. (6 – 2) involves the following steps:

1. flip the window function in x and y (equivalent to a 180° rotation)
2. shift the window such that it is centered on the pixel being processed
3. multiply the window weights and the corresponding original image pixels
4. add the weighted pixels and save as the output pixel
5. repeat steps 2 through 4 until all pixels have been processed

The first step, flipping of the window, is often forgotten because many window functions are symmetric. However, for asymmetric window functions, it is important in determining the result.

Low-pass and high-pass filters (LPF, HPF)

Equation (6 – 1) defines a complementary pair of filters, the sum of which equals a unit filter. For example, a 1 x 3 LP-filter has the weights [1/3 1/3 1/3], and its

---

[2] If W is an odd number, then $W/2$ is rounded down, e.g. for W equal 5, $W/2$ equals 2.

complementary HP-filter is [-1/3 2/3 -1/3]. Their sum is a filter with the weights
[0 1 0]. Examples of 1 x 3 and 1 x 7 filters applied to a one-dimensional signal
are shown in Fig. 6-3 and Fig. 6-4. The LP-filters preserve the local mean (the
sum of their weights is one) and smooth the input signal; the larger the window,
the more the smoothing. The HP-filters remove the local mean (the sum of their
weights is zero) and produce an output which is a measure of the deviation of the
input signal from the local mean. These same characteristics pertain to two-
dimensional LP- and HP-filters (Table 6-2).

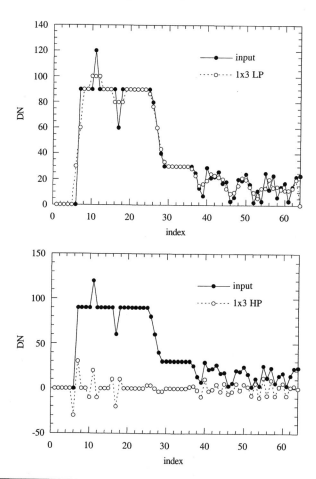

*FIGURE 6-3. 1–D signal processed with 1 x 3 LP- and HP-filters. The LP result
approximates the original signal, but is somewhat smoother. The HP result has zero
mean and zero-crossings wherever there is step or pulse in the original signal.*

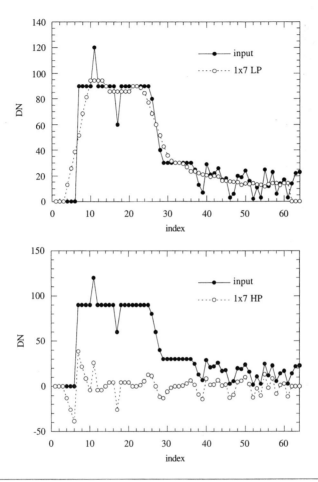

FIGURE 6-4. *1–D signal processed with 1 x 7 LP- and HP-filters. The LP result is smoother than that for the 1 x 3 LP-filter, and the HP result has a broader response around the zero-crossings.*

High-boost filters (HBF)

The basic LP- and HP-filters can be additively combined to form a wide variety of more complex filters. For example, a *high-boost filter* can be created by adding a weighted HP version of an image to the original,

$$HB(x, y;K) = original(x, y) + K \cdot HP(x, y) \qquad (6-5)$$

TABLE 6-2. Examples of simple box filters, which have uniform weights in the LP-filter and the complementary weights in the HP-filter. The normalizing factor is required to preserve the relation in Eq. (6 – 1).

size	LPF	HPF
3 x 3	$1/9 \cdot \begin{bmatrix} +1 & +1 & +1 \\ +1 & +1 & +1 \\ +1 & +1 & +1 \end{bmatrix}$	$1/9 \cdot \begin{bmatrix} -1 & -1 & -1 \\ -1 & +8 & -1 \\ -1 & -1 & -1 \end{bmatrix}$
5 x 5	$1/25 \cdot \begin{bmatrix} +1 & +1 & +1 & +1 & +1 \\ +1 & +1 & +1 & +1 & +1 \\ +1 & +1 & +1 & +1 & +1 \\ +1 & +1 & +1 & +1 & +1 \\ +1 & +1 & +1 & +1 & +1 \end{bmatrix}$	$1/25 \cdot \begin{bmatrix} -1 & -1 & -1 & -1 & -1 \\ -1 & -1 & -1 & -1 & -1 \\ -1 & -1 & +24 & -1 & -1 \\ -1 & -1 & -1 & -1 & -1 \\ -1 & -1 & -1 & -1 & -1 \end{bmatrix}$

Example HB box filters are given in Table 6-3; the sum of their weights is one, which means that the output image will have the same mean *DN* as the input image. Image examples are shown in Fig. 6-5 for different values of *K*. The amount of *edge enhancement* is proportional to *K*. In this process, we are reinforcing ("boosting") the high frequency components of the original image, relative to the low frequency components (Wallis, 1976; Lee, 1980). The parametric nature of the filter allows "tuning" of *K* to achieve the degree of enhancement desired and would be ideally implemented in an interactive mode.

TABLE 6-3. Example 3 x 3 HB box filters for different values of K.

$K = 1$	$K = 2$	$K = 3$
$1/9 \cdot \begin{bmatrix} -1 & -1 & -1 \\ -1 & +17 & -1 \\ -1 & -1 & -1 \end{bmatrix}$	$1/9 \cdot \begin{bmatrix} -2 & -2 & -2 \\ -2 & +25 & -2 \\ -2 & -2 & -2 \end{bmatrix}$	$1/9 \cdot \begin{bmatrix} -3 & -3 & -3 \\ -3 & +33 & -3 \\ -3 & -3 & -3 \end{bmatrix}$

Directional filters

Directional enhancement of image features is a useful application of spatial filter-ing. Some directional filters are shown in Table 6-4. These are all variations on the notion of a discrete derivative, which is a type of HP-filter. Fig. 6-6 depicts the results of preferential processing for oriented-features. Note how the diagonal

FIGURE 6-5.   Example application of 3 x 3 HB-filters from Table 6-3. Each image is stretched or saturated to [0,255], as necessary.

filter accentuates horizontal *and* vertical edges (in addition to those at –45°), because they both have components in the diagonal direction. Considerable care obviously must be exercised when interpreting spatially-filtered images, particularly directional enhancements, because of their abstract nature.

The border region

In spatial filtering, it is desirable to have the size of the output and input images equal, because that allows further algebraic operations such as discussed at the beginning of this chapter. If a W x W window (W being an odd integer) is used for a local spatial filter, the *border region* includes the first and last $W/2$ (truncated to an integer) rows and columns of the input image (Fig. 6-7). Output pixels within

TABLE 6-4. Example directional filters. Features in the indicated direction will be enhanced by the filter. For the azimuthal filter, the angle α is measured counterclockwise from the horizontal axis; features in the directions $\alpha \pm 90°$ will be enhanced.

vertical	horizontal	diagonal		azimuthal
$\begin{bmatrix} -1 & +1 \end{bmatrix}$	$\begin{bmatrix} -1 \\ +1 \end{bmatrix}$	$\begin{bmatrix} -1 & 0 \\ 0 & +1 \end{bmatrix}$	$\begin{bmatrix} 0 & -1 \\ +1 & 0 \end{bmatrix}$	$\begin{bmatrix} \sin\alpha & 0 \\ -\sin\alpha-\cos\alpha & \cos\alpha \end{bmatrix}$
$\begin{bmatrix} -1 & +2 & -1 \end{bmatrix}$	$\begin{bmatrix} -1 \\ +2 \\ -1 \end{bmatrix}$	$\begin{bmatrix} -1 & 0 & 0 \\ 0 & +2 & 0 \\ 0 & 0 & -1 \end{bmatrix}$	$\begin{bmatrix} 0 & 0 & -1 \\ 0 & +2 & 0 \\ -1 & 0 & 0 \end{bmatrix}$	

the border region cannot be directly calculated, and since the window cannot extend beyond the borders of the original image, some "trick" must be used to maintain the size of the image from input to output. Any number of techniques can be used to fill this area and make the output image the same size as the input image:

- Repeat the nearest valid output pixel in each border pixel.
- Reflect the input pixels in the border area outward to effectively increase the size of the input image. The output image is then the same size as the original input image.
- Reduce the size of the window in the border area.
- Set the border pixels to zero or to the mean *DN* of the output image.
- Wrap the window around to the opposite side of the image and include those pixels in the calculation. This *circular convolution* is equivalent to the discrete Fourier implementation of convolution.

Each approach causes a different type of artifact around the border. In general, the first and second are most effective in maintaining image size without introducing severe artifacts.

Characteristics of filtered images

The LP-component of an image is statistically *nonstationary*, i.e. its properties (local mean and variance) change from point to point, while the HP-component can be modeled statistically as resulting from a *stationary*, zero mean Gaussian process (Hunt and Cannon, 1976). One sees this empirically in that the histograms of HP-images universally exhibit a Gaussian-like shape with zero mean.

FIGURE 6-6. Examples of directional enhancement using derivative filters. The direction of the derivative is indicated by the arrows. In some commercial art image processing programs, this effect is called "embossing" or "relief" because of the 3–D impression conveyed by the filtering.

The variance of a HP-image is typically much less than that of the original image. The histogram of a LP-image resembles that of the original image, with a slightly reduced *DN* range (Fig. 6-8).

The box-filter algorithm

The simplest LP box filter can be programmed in a very efficient *recursive* form. Consider a 3 x 3 linear convolution filter, as shown in Fig. 6-9. A straightforward

FIGURE 6-7. The border region for a 3 x 3 filter. Some "trick" must be used to calculate the output pixels at the shaded locations.

calculation of the average at each pixel would require eight additions.[3] However, if we save and update the sum of the input pixels in each of the three columns of the window (C_1, C_2, C_3), the eight additions need to be computed only once at the beginning of an image line. The next output pixel is calculated as,

$$\text{output pixel} = C_2 + C_3 + C_4$$
$$= C_1 + C_2 + C_3 - C_1 + C_4 \qquad , \qquad (6-6)$$
$$= \text{previous output pixel} - C_1 + C_4$$

an algorithm that can obviously be extended to all pixels in the line. Thus, except for the first output pixel of each line, only three additions and one subtraction are needed to calculate each output pixel for a 3 x 3 filter, yielding a computational advantage of two for the recursive algorithm over the direct calculation. The advantage obviously increases with the size of the window, and the recursion algorithm also can be applied in the vertical direction by maintaining three full lines of data in memory for the previously calculated output image line, the oldest input image line and the newest input image line (McDonnell, 1981). With a full bidirectional implementation of the recursion algorithm, the number of computations required for each output pixel is *independent of the size of the filter window.*

The box-filter algorithm can be applied along image rows only if the filter weights are constant within each row of the window. In that case, it is unquestionably the fastest algorithm for convolution. Otherwise, Fourier domain

---

[3] We are ignoring the multiplication by the filter weights. For the LP box filter, this multiplication can be done after summing the pixels in the window, and is one additional operation.

FIGURE 6-8. Histograms of a TM image, and its LP- and HP-components. A 5 x 5 convolution window was used to generate the two components. Note the larger amplitudes of the LP and HP histograms because of their narrower widths and that the HP histogram has a near-zero mean, with both positive and negative DNs.

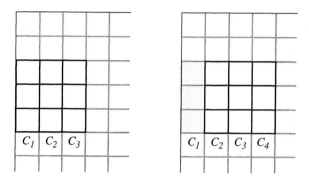

FIGURE 6-9. *Depiction of the box-filter algorithm applied to neighboring regions along an image row. The weighted average of pixels in the window column j is C_j.*

implementation of the convolution can be a competitive alternative (see Sect. 6.4.4), particularly for large windows.

Cascaded linear filters

A series of filters applied sequentially to an image can be replaced by a single, net filter that is equal to the convolution of the individual filters,

$$g = (f * w_1) * w_2 = f * (w_1 * w_2) = f * w_{net} \qquad (6-7)$$

where,

$$w_{net} = w_1 * w_2. \qquad (6-8)$$

Suppose that w_1 and w_2 are each $W \times W$ pixels in size. The net window function w_{net} is then $(2W - 1) \times (2W - 1)$; examples are given in Table 6-5. Any number of cascaded filters may be replaced by a single filter, as long as there are no nonlinear operations (such as DN thresholding) in the processing chain. This property of linear filters was used in Chapter 3 to "build" the net sensor spatial response from the individual component responses.

6.3.2 Statistical Filters

Statistical filters measure a local statistical property of an image, and use it for the pixel values of the output image. Statistical measures calculated over small neighborhoods have low statistical significance because of the small sample size,

TABLE 6-5. Example cascaded filters and their equivalent net filter.

filter 1	filter 2	net filter
$1/9 \cdot \begin{bmatrix} +1 & +1 & +1 \\ +1 & +1 & +1 \\ +1 & +1 & +1 \end{bmatrix}$	$1/9 \cdot \begin{bmatrix} +1 & +1 & +1 \\ +1 & +1 & +1 \\ +1 & +1 & +1 \end{bmatrix}$	$1/81 \cdot \begin{bmatrix} +1 & +2 & +3 & +2 & +1 \\ +2 & +4 & +6 & +4 & +2 \\ +3 & +6 & +9 & +6 & +3 \\ +2 & +4 & +6 & +4 & +2 \\ +1 & +2 & +3 & +2 & +1 \end{bmatrix}$
$1/3 \cdot \begin{bmatrix} +1 & +1 & +1 \end{bmatrix}$	$1/3 \cdot \begin{bmatrix} +1 \\ +1 \\ +1 \end{bmatrix}$	$1/9 \cdot \begin{bmatrix} +1 & +1 & +1 \\ +1 & +1 & +1 \\ +1 & +1 & +1 \end{bmatrix}$

but are nevertheless useful for tasks such as noise reduction or texture feature extraction.

Morphological filters

The *minimum filter* creates an output image in which the value of each pixel is the local minimum *DN* of the input image. The *maximum filter* creates an output image from the local maximum *DN*. These two filters are similar to *dilation* and *erosion filters*, which are applied to binary images. If a minimum filter is applied to a binary image of dark and light objects, the result–expansion of the dark objects–is the same as that of a dilation filter. A maximum filter applied to a binary image is equivalent to an erosion filter. If the two morphological operations are cascaded, an *opening* or *closing* operation is obtained,

$$\text{opening(binary image)} = \text{dilation[erosion(binary image)]} \qquad (6-9)$$

$$\text{closing(binary image)} = \text{erosion[dilation(binary image)]} \ . \qquad (6-10)$$

This type of image processing is illustrated in Fig. 6-10. It has application to spatial segmentation and noise removal.

Dilation and erosion filters are examples of *morphological image processing* (Serra, 1982; Giardina and Dougherty, 1988). The window shape is important because it affects the changes induced in binary objects; in addition to being simply square or rectangular, it can be a plus shape, a diagonal shape, or in fact, any desired pattern. In morphological image processing, the shaped window is called a *structuring element* and can be designed to perform pattern matching or *original*   *local minimum*   *local maximum*

thresholded original   *local minimum (dilation)*   *local maximum (erosion)*

closing   *opening*

FIGURE 6-10. Examples of minimum and maximum filter processing of a TM band 4 image of Dulles International Airport, Virginia. The lower two rows illustrate morphological filters applied to a binary version of the original. A 3 x 3 pixel window is used in all cases. The terms "erosion" and "dilation" always refer to the dark objects in a binary image, independent of its content. In our example, the dilation operation "erodes" the airport, and the erosion operation "dilates" the airport!

modification of particular shapes. A good set of examples is presented in (Schalkoff, 1989).

The *median filter* is a particularly useful statistical filter. If the pixel *DN*s within the window are sorted into decreasing or increasing order, the output of the median filter is the *DN* of the pixel at the middle of the list (the number of pixels must be odd). The median operation has the effect of excluding pixels that do not fit the "typical" statistics of the local neighborhood, i.e. outliers. Isolated noise pixels can therefore be removed with a median filter. One can see this by generating a sequence of numbers with one or more outliers, for example,

$$10, 12, 9, 11, 21, 12, 10, \text{ and } 10 \text{ .}$$

If the median of five points is calculated in a moving window, the result is (excluding the two border points on either end),

$$. \quad . \quad , 11, 12, 11, 11, \quad . \quad .$$

and the outlier value, 21, has magically disappeared! If we apply a median to the data of Fig. 6-3, we see that it very neatly removes the two spikes in the data, with minimal smoothing of other structure (Fig. 6-11). Efficient data sorting algorithms, such as HEAPSORT and QUICKSORT (Press *et al.*, 1986), can speed median filter processing for large windows.

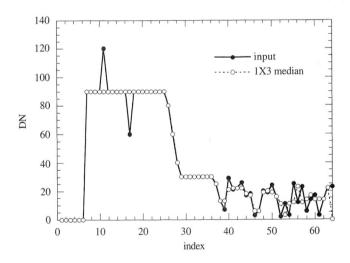

FIGURE 6-11. Application of the median filter to the previously used 1-D dataset.

6.3.3 Gradient Filters

The detection of significant *DN* changes from one pixel to another is a common problem in image processing. Such changes usually indicate a physical boundary in the scene, such as a coastline, a paved road or the edge of a shadow. Although many different approaches to this problem have been studied for many years (Davis, 1975), a combination of high-pass spatial filtering and *DN* thresholding provides a simple and effective technique that is widely used.

The directional high-pass filters of Table 6-4 produce images whose *DN*s are proportional to the difference between neighboring pixel *DN*s in a given direction, i.e. they calculate the *directional gradient*. An isotropic gradient may be calculated by filtering the image in two orthogonal directions, e.g., horizontally and vertically, and combining the results in a vector calculation at every pixel (Fig. 6-12). The *magnitude* of the local image gradient is given by the length of the composite vector,

$$\text{gradient magnitude} = |g| = \sqrt{g_x^2 + g_y^2} \qquad (6-11)$$

The *direction* of the local gradient is given by the angle between the composite vector and the coordinate axis,

$$\text{gradient angle} = \phi = \text{atan}\left(\frac{g_y}{g_x}\right). \qquad (6-12)$$

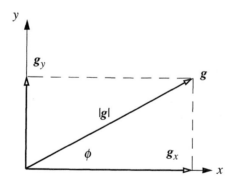

FIGURE 6-12. *Vector geometry for calculating image gradients. The x-derivative of the image at the given pixel is* g_x *and the y-derivative is* g_y. *Their vector sum is* **g**.

Numerous local gradient filters have been proposed (Table 6-6). There often is little visual difference in their results, however (Fig. 6-13). The edge boundaries produced by the 3 x 3 filters are not as sharp as those produced by the Roberts filter. Another difference is that the output of the Roberts filter is shifted by one-half pixel relative to the filters with odd sizes. No matter what filter is used, the overall gradient calculation is nonlinear because of the vector magnitude calculation in Eq. (6 – 11).

TABLE 6-6. Example local gradient filters. See also (Robinson, 1977).

filter	horizontal component	vertical component
Roberts	$\begin{bmatrix} 0 & +1 \\ -1 & 0 \end{bmatrix}$	$\begin{bmatrix} +1 & 0 \\ 0 & -1 \end{bmatrix}$
Sobel	$\begin{bmatrix} +1 & +2 & +1 \\ 0 & 0 & 0 \\ -1 & -2 & -1 \end{bmatrix}$	$\begin{bmatrix} -1 & 0 & +1 \\ -2 & 0 & +2 \\ -1 & 0 & +1 \end{bmatrix}$
Prewitt	$\begin{bmatrix} +1 & +1 & +1 \\ +1 & -2 & +1 \\ -1 & -1 & -1 \end{bmatrix}$	$\begin{bmatrix} -1 & +1 & +1 \\ -1 & -2 & +1 \\ -1 & +1 & +1 \end{bmatrix}$

The *detection* of edges is a binary classification problem that may be addressed with a *DN* threshold applied to the gradient magnitude. A threshold that is too low results in many isolated pixels classified as edge pixels and thick, poorly defined edge boundaries, while a threshold that is too high results in thin, broken segments (Fig. 6-14). These problems arise because we are using only local information in the gradient filters. The use of scale-space filtering and zero-crossing mapping, as described later in this chapter, greatly increases the robustness of edge detection and the spatial continuity of edge maps.

6.4   Fourier Transforms

Fourier theory dates from the Eighteenth century and has applications in almost all areas of science and engineering and even the fine arts, such as stock market analysis! It is a general framework for analysis of signals in one or more original   Prewitt

Roberts   Sobel

FIGURE 6-13. Comparison of the gradient magnitude images produced by common gradient filters.

dimensions as linear combinations of basic sinusoidal functions. We will introduce the topic in the context of synthesis of 1-D signals and 2-D images.

6.4.1 Fourier Synthesis

In Sect. 6.2 we introduced the idea that an image could be represented by a sum of two components with different spatial scales. The Fourier transform is an extension of that idea to *many* scales. We will use a 1–D signal example to illustrate and then extend to two dimensions. In Fig. 6-15 we have a 1–D periodic, infinite square wave signal with a spatial period of 100 units. It can be shown that this signal is composed of an infinite set of sine wave signals of different frequencies, amplitudes and phases (see Fig. 4-15). The first component actually has zero frequency because it represents the mean amplitude of the signal, and is

gradient magnitude $DN_T = 60$

$DN_T = 90$ $DN_T = 120$

FIGURE 6-14. Edge maps produced by thresholding the Roberts gradient magnitude at different levels.

sometimes called the *DC* term.[4] The lowest nonzero frequency component has the same period as the square wave and is known as the *fundamental*, the next highest frequency component, the *third harmonic*, is three times the frequency of the fundamental, and so on. The relative strengths of these components are 1, 1/3, and 1/5 when they are added together to synthesize the original square wave. As each additional higher frequency component is included, the sum approaches a square wave, with the edges of the pulses becoming sharper and the tops and bottoms becoming flatter. We need to add an infinite number of sine waves,

---

[4] DC is an acronym for Direct Current and is used in physics and electrical engineering to describe a constant current from a stable source such as a battery. Its use has persisted to describe the constant mean level of any signal.

however, to retrieve the square wave exactly. The full sum is termed the *Fourier Series* of the square wave.

In Fig. 6-16 and Fig. 6-17 we extend this idea to 2–D, discrete images. Because the images are discrete (arrays of digital numbers), the Fourier series is a finite sum of sines and cosines. If all the terms are included, the original image is obtained. The examples show how larger partial sums provide better approximations to the original image. Each partial sum amounts to a LP version of the image, and the error maps are the complementary HP versions. They satisfy the relation of Eq. (6 – 1), namely,

$$
\begin{aligned}
\text{image} &= \text{partial sum} + \text{error} \\
&= LP + HP \, .
\end{aligned}
\tag{6-13}
$$

6.4.2 Discrete Fourier Transforms in 2–D

The 2–D discrete Fourier series is defined mathematically as,

$$
f_{mn} = \frac{1}{N_x N_y} \sum_{k=0}^{N_x-1} \sum_{l=0}^{N_y-1} F_{kl} \, e^{j2\pi\left(\frac{mk}{N_x} + \frac{nl}{N_y}\right)}
\tag{6-14}
$$

where F_{kl} is the complex amplitude of component kl. The complex exponential term can be written as,[5]

$$
e^{j2\pi\left(\frac{mk}{N_x} + \frac{nl}{N_y}\right)} = \cos 2\pi\left(\frac{mk}{N_x} + \frac{nl}{N_y}\right) + j\sin 2\pi\left(\frac{mk}{N_x} + \frac{nl}{N_y}\right)
\tag{6-15}
$$

The coefficient j is the imaginary number $\sqrt{-1}$.

Equation (6 – 14) thus represents the *superposition of $N_x N_y$ cosine and sine waves to synthesize the original image f_{mn}*. The inverse of Eq. (6 – 14) is defined as,

$$
F_{kl} = \sum_{m=0}^{N_x-1} \sum_{n=0}^{N_y-1} f_{mn} \, e^{-j2\pi\left(\frac{mk}{N_x} + \frac{nl}{N_y}\right)}
\tag{6-16}
$$

which *yields the Fourier coefficients, given the image*. Equation (6 – 16) is the *Discrete Fourier Transform (DFT)* of f_{mn} and Eq. (6 – 14) is the *inverse DFT of F_{kl}*.

When we do the 2–D Fourier transform of a digital image, we implicitly assume that the image is replicated infinitely in all directions (Fig. 6-18), by

---

[5]   This fundamental relation is known as *Euler's Theorem.*

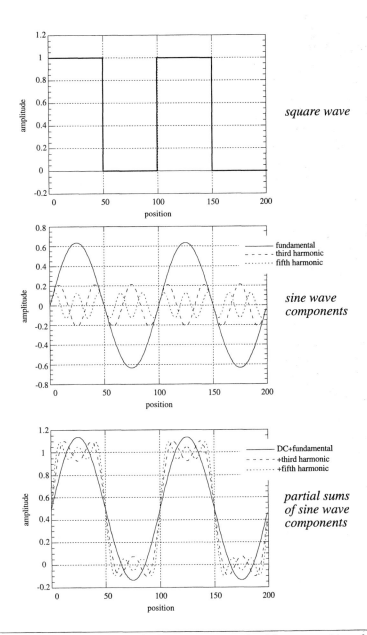

FIGURE 6-15. *Fourier synthesis of a 1–D square wave signal by superposition of sine wave signals. Even with only the DC term and three sinusoids, the synthesized signal is a fair approximation of the square wave.*

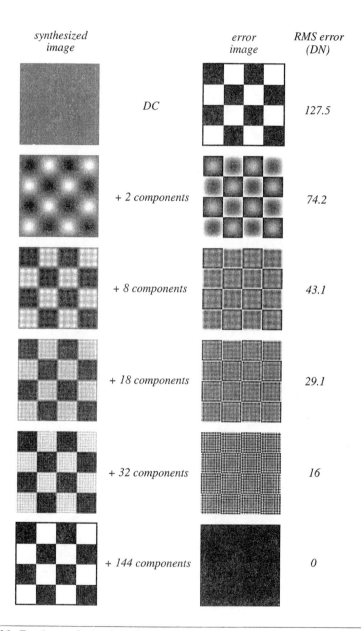

FIGURE 6-16. Fourier synthesis of a 2–D square wave. The error images are stretched to [0,255]. Only the nonzero Fourier components are counted in the partial sums.

FIGURE 6-17. Fourier synthesis of a TM image. The error images are stretched to
[0,255]. The 64 x 64 image has 4096 complex Fourier coefficients, but because of
symmetry, only 2048 are unique.

virtue of the periodicity of the component cosines and sines.[6] Likewise, its Fourier transform is infinitely periodic. It is this periodic representation that leads to the wraparound effect when using Fourier transforms to implement spatial convolution. The Fourier array, as produced by Eq. (6 – 16), is often reordered to the so-called *natural order*, which arises from optical Fourier transforms (Fig. 6-19).

The spatial frequency units for image processing, when we are not concerned with absolute quantities, are *cycles/pixel*. The spatial frequency *intervals* along each axis are given by,

$$\Delta u = 1/N_x \quad , \quad \Delta v = 1/N_y \quad \text{(cycles/pixel)} \tag{6-17}$$

and the spatial frequency *coordinates* are given by,

$$u = k\Delta u = k/N_x \quad , \quad v = l\Delta v = l/N_y . \text{(cycles/pixel)} \tag{6-18}$$

If absolute units are necessary, the spatial frequency intervals are divided by the pixel intervals Δx and Δy, in units of length, resulting in

$$\Delta u = \frac{1}{N_x \Delta x} \quad , \quad \Delta v = \frac{1}{N_y \Delta y} \quad \text{(cycles/unit length).} \tag{6-19}$$

6.4.3 The Fourier Components

The Fourier transform (Eq. (6 – 16)) produces *real* and *imaginary* components. These are related by the complex number relationship,

$$F_{kl} = Re(F_{kl}) + j \cdot Im(F_{kl}) \tag{6-20}$$

where j is $\sqrt{-1}$. A complex number can equivalently be written in terms of *amplitude*, A_{kl}, and *phase*, ϕ_{kl},

$$F_{kl} = A_{kl} \cdot e^{-j \cdot \phi_{kl}} \tag{6-21}$$

where,

$$A_{kl} = |F_{kl}| = \sqrt{[Re(F_{kl})]^2 + [Im(F_{kl})]^2} \tag{6-22}$$

and

$$\phi_{kl} = \text{atan}[Im(F_{kl})/Re(F_{kl})] \quad . \tag{6-23}$$

---

[6] This is the fundamental connection between the *continuous* Fourier transform and the *discrete* Fourier transform.

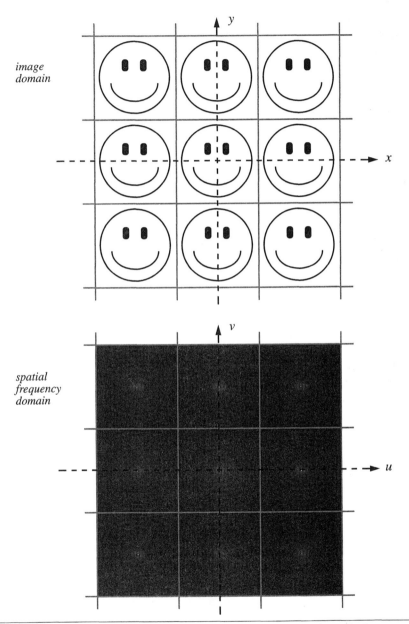

FIGURE 6-18. The implied periodicity of the discrete Fourier transform. Each square above is a copy of the original image and each square below is a copy of the Fourier transform of the image. The replication is known as the "postage stamp" representation!

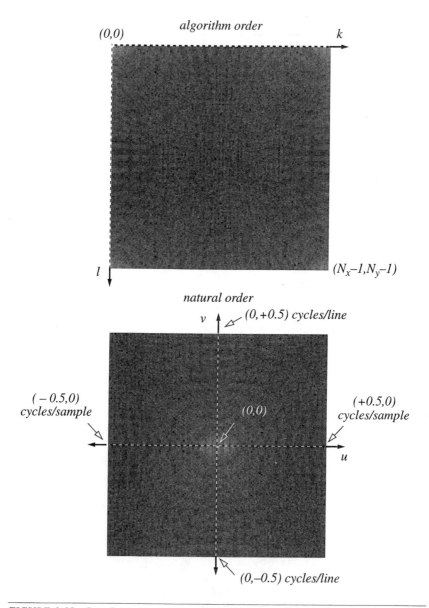

FIGURE 6-19. *Coordinate geometry in the Fourier plane. The upper display shows the spectrum as produced by the FFT algorithm. Because of the periodicity of the array, we can shift the origin, as shown below. The directions of k and l, and u and v are consistent with the discrete and natural coordinate systems described in Chapter 4.*

The various components of the Fourier transform are illustrated in Fig. 6-20.[7]

The phase component is critical to the spatial structure of the image (Oppenheim and Lim, 1981). An impressive demonstration of this fact can be made by setting the amplitude component to a constant value, and doing an inverse Fourier transform of the modified spectrum. The result shows that the phase component carries information about the relative positions of features in the image. Conversely, if the phase part is set to a constant value (zero), with the original amplitude component retained, and the inverse Fourier transform performed, the result is unintelligible (Fig. 6-21). Thus, while the phase component is often ignored in Fourier filter design, it should be treated with respect!

6.4.4 Filtering with the Fourier Transform

The major application of the Fourier transform to filtering is in implementation of Eq. (6 – 2). Taking the Fourier transform of both sides of that equation, we obtain the relatively simple product relation in the Fourier domain,

$$\mathcal{F}[g_{ij}] = G_{kl} = \mathcal{F}[f * w]$$
$$= F_{kl}W_{kl}$$

$$(6-24)$$

The arrays F and G are the *spatial frequency spectra* of the input and output images, respectively. W is the *transfer function* of the filter. To use Fourier transforms for a spatial domain convolution, we must take the Fourier transform of the original image and the window weighting function, multiply the image spectra and the transfer function, which yields the filtered image spectrum, and then take the inverse of that product to obtain the filtered image in the spatial domain. This sequence is depicted in Fig. 6-22. Since the spatial window is normally much smaller than the image, it must be "padded" to the same size before taking its Fourier transform to insure that the Fourier components are at the same locations for both F and W (Eq. (6 – 17)). This padding is accomplished by surrounding the window function with zeros. If one filter is to be applied to many different images, its transfer function can be calculated once and used directly in the multiplication step, thus avoiding one of the Fourier transform operations.

The amplitude and phase components of Eq. (6 – 24) are given by,

$$|G_{kl}| = |F_{kl}||W_{kl}|$$
$$\phi(G_{kl}) = \phi(F_{kl}) + \phi(W_{kl})$$

$$(6-25)$$

---

[7] The amplitude component is difficult to display as a greyscale image because it has a large dynamic range, dominated by the DC value. One way to handle this is to set the DC component to zero, and then take the logarithm of the result, which reduces the dynamic range.

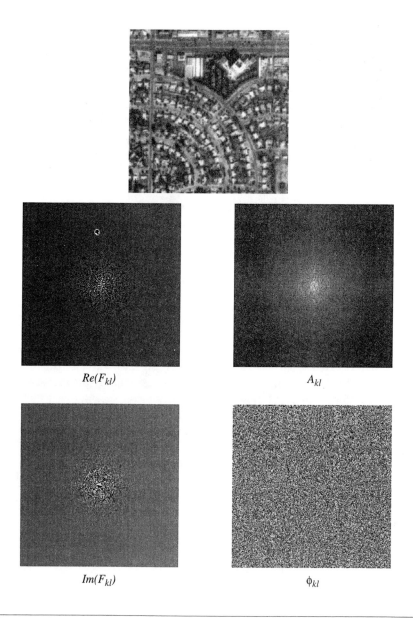

$Re(F_{kl})$ A_{kl}

$Im(F_{kl})$ ϕ_{kl}

FIGURE 6-20. The Fourier components of an image. To completely specify the transform, only the real and imaginary, or amplitude and phase components are needed. The natural coordinate system is used in this figure.

phase only ($A_{kl} = 1$)   *amplitude only ($\phi_{kl} = 0$)*

FIGURE 6-21. Evidence for the importance of spatial phase information. The spectrum of the image in Fig. 6-20 was modified as indicated and inverse Fourier transformed to the spatial domain.

These equations describe how the filter W affects the modulation and spatial phase, respectively, of the input digital image spectrum F.

A general guide for choosing spatial or Fourier domain processing is to use spatial domain convolution if the window is 7 x 7 or smaller and Fourier domain filtering otherwise (Pratt, 1991). Of course this "rule" depends also on the size of the image, the speed of the particular algorithms used, and the availability of special purpose hardware for performing convolutions or FFTs. The most efficient software implementation of Eq. (6 – 14) is the *Fast Fourier Transform (FFT)* (Brigham, 1988). Common FFT algorithms require that the size of the input array is a power of two; less efficient algorithms can transform images of other sizes. The choice in remote sensing is usually in favor of spatial domain convolution because of the large size of remote-sensing images, although the steadily increasing computation speed of computers makes the Fourier domain approach worth considering in some cases. As we will see in Chapter 7, the Fourier approach provides a unique view of the data that is quite useful for analysis and removal of periodic noise components.

Transfer functions

One of the most significant advantages of Eq. (6 – 24) over Eq. (6 – 2) is that it allows us to view filters as multiplicative "masks" in the spatial frequency domain. For example, an "ideal" amplitude LP-filter has the characteristics,

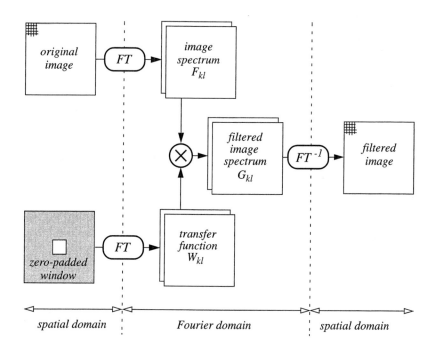

FIGURE 6-22. A filtering algorithm that uses the Fourier transform to compute a spatial domain convolution. The Fourier domain arrays are shown as doubles because the data are complex in that domain.

$$
\begin{aligned}
|W_{kl}| &= 1, |k| \le k_c \, , \, |l| \le l_c \\
|W_{kl}| &= 0, |k| > k_c \, , \, |l| > l_c
\end{aligned}
\qquad (6-26)
$$

where k_c and l_c are the effective "cutoff" frequencies for the LPF. The ideal LPF is a binary mask in the Fourier domain that transmits frequency components below the cutoff frequency unchanged, and does not transmit frequency components above the cutoff frequency. The amplitude filter, $|W_{kl}|$, is called the *Modulation Transfer Function (MTF)*. The MTFs of some of the box filters described earlier are shown in Fig. 6-23.

When filters are viewed in the frequency domain, two potential problems with the simple box filters described earlier are seen–their lack of rotational symmetry and inconsistent behavior at higher frequencies. The latter is an indication of spatial *phase reversal*, which can cause artifacts for small, periodic targets (Castleman, 1996). For simple visual enhancement, these problems may not matter, but

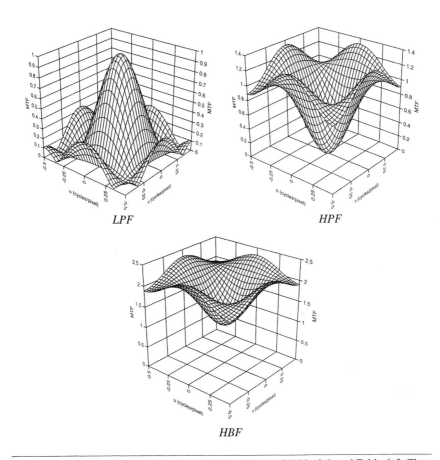

FIGURE 6-23. The 2-D MTFs for the 3 x 3 box filters of Table 6-2 and Table 6-3. The HBF is for a K value of one. The MTF of the 3 x 3 LPF is considerably different from that of an ideal LPF. Note also how each of these filters shows significantly different responses in different directions, i.e. they are not rotationally symmetric.

for modeling and more precise work, it is desirable to use filters that have an equal response in all directions (unless one is specifically modeling an asymmetric response). If a Gaussian function is used within the square window, the situation is improved. For example, a 3 x 3 Gaussian LPF, with a 1/e radius of 1.5 pixels has the weights,

$$w_g = \begin{bmatrix} +0.079 & +0.123 & +0.079 \\ +0.123 & +0.192 & +0.123 \\ +0.079 & +0.123 & +0.079 \end{bmatrix}$$ (6 – 27)

and the MTF shown in Fig. 6-24. The Gaussian is still spatially truncated by the 3 x 3 window, but the effect in the frequency domain is much less than for a box filter. A larger window would reduce the remaining rotational asymmetry. HP- and HB-filters corresponding to Eq. (6 – 27) can be easily derived.

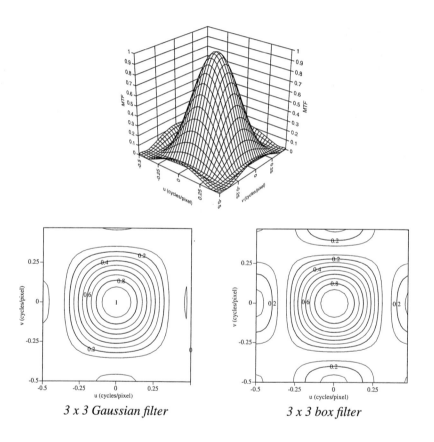

3 x 3 Gaussian filter   *3 x 3 box filter*

FIGURE 6-24. *The MTF of a Gaussian filter with a spatial radius of 1.5 pixels and truncated to 3 x 3 pixels. The contour map is compared to that for a box filter at the bottom.*

6.4.5 The Power Spectrum

Applications of the Fourier transform also exist in image pattern analysis and recognition. In particular, the *power spectrum* (the square of the Fourier amplitude function) is a useful tool. Application of the power spectrum for fractal analysis of a TM image was presented in Chapter 4, including some example spectra (Fig. 4-26). Additional example power spectra for 256 x 256 aerial photographic images are shown in Fig. 6-25 and Fig. 6-26; more examples can be found in (Jensen, 1996) and (Schott, 1996).

desert

fields

FIGURE 6-25. *Two aerial photographs with little spatial structure and their respective power spectra.*

From these examples and knowledge of the Fourier transform we can make the correspondances of Table 6-7. Because the spatial frequency power spectrum *localizes* information about *global* patterns in the spatial domain, it is useful as a

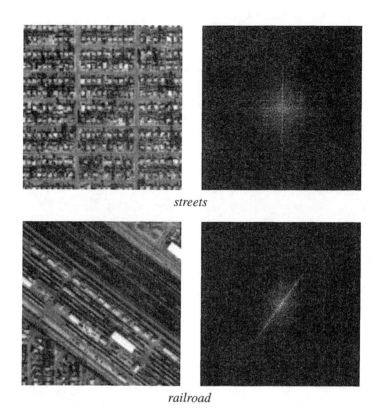

streets

railroad

FIGURE 6-26. *Two aerial photographs with strong spatial structure and their respective power spectra.*

diagnostic tool for global periodic noise or as a pattern recognition tool for global spatial patterns. In Chapter 7, we will illustrate its use in designing noise filters.

6.5 Scale-Space Transforms

In many cases, we would like to extract the spatial information from an image over a range of scales, from fine details in local areas to large features that extend across the image. The human vision system does a remarkable job at this task, without even thinking about it! Implementation of a similar capability in computer algorithms is a challenging task. A number of approaches are promising,

TABLE 6-7. Descriptive relationships between the spatial and spatial frequency domains.

spatial description	spatial frequency description
periodic patterns	high amplitude "spikes," localized at the frequencies of the patterns
linear, quasi-periodic features	high amplitude "line" through DC, oriented orthogonal to the spatial patterns
nonlinear, aperiodic features	high amplitude "cloud," primarily at lower frequencies

however, and we have grouped them within the category *scale-space filtering*. The algorithms generally behave as filters, but are applied repeatedly on scaled versions of the image (resolution pyramids), or the filter itself is scaled (zero-crossing filters). Some of the ideas, such as the scale-space LoG filter and zero-crossings, originated in human vision system models (Marr and Hildreth, 1980; Marr, 1982; Levine, 1985).

6.5.1 Image Resolution Pyramids

Image resolution pyramids are a way to efficiently include global, intermediate and local scales in an analysis. A pyramid can be defined as,

$$i_L = \text{REDUCE}(i_{L-1}) \qquad (6-28)$$

where i_L is the image at level L and REDUCE is any operation on the image at level *L-1* that reduces its size for level *L*. For example, REDUCE can be simply an average over 2 x 2 pixel neighborhoods, with an equal down-sampling[8] along rows and columns, producing a *box pyramid* (Fig. 6-27). The linear size of the image at level *L* is related to the size at level *L-1* by,

$$N_L = \frac{N_{L-1}}{2^L}. \qquad (6-29)$$

One popular REDUCE operator was proposed in (Burt, 1981) and (Burt and Adelson, 1983). The weighting function in this case is separable,

$$w_{mn} = w1_m \cdot w2_n \qquad (6-30)$$

---

[8] *Down-sampling* means sampling of every other row and column of an image, discarding the intermediate pixels. Conversely, *up-sampling* means insertion of rows and columns of zero *DN* between existing rows and columns.

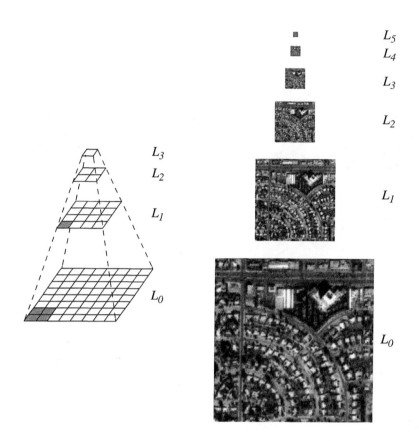

FIGURE 6-27. Construction of a 2 x 2 box pyramid and an example of six levels, starting with a 256 x 256 image. The grey area in level 0 indicates the four pixels that are averaged to calculate the lower left pixel in level 1.

where both 1-D functions are parametric,

$$w = \begin{bmatrix} 0.25 - a/2, & 0.25, & a, & 0.25, & 0.25 - a/2 \end{bmatrix}. \qquad (6-31)$$

A Gaussian-like function is generated with a equal to 0.4 in both directions. The 2-D weighting function is then,

$$
w_{mn} = \begin{bmatrix}
0.0025 & 0.0125 & 0.0200 & 0.0125 & 0.0025 \\
0.0125 & 0.0625 & 0.1000 & 0.0625 & 0.0125 \\
0.0200 & 0.1000 & 0.1600 & 0.1000 & 0.0200 \\
0.0125 & 0.0625 & 0.1000 & 0.0625 & 0.0125 \\
0.0025 & 0.0125 & 0.0200 & 0.0125 & 0.0025
\end{bmatrix} . \tag{6-32}
$$

In the REDUCE operation the image is convolved with the weighting function, but only every other calculated pixel and line are used in the next level. Thus, a full convolution is not necessary, making the algorithm particularly efficient. If the weights in Eq. (6 – 32) are used, a *Gaussian pyramid* results (Burt, 1981), as shown in Fig. 6-28.

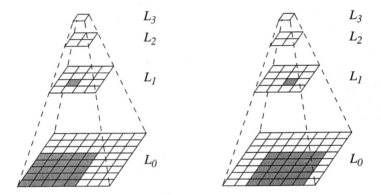

FIGURE 6-28. Gaussian pyramid construction with the 5 x 5 pixel weighting function of Eq. (6 – 32). On the left, the weighted average of 25 pixels in level 0 gives the shaded pixel in level 1. On the right, the weighting function is moved two pixels along a row and the weighted average of the corresponding pixels gives the next pixel in level 1. This process is repeated across the image in level 0, and then performed on level 1 to calculate level 2, and so forth. In this way, the linear size of the image is reduced by two from level to level. This is equivalent to a full convolution of level 0 with the weighting function, followed by a down-sample, but avoids unnecessary calculations. The border pixels require special attention as discussed earlier.

The REDUCE operation with this filter is depicted in Fig. 6-29. The corresponding EXPAND operation consists of expansion of the image at level L by insertion of one row and column of zeros between existing rows and columns, and convolution of the resulting array with the filter in Eq. (6 – 32), to yield the image at level L-1.

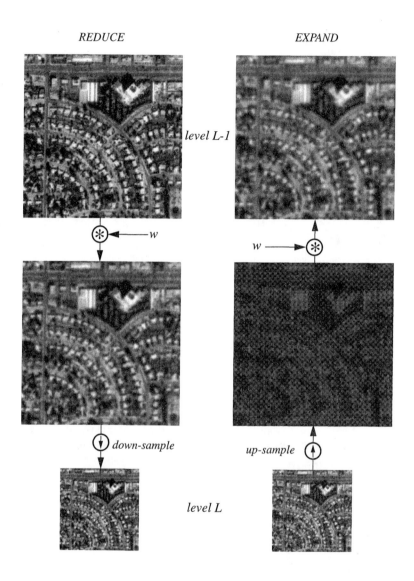

REDUCE   *EXPAND*

level L-1

down-sample   *up-sample*

level L

FIGURE 6-29. The REDUCE and EXPAND procedures as defined in (Burt and Adelson, 1983). Any spatial filter can be used, but these examples use the Gaussian window of Eq. (6 – 32).

The process used to create and reconstruct one level of the Gaussian pyramid from another level is shown in Fig. 6-30. Also shown is level 0 of the *Laplacian pyramid*, constructed by subtracting the up-sampled, reconstructed Gaussian level 1 image from the level 0 image. The Laplacian pyramid is useful for image coding and compression (Burt and Adelson, 1983) and for finding contrast edges at different scales as discussed in the next section. The box and Gaussian pyramids are compared in Fig. 6-31. The Gaussian algorithm avoids the discontinuities characteristic of aliasing due to undersampling (Chapter 3) and produces a smoother image at each level.

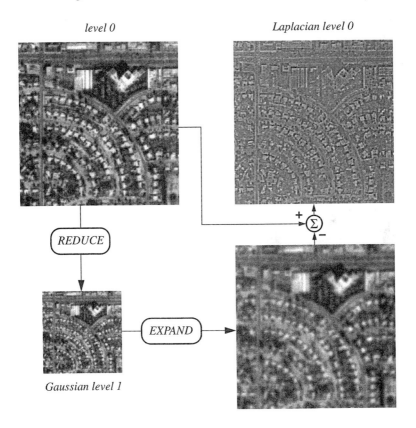

FIGURE 6-30. *The construction of level 1 of the Gaussian pyramid and level 0 of the Laplacian pyramid. Level 2 would be constructed by the same process, starting at level 1. The Laplacian at level 1 would be calculated as the difference between the level 1 image and the EXPANDed Gaussian level 2 image.*

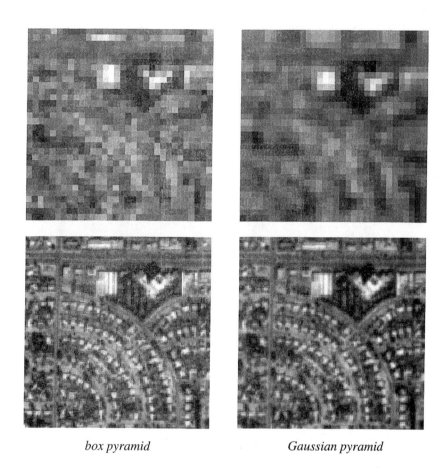

box pyramid   *Gaussian pyramid*

FIGURE 6-31. Level 3 and level 1 images compared for the box and Gaussian
pyramids. The level 3 images are magnified to the level 1 scale by pixel replication.

Because of the down-sampling operation, the Gaussian pyramid algorithm is
equivalent to successive convolutions of the original image with a weight
window that expands by a factor of two at each pyramid level (Fig. 6-32). In
Fig. 6-33, the filtered images at the first three levels are shown, without down-
sampling. The down-sampling used in construction of the pyramid allows the use
of a single weight function (Eq. (6 – 32)) *directly on the image at each level*,
thereby avoiding unnecessary convolution calculations.

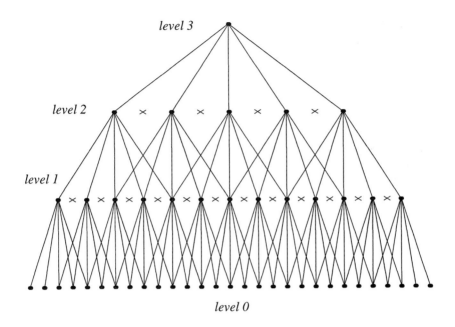

FIGURE 6-32. *The links between a pixel in level 3 of the Gaussian pyramid and pixels at lower levels. The x-marks indicate pixels that are not included at each level because of downsampling. The effective convolution window size at the original level 0 for a pyramid that reduces by two at each level L is* $4(2^L - 1) + 1$ *(Burt, 1981).*

6.5.2 Zero-Crossing Filters

A significant limitation of the local edge gradient filters described earlier in this chapter is that they only use information from a local neighborhood about each pixel. Large-scale edges, extending over many pixels, are not found explicitly, but only by connecting individual "edge pixels" produced by the gradient operation. The pyramid representation provides a way to access multiple image scales with a single-size filter.

In the original descriptions of the Gaussian pyramid (Burt and Adelson, 1983), a Laplacian pyramid was also calculated as the difference between level k and $k+1$ of the Gaussian pyramid (Fig. 6-30). The name comes from the Laplacian second derivative operator (Castleman, 1996), a connection that will be made in the following. Just as the Gaussian pyramid represents a series of LPFs with different bandpasses, the Laplacian pyramid is a series of *Band-Pass Filters (BPFs)* over different frequency regions.

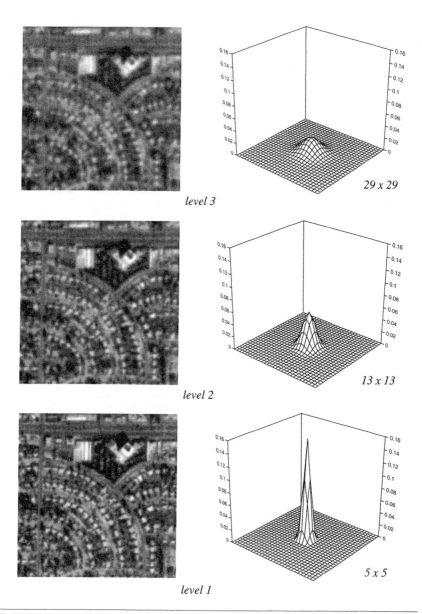

level 3

29 x 29

level 2

13 x 13

level 1

5 x 5

FIGURE 6-33. Levels 1 through 3 of the Gaussian pyramid, without down-sampling. The right-hand column shows the effective weight function at level 0. Note the weight amplitudes decrease as the spatial size increases, in order to maintain image normalization.

Laplacian-of-Gaussian (LoG) filters

Suppose we have a 1-D function and want to find the location of "significant" changes in that function. If we calculate its second derivative (Fig. 6-34), we see that it crosses the zero ordinate at the locations of such changes. These locations indicate changes in the second derivative, from either negative to positive, or positive to negative, and are therefore called *zero-crossings*. Now, suppose we convolve the original function with a smoothing Gaussian five points wide and again calculate the second derivative. Notice that the zero-crossings for the more significant changes remain in approximately the same locations as before, and some previous zero-crossings from lesser changes have disappeared.

Starting with a unit area, Gaussian function with zero mean and standard deviation σ,

$$g(x) = \frac{1}{\sigma\sqrt{2\pi}}e^{-x^2/2\sigma^2} \, , \tag{6-33}$$

the first derivative is,

$$g'(x) = -\left(\frac{x}{\sigma^2}\right)g(x) \tag{6-34}$$

and the second derivative, or the *Laplacian-of-Gaussian (LoG)*, is,

$$g''(x) = \left(\frac{x^2 - \sigma^2}{\sigma^4}\right)g(x) \, . \tag{6-35}$$

These functions are plotted in Fig. 6-35. Now, using the principles of linear systems theory, it is not difficult to show that, if a function f is convolved with g,

$$s(x) = f(x) * w(x) \, , \tag{6-36}$$

then the second derivative of s is given by,

$$s''(x) = f''(x) * g(x) = f(x) * g''(x) \, . \tag{6-37}$$

This means that the second derivative of a function convolved with a Gaussian is equal to the function convolved with the second derivative of the Gaussian. The implication is that we can generate the LoG filter (Eq. (6 – 35)) once, and use it on different functions to find their second derivatives.

The Laplacian of a 2-D, rotationally-symmetric Gaussian is nearly the same form as Eq. (6 – 35) (Castleman, 1996). An additional factor of two occurs in the second term because of the 2-D context,

$$g''(r) = \left(\frac{r^2 - 2\sigma^2}{\sigma^4}\right)g(r) \tag{6-38}$$

FIGURE 6-34. Example of the relation between slope changes in a function and zero-crossings in its second derivative (marked with small circles). The more significant zero-crossings are retained even if the function is smoothed with a low-pass filter, but some of the less significant zero-crossings are lost. Note the similarities to Fig. 6-3 and Fig. 6-4.

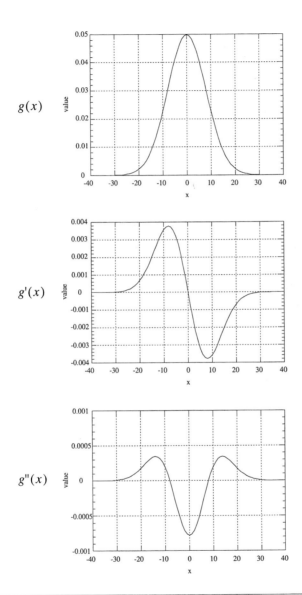

FIGURE 6-35. *First and second derivatives of the 1-D Gaussian function. The first derivative has a zero-crossing at the location of the Gaussian's maximum, and the second derivative has zero-crossings at the locations of the Gaussian's maximum gradient, which are at x equal $\pm\,\sigma$ (eight in this example).*

where,

$$g(r) = \frac{1}{\sigma^2 2\pi} e^{-r^2/2\sigma^2} \, . \qquad\qquad (6-39)$$

A special algorithm can be used to find zero-crossings in filtered images. However, a simple series of standard algorithms works as well. To produce a zero-crossing map, we first threshold the filtered image at a *DN* of zero (which is the mean). This segments the image into negative and positive regions. Then a Roberts gradient filter (Table 6-6) is applied, and the gradient magnitude is thresholded to find the transition pixels between negative and positive regions, i.e. the zero-crossings. The process is depicted in Fig. 6-36.

Laplacian-filtered image   *thresholded at zero*

thresholded Roberts gradient

FIGURE 6-36. *The process used here to find zero-crossings. The Laplacian image is thresholded at zero DN and a Roberts gradient filter is applied. The gradient magnitude is then thresholded at a DN of one to form the binary zero-crossing map.*

Difference-of-Gaussians (DoG) filters

The DoG filter is created by subtracting two Gaussian functions of different widths. The volume under each Gaussian is first normalized to one, so that their difference has a mean of zero,

$$DoG_{mn}(\sigma_1, \sigma_2) = \frac{1}{2\pi}\left[\frac{1}{\sigma_1^2}e^{-\left(\frac{m^2+n^2}{2\sigma_1^2}\right)} - \frac{1}{\sigma_2^2}e^{-\left(\frac{m^2+n^2}{2\sigma_2^2}\right)}\right] \qquad (6-40)$$

The shape of the resulting filter is similar to the LoG filter, and zero-crossings are found in the same way.

Assuming that each Gaussian is normalized as mentioned, there are two ways to alter the DoG filter characteristics. One is to hold the smaller Gaussian fixed, and vary the size of the larger (subtracted) Gaussian. The resulting filters maintain high spatial resolution in the zero-crossings, but progressively ignore smaller and smaller features as the ratio, R_g, decreases,

$$R_g = \sigma_2/\sigma_1. \qquad (6-41)$$

Cross-sections of the DoG filter for different values of R_g are shown in Fig. 6-37. Example zero-crossing maps are shown in Fig. 6-38 for the same aerial photograph used previously.

The second way to change the DoG filter characteristics is to hold R_g constant, while allowing the overall size of the filter to vary (Fig. 6-39). In this case, the resulting zero-crossing maps lose resolution as the filter size increases. Larger filters tend to map larger features with their zero-crossings. Three DoG filters with a range of parameters were used to find the zero-crossings for our test image. The results illustrate the general parametric behavior of this class of scale-space filter (Fig. 6-40).

Finally, a comparison of the zero-crossings produced by a DoG(1,2) filter and a Roberts gradient edge map is made in Fig. 6-41. The zero-crossing map is fully connected and only one pixel wide, but does not indicate the *slope* of the filtered image at the zero-crossing, i.e. the *strength* of the local gradient. Therefore, low contrast features, such as outside the lake boundary, and high contrast features are both mapped by the zero-crossings. It is possible to combine the gradient information at the zero-crossings with the thresholding step (see Fig. 6-36) to produce a more selective zero-crossing map.

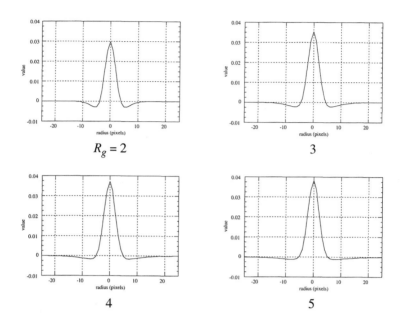

FIGURE 6-37. Profiles of the DoG filter for various size ratios between the two
Gaussian functions in Eq. (6 – 40). These filters were used to produce the zero-
crossing maps in Fig. 6-38. Note the locations of the zero-crossings are a weak
function of R_g.

6.5.3 Wavelet Transforms

Resolution pyramids and scale-space filtering provide a foundation for the relatively newer *wavelet transform*, which has attracted a great deal of attention in
the last few years. Their introduction in the literature was in 1984 (Grossman and
Morlet, 1984); other background papers of interest include (Daubechies, 1988;
Cohen, 1989; Mallat, 1989; Mallat, 1991) and (Shensa, 1992). We want to put
our discussion in the context of image resolution pyramid representations, which
are only a subset of the entire subject of wavelets. The clear presentation in
(Castleman, 1996), which is oriented to image processing, will be largely
followed here.

Wavelet theory provides a general mathematical framework for decomposition of an image into components at different scales and with different resolutions. Just as the 2-D discrete Fourier transform expands an image into a
weighted sum of global cosine and sine functions, the 2-D discrete wavelet transform expands the image into a sum of four components at each resolution level

FIGURE 6-38. *Zero-crossing maps for different size ratios in the DoG filter. The smaller Gaussian has a sigma of two. The contrast is reversed from that in Fig. 6-36.*

(Fig. 6-42). The wavelet transform operation is separable, consisting of two 1-D operations along rows and columns (Fig. 6-43).

A multiscale wavelet pyramid is constructed in a series of steps similar to those for the Gaussian and Laplacian pyramids. Each step is a convolutional filtering operation, followed by a down-sampling by two. It is therefore a type of REDUCE operation as described for the Gaussian and Laplacian pyramids. A major difference in wavelet decompostion is, however, that four components are calculated from the different possible combinations of row and column filtering. These four components may be most simply thought of as isotropic low-pass and high-pass components, plus horizontal and vertical high-pass components.

Much of the important content of wavelet theory is related to determining the particular window functions to use in the convolutions. Many functions are

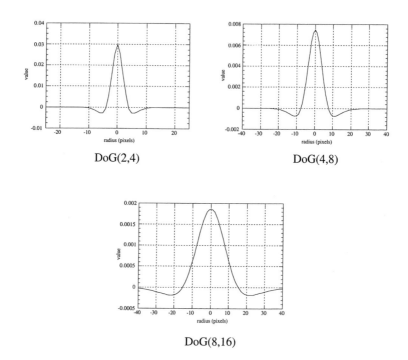

DoG(2,4)   DoG(4,8)

DoG(8,16)

FIGURE 6-39. *Profiles of the DoG filter for different overall sizes, but the same size ratio between the two Gaussians. These filters were used to produce the zero-crossing maps in Fig. 6-40. The locations of the zero-crossings here are a strong function of overall size.*

possible, some looking quite unusual, with multiple peaks and asymmetry (Castleman, 1996). The *biorthogonal filters* are either even or odd symmetric; they also may be any length as long as the low-pass w^L and high-pass w^H filters are consistent. The constraint between them is,

$$w_m^H = (-1)^{1-m} w_{1-m}^L \qquad (6-42)$$

where m is the index of the weight location in the high-pass filter w^H. In signal processing, these types of dual filters are called *quadrature mirror filters*.

Example wavelet components obtained with the following 1-D convolution filters,

DoG(2,4)

DoG(4,8)

DoG(8,16)

FIGURE 6-40. *Zero-crossing maps for constant size ratio, but different overall DoG filter size. The size ratio of the two Gaussians is two in all cases. Note that some edges are mapped at more than one scale; these are "significant" extended edges with high contrast.*

thresholded Roberts gradient   *zero-crossing map*

FIGURE 6-41. *Comparison of the Roberts thresholded gradient edge map with a zero-crossing map for the Lake Anna image.*

$$w^L = \begin{bmatrix} -0.05 & +0.25 & +0.6 & +0.25 & -0.05 \end{bmatrix}$$

$$w^H = \begin{bmatrix} +0.05 & +0.25 & -0.6 & +0.25 & +0.05 \end{bmatrix}$$

$$(6-43)$$

are shown in Fig. 6-44; these filters are collectively called a "Laplacian analysis filter" in (Castleman, 1996), and are analogous to Burt's Gaussian window function (Eq. (6 – 32)). Each level in the wavelet pyramid contains a low-pass filtered version and three high-pass filtered versions of the original image. The original image can be reconstructed, just as in the Gaussian and Laplacian pyramids, by an inverse wavelet transform consisting of upsampling and convolution.

Wavelet representations have been used for mapping high frequency features (points, lines and edges) for automated registration of two images and as a scale-space framework for fusion of images from sensors with different GIFOVs (Chapter 8). Although the mathematics of wavelet transforms can be quite complex, the transform's effect is the same as that of the Gaussian and Laplacian pyramid generation scheme; a multiresolution, scale-space representation of the image is created.

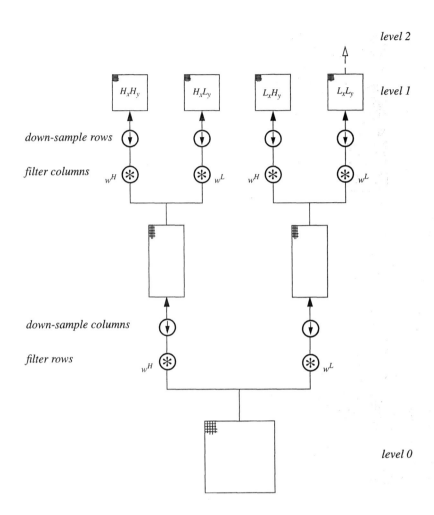

FIGURE 6-42. Wavelet decomposition from one pyramid level to the next. The letter L means low-pass and H means high-pass. Thus, L_xL_y means a low-pass filtered version that has been filtered in x (along rows) and y (along columns) and down-sampled by two. Likewise, H_xL_y means the image from the previous level is first high-pass filtered and down-sampled in x and then low-pass filtered and down-sampled in y. The L_xL_y component at each level is used as input to the calculations for the next level. This processing architecture is known as a filter bank.

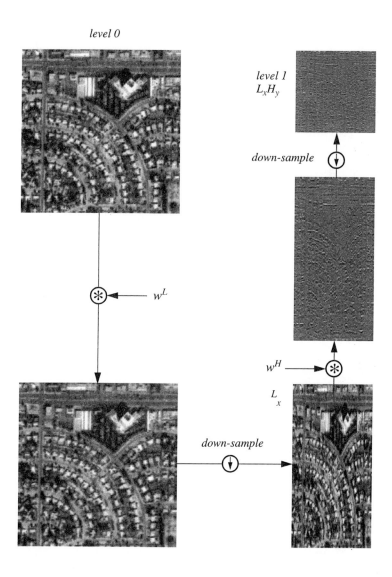

FIGURE 6-43. Calculation of one of the four wavelet components in level 1. The filtering and down-sampling combinations are similar to the REDUCE operation described earlier, except that here they are done one direction at a time.

FIGURE 6-44. *The four wavelet transform components of level 1 produced by the filter bank in Fig. 6-42. To create level 2, the upper left image, L_xL_y, is processed by the wavelet transform to produce a similar set of four images that are smaller by a factor of two in each direction.*

6.6   Summary

A variety of spatial transforms were discussed, including convolution, Fourier transforms, various types of filters and scale-space filtering. The important points of this chapter are:

- Convolution and Fourier transform filtering are equivalent global processing techniques, except in the border region.

- A wide range of processing can be performed with small neighborhood windows, including noise removal and edge detection.

- Scale-space filters allow access to image features according to their size, which is not possible with linear convolution or Fourier filters.

- The resolution pyramid provides a unified description of scale-space filters. Gaussian, Laplacian and wavelet pyramids are examples.

In Chapters 7, 8 and 9 we will see some specific applications of spatial transforms to solve image processing problems in remote sensing.

6.7   Exercises

6–1.   Determine the 2-D weighting functions for the Burt pyramid for values of a equal to 0.6, 0.5 and 0.3.

6–2.   Suppose you calculate the local mean and standard deviation in a 3 x 3 pixel window, moved across an image. What is the statistical uncertainty in the mean values? What if the window is 5 x 5 pixels?

6–3.   Show that the first step of a convolution, the flipping of the window function, is included in Eq. $(6 - 4)$.

6–4.   Explain Table 6-5 in terms of what each filter does to the image *DN*s.

CHAPTER 7

Correction and Calibration

7.1   Introduction

Remote-sensing imagery generally requires correction of systematic defects or undesirable sensor characteristics before performing reliable data analysis. Typical corrections include reduction of detector striping or other noise and geometric rectification. Relative *DN*-to-radiance radiometric calibration (e.g. EOS level 1B, Table 1-7) is necessary for analysis of multitemporal or multisensor images. Absolute *DN*-to-reflectance calibration (e.g. EOS level 2, Table 1-7) must be done to compare remote sensing measurements to laboratory or field reflectance data. Since many users purchase level 1 data for cost reasons, they must do corrections and calibration as needed for their application. This situation allows for custom algorithm development, but can make any comparison of the results from different projects difficult.

7.2   Noise Reduction

The basic types of image noise were introduced in Chapter 4. If the noise is severe enough to degrade the quality of the image, i.e. to impair extraction of information, an attempt at noise suppression is warranted. Analysis of the image and noise structure is necessary before applying a noise suppression algorithm. In some cases, sensor calibration data and even test images may exist that are sufficiently comprehensive to estimate noise parameters. Unfortunately, however, such data are often incomplete or the noise is unexpected (for example, interference from other equipment or external signal sources). We must then learn as much as we can about the noise from the noisy image itself; some examples are given in the following sections. Periodic noise is relatively easier to diagnose and characterize from the noisy imagery than is random noise. We have attempted to neatly categorize image noise in the following discussion, but many variations exist and may occur simultaneously in the same image.

7.2.1 Global Noise

Global noise is characterized by a random variation in *DN* at every pixel. Low-pass spatial filters can reduce the variance of such noise, particularly if it is uncorrelated from pixel-to-pixel, by averaging several neighboring pixels. Unfortunately, the variance of the noiseless part of the image, i.e. the signal, will also be reduced, albeit by a lesser amount because of the intrinsic spatial correlation present in the signal (Chapter 4). More complex algorithms that simultaneously preserve image sharpness and suppress noise are known as *edge-preserving smoothing* algorithms (Chin and Yeh, 1983; Abramson and Schowengerdt, 1993) and are discussed in the following sections.

Sigma filter

The fundamental conflict in spatial noise smoothing is the trade-off between noise averaging and signal averaging. A way is needed to separate the two components *before* averaging. In the sigma filter (Lee, 1983), they are separated by contrast. A moving-average window is used, as described in Chapter 4. At each output pixel, only those input pixels having *DN*s within a specified threshold of the center pixel DN_c are averaged (Fig. 7-1). A fixed threshold, Δ, can be set in terms of the *global* noise *DN* standard deviation, σ,

$$\Delta = k\sigma, \tag{7-1}$$

resulting in an acceptable range about the center pixel DN_c of,

$$DN_c \pm k\sigma . \qquad (7-2)$$

Lee suggested a k value of two, since that corresponds to 95.5% of the data from a Gaussian distribution (the noise is assumed additive and Gaussian distributed).

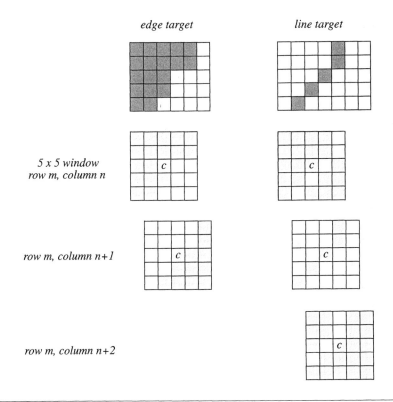

FIGURE 7-1. The behavior of the sigma filter near edges and lines. The image data is noisy with a standard deviation of σ, but the difference between the mean DNs of the target and background is greater than $2k\sigma$. Therefore, the 5 x 5 sigma filter calculates the average of only the light gray areas as the window moves across the target, avoiding averaging of the target and background. The "pool" of pixels from which the filter calculates an average can take any irregular, even disconnected, shape within the 5 x 5 window. If the target-to-background contrast is less than $2k\sigma$, there will be some averaging of the target and background.

If the noise standard deviation is signal-dependent, the threshold can be made adaptive by using the *local* standard deviation,

$$\Delta_{local} = k\sigma_{local}. \tag{7-3}$$

However, the filter then becomes sensitive to higher contrast signal features, and they will be smoothed accordingly.

Nagao-Matsuyama filter

The sigma filter has no directional sensitivity. Many spatial features of interest are linear, at least over local regions. The Nagao-Matsuyama filter is designed to adapt to linear features at different orientations (Nagao and Matsuyama, 1979). A 5 x 5 window is used, and at each location of the window, the *DN* mean and variance in each of nine subwindows is calculated (Fig. 7-2). The mean of the subwindow with the lowest varaiance is used for the output *DN*. Thus, the filter finds the most homogeneous local region (of the nine) and uses its average for the output.

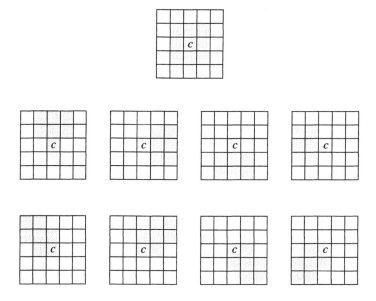

FIGURE 7-2. *The nine subwindows used to calculate local DN variance in the Nagao-Matsuyama algorithm. The pixels in the subwindow with the lowest DN variance are averaged for the output pixel at c.*

To illustrate these methods for reducing global noise, a Side-Looking Airborne Radar (SLAR) image will be used (Fig. 7-3). The image has the global, random speckle noise associated with synthetic aperture radar data. A 5 x 5 LPF smooths the noise, but also smooths the signal component of the image by an unacceptable amount. The global sigma filter achieves considerable smoothing of the noise, with much less signal smoothing. The small bright features and street patterns in the city are generally better preserved by the sigma filter. The sigma filter can average any number of pixels, and may not average at all for high contrast point-like objects against a background. The Nagao-Matsuyama filter is better than the sigma filter at retaining linear features through the smoothing process. Small details in the city and the rows of spots at the top are lost in the seven-pixel averaging, but linear features such as the airport runways at the lower right are better preserved than with the sigma filter.

7.2.2 Local Noise

Local noise refers to individual bad pixels and lines, typically caused by data transmission loss, sudden detector saturation or other intermittent electronic problems. The pixels affected by the noise usually have zero *DN*, indicating data loss, or maximum *DN*, indicating saturation.[1] An example is shown in Fig. 7-4. A 3 x 1 median filter removes the noise, but also alters many other pixels, as shown in Fig. 7-5. We would like to design an approach that makes use of the models of Chapter 4 to reduce the number of modified pixels, while still removing most of the noise.

Detection by spatial correlation

The fact that images are spatially correlated is useful in designing correction algorithms for local noise. Because of the spatial correlation in images, we expect neighboring pixels to have similar values. As shown in Chapter 4, the *DN* semivariogram provides a measure of local pixel-to-pixel correlation. The   quantity,

$$\Delta_{ij} = \left(DN_{i,\,j} - DN_{i-1,\,j} \right)^2 , \qquad (7-4)$$

the squared difference between a pixel in row *i* and the pixel immediately above in row *i-1*, is the similarity measure used in the semivariogram and is shown in Fig. 7-6 as calculated for our test image. If Eq. (7 – 4) were summed over all pixels, we would essentially have $\gamma(1)$ in the vertical direction. Now, we would like to have a local noise removal algorithm that could control the number of

---

[1] Local pixel noise is called "salt noise" if the bad pixels are bright, "pepper noise" if the bad pixels are dark, and "salt and pepper noise" if both cases are present!

original   *5 x 5 LPF*

5 x 5 sigma filter, k = 2   *Nagao-Matsuyama filter*

FIGURE 7-3. Speckle noise filtering of a SLAR image of Deming, New Mexico, acquired in the X-band with ḤH polarization on July 1, 1991, from 22,000 feet altitude. The GSI is about 12 x 12 meters. Metallic objects such as cars, metal roofs and power lines show a high signal ("return"). The two rows of periodic spots at the top are probably power line support structures. (The SLAR image is from the U. S. Geological Survey's Side-Looking Airborne Radar (SLAR) Acquisition Program 1980-1991 and is available on CD-ROM. The processed images are courtesy of Justin Paola, Oasis Research Center.)

pixels that are changed. It is not difficult to provide an estimate of the maximum number of bad lines expected, for example. The *Cumulative Distribution Function (CDF)* of Δ gives us a way to convert that specification to a meaningful threshold based on the data statistics. The $CDF(\Delta)$ is plotted in Fig. 7-6, and

local line noise

3 x 1 median filtered

FIGURE 7-4. *Local line noise in an MSS image. A 3 x 1 median filter removes the noise, but also changes many good pixels.*

FIGURE 7-5. *The difference between the noisy image and the median-filtered image. Values are scaled [0,255].*

three thresholds are shown, corresponding to the percentage of pixels that would be corrected. Using a specification of 2%, the Δ value of .0147 is obtained from the *CDF*. This number is used, in turn, to threshold the Δ image into a binary *noise pixel mask* (Fig. 7-7).

The algorithm proposed is to simply retain all original pixels that are not in the noise pixel mask, and replace those pixels within the mask by the corresponding pixels in the median-filtered image (Fig. 7-4). The result shows a high degree of noise reduction, with relatively few noiseless pixels changed (Fig. 7-8). In terms of fraction of the image, we have changed about 0.9% pixels with the selective algorithm, compared to 41.8% with the median filter.[2]

Detection by spectral correlation

As mentioned in Chapter 5, the *PCT* can isolate uncorrelated image features in the higher-order *PC*s. This is not only useful for change detection in multitemporal images, but also for isolation of uncorrelated noise in multispectral images. As

---

[2] The number is less than the 2% specification, because about half the pixels replaced by the local median were actually unchanged. Therefore, it is advisable to use a larger noise pixel fraction than actually expected.

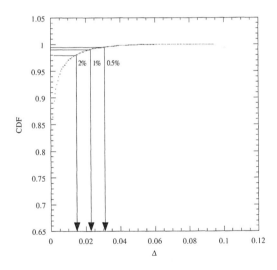

FIGURE 7-6. The squared-difference image Δ_{ij} between each pixel and its immediate neighbor in the preceding line, and the CDF(Δ) of that image. From the CDF, we can see that a threshold lower than about 98% would greatly increase the number of candidate noise pixels.

FIGURE 7-7. The Δ_{ij} image, thresholded so that 2% of the pixels are labeled candidate noise pixels. This noise mask includes only those pixels where the local spatial correlation is below the specified threshold.

an example, TM bands 2, 3 and 4 from the Marana agricultural scene (Plate 9-1) will be used. If a *PCT* is performed, some local line noise is clearly isolated into PC_3 because the noise occurs only in band 2 (Fig. 7-9). It can then be removed if desired, by setting PC_3 equal to a constant value (e.g. its mean *DN*), followed by an inverse *PCT*. That would also, of course, remove any image information in PC_3. Alternatively, the PC_3 image could be used to localize the bad pixels by thresholding, and a local spatial correction could be performed only on those pixels in the original image bands, similar to the process described above.

Some of the research papers on techniques for reducing global and local random image noise are noted in Table 7-1. Comparisons of various algorithms are provided in a number of papers, including (Chin and Yeh, 1983), (Mastin, 1985), and (Abramson and Schowengerdt, 1993). As in most noise filtering, no single algorithm has been proven to be superior in all cases. Most procedures require specification of image-dependent parameters, and in practice, interactive "tuning" of those parameters to achieve satisfactory results.

median-filtered with noise mask

removed noise

FIGURE 7-8. *The processed image using the candidate noise pixel mask and the difference with the original noisy image. This algorithm is much more selective than the global median filter (compare to Fig. 7-5). Consequently, the narrow roads between the fields are better preserved.*

FIGURE 7-9. Application of the PCT to noise isolation in multispectral images. The noise can then be removed by setting PC_3 equal to a constant, followed by an inverse PCT. The local noise in this example has a maximum range of about $\pm 10DN$. On inspection of the original three bands, the noise is seen to occur only in band 2, explaining why it is well isolated by the PC transformation.

TABLE 7-1. Some research papers on reducing random image noise.

sensor	type of noise	reference
aerial photo	global	(Nagao and Matsuyama, 1979)
aerial photo	global	(Lee, 1980)
Seasat SAR	global	(Lee, 1981)
aerial photo	global, local	(Lee, 1983)
Viking Mars Orbiter	global, local	(Eliason and McEwen, 1990)
TM with simulated noise	local	(Centeno and Haertel, 1995)

7.2.3 Periodic Noise

Global periodic noise (commonly called *coherent* noise) is a spurious, repetitive pattern that has consistent characteristics throughout an image. One source is electronic interference from data transmission or reception equipment; another is calibration differences among detectors in whiskbroom or pushbroom scanners (Chapter 3). The consistent periodicity of the noise leads to well-defined spikes in the Fourier transform of the noisy image (Chapter 6). If the noise spikes are at a sufficient distance from the image spectrum (i.e. the noise is relatively high frequency), they may be removed by simply setting their Fourier amplitude to zero. The filtered spectrum may then be inverse Fourier transformed to yield a noise-free image. Spatial domain convolution filters can be designed that produce the same result; large spatial windows are required to achieve localized frequency domain filtering (Pan and Chang, 1992).

If the frequency of the noise falls within the image spectrum, then valid image structure will also be removed if the Fourier domain noise spikes are simply set to zero. Although this is usually not a serious problem, in practice it can be partially ameliorated by replacing the noise spikes in the Fourier domain with values interpolated from surrounding noise-free areas. Both the real and imaginary components of the spectrum must be interpolated.

Local periodic noise is periodic, but the noise amplitude, phase or frequency varies across the image. One approach to this problem is to estimate the local amount of noise and remove only that amount at each pixel. For example, the global noise pattern may be obtained by isolating the Fourier domain noise spikes and inverse Fourier transforming them to the spatial domain. The local spatial correlation between this global noise pattern and the noisy image is then calculated for every image pixel (see Chapter 8 for a discussion of spatial correlation). At each pixel of the noisy image, a weighted fraction of the noise pattern is then subtracted; the weight is proportional to the magnitude of the local correlation. In areas where the noise component dominates, a correspondingly larger fraction of the noise pattern is removed and conversely, in areas where the noise is weak and the image dominates, a smaller fraction of the noise pattern is removed. A simplified version of this procedure, implemented entirely in the spatial domain, was used in (Chavez, 1975) to remove the single frequency, variable amplitude noise in Mariner 9 images of Mars.

Destriping

Unequal detector sensitivities and other electronic factors cause line-to-line striping in Landsat MSS and TM imagery (Chapter 4). In the MSS with six detectors/scan, the striping has a periodicity of six lines, and in TM with 16 detectors/scan, a periodicity of sixteen lines. In SPOT imagery, there is no periodicity since the scene is acquired with 6000 detectors cross-track. However, other types of periodic noise have been observed in SPOT data, probably arising from read-out of the CCD arrays. A variety of techniques have been applied to correction of striping and coherent periodic noise (Table 7-2).

TABLE 7-2. *Some of the published work on sensor-specific periodic noise reduction.*

sensor	type of noise	reference
AVHRR	coherent	(Simpson and Yhann, 1994)
AVIRIS	coherent	(Curran and Dungan, 1989) (Rose, 1989)
Fuyo-1	striping, periodic	(Filho *et al.*, 1996)
GOES	striping	(Weinreb *et al.*, 1989)
MSS	striping striping striping within-line, coherent striping striping striping	(Chavez, 1975) (Horn and Woodham, 1979) (Richards, 1985) (Tilton *et al.*, 1985) (Srinivasan *et al.*, 1988) (Wegener, 1990) (Pan and Chang, 1992)
SPOT	near in-track striping cross-track striping 2 x 2 "chessboard"	(Quarmby, 1987) (Büttner and Kapovits, 1990) (Westin, 1990)
TIMS	scanline, periodic	(Hummer-Miller, 1990)
TM	striping banding banding banding banding	(Poros and Peterson, 1985) (Fusco *et al.*, 1986) (Srinivasan *et al.*, 1988) (Crippen, 1989) (Helder *et al.*, 1992)
VHRR	aperiodic striping	(Algazi and Ford, 1981)

Destriping should be done before any geometric corrections, while the data array is still aligned to the scan direction. The models discussed in Chapter 4 can then be applied. If only geometrically-processed data are available, then the data array will not be orthogonal to the scan direction, i.e. each original scan line will cross several of the data array lines because of the rotational geometric corrections that are applied (see Fig. 7-28). More complicated correction schemes are then necessary.

Striping is not always the same across an image. Thus, the model with fixed characteristics described in Chapter 4 is not completely correct. The degree of correction achieved with that model, however, is often adequate and more complex models are unnecessary.

Global, linear detector matching

Over a large image, the data from each individual detector should have nearly the same *DN* mean and standard deviation. This realization provides a simple rationale for destriping. One of the detectors is chosen as a reference and its global mean *DN* and standard deviation are calculated. The global *DN* means and standard deviations in each of the other detectors (five for MSS, 15 for TM) are then linearly adjusted to equal the values for the reference detector. The transformation for detector *i*,

$$DN_i^{new} = \frac{\sigma_{ref}}{\sigma_i}(DN_i - \mu_i) + \mu_{ref} \qquad (7-5)$$

is applied to all pixels from that detector. After this transformation, each detector will produce the same global *DN* mean and standard deviation as that produced by the reference detector.[3] In effect, a linear gain and bias correction has been made to equalize the sensitivities of all detectors.

Nonlinear detector matching

In some cases, a linear *DN* correction is not sufficient, and a nonlinear correction may be indicated. An algorithm for matching two image *DN* distributions was presented in Chapter 5 in the context of contrast matching. Extending this scheme to the destriping problem, we select one detector (or the average of all detectors) as the reference, calculate the cumulative distribution functions $CDF_{ref}(DN_{ref})$ and $CDF_i(DN_i)$, and do the following transformation,

$$DN_i^{new} = CDF_{ref}^{-1}[CDF_i(DN_i)] . \qquad (7-6)$$

The result is that the *CDF* of the data from each detector matches that from the reference detector, and nonlinearities in the detector calibrations are accounted for by Eq. (7 – 2) (Horn and Woodham, 1979).

---

[3] The global *DN* mean and standard deviation, averaged over all detectors, can also be used. That would perhaps be a little more robust, since the possibility of picking a reference detector with some peculiarity would be avoided.

Statistical modification

The use of a deterministic truncation rule in converting floating point calculated DNs to integer DNs can result in residual striping. One solution is to use the fractional part of the floating point DN as a probability measure controlling whether to round up or down in the conversion to integer. Thus, any systematic round-off striping is converted to a random variation, which is less visually apparent and perhaps more statistically precise (Bernstein *et al.*, 1984; Richards, 1993).

Spatial filtering approaches

Fourier domain filtering can be effective for destriping of relatively small images. The approach is to examine the amplitude or power spectrum of the image to locate the noise frequency components, design a blocking ("notch") filter to remove them, apply the filter to the noisy image spectrum in the frequency domain, and finally calculate an inverse Fourier transform to obtain the corrected image (Fig. 7-10).[4] This approach to removing periodic noise from images dates from the earliest spacecraft missions (Rindfleisch *et al.*, 1971) and has been applied to MSS (Pan and Chang, 1992), TM (Srinivasan *et al.*, 1988) and Fuyo-1 imagery (Filho *et al.*, 1996). A concise introduction to Fourier domain filtering of spatial data is given in (Pan, 1989).

A 256-by-256 image block from an early TM band 1 image over the Pacific Ocean is shown in Fig. 7-11. The signal level is low, so the striping noise is readily apparent, both in the image and in the amplitude spectrum, where it appears as the column through zero frequency. A filter that blocks that column, except at low frequencies, appears to remove the striping noise and retain the lower, image frequency components. If we also block the pair of high frequency columns on either side of the center noise spike, a more subtle, within-line coherent noise is also removed. The noise components for the two cases are shown in Fig. 7-12.

A practical question that may arise is, can the same filter remove noise from another part of the image? A second image block over downtown San Francisco (Fig. 7-13) was selected from the same scene. The amplitude spectrum of this block shows a more complex structure, as expected, but with similar noise components. The filter derived previously removes this noise, without appearing to harm the image. The noise pattern looks entirely different than that for the ocean block (Fig. 7-12); the frequency components are the same, which is why the same notch filter works, but their amplitudes are different.

---

[4]   A Gaussian amplitude notch filter, centered at frequencies ($\pm u_0$, $\pm v_0$) is given by,

$$W(u, v) = 1 - e^{-[\sqrt{(u-u_0)^2/\sigma_u^2 + (v-v_0)^2/\sigma_v^2}]} \quad .$$

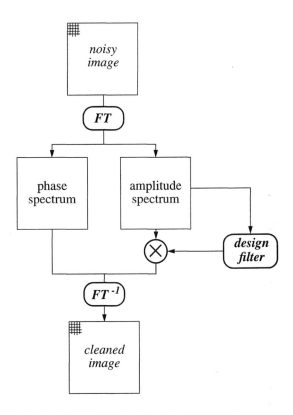

FIGURE 7-10. Algorithm flow diagram for Fourier amplitude filtering. The "design filter" step can be manual or semi-automated, as described in the text.

A more automated and conservative approach to designing the noise filter is possible for some images. When the frequency content of the scene is low (such as for the example ocean image segment) and the noise is primarily at higher frequencies, we can automatically find a noise filter with the following steps:

- Apply a "soft" high-pass filter, such as an inverted Gaussian, to the amplitude spectrum of the image. This reduces the amplitude of the lower frequency components containing the scene content and removes the zero frequency component.

- Threshold the result at a reasonable value.

- Invert and scale the result to zero for the noise components and one for all others, thus creating a noise mask.

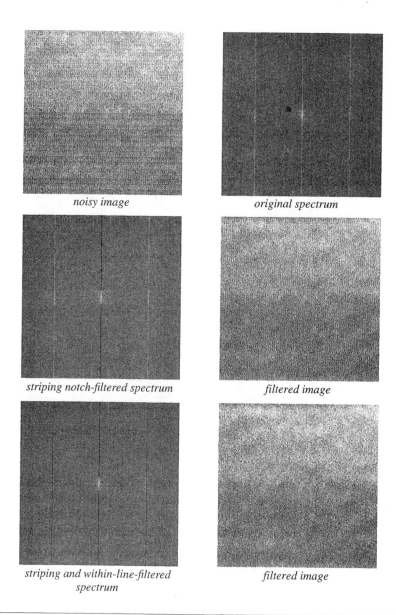

noisy image

original spectrum

striping notch-filtered spectrum

filtered image

striping and within-line-filtered
spectrum

filtered image

FIGURE 7-11. *Fourier amplitude filtering to remove striping is illustrated for a nearly uniform ocean area from an early TM image acquired on December 31, 1982. Blocking the primary striping noise component results in a much cleaner image and blocking the three periodic noise components results in further correction.*

striping noise   *striping and within-line noise*

FIGURE 7-12. The noise components removed by the two filters in Fig. 7-11. Both patterns are strongly contrast enhanced; the noise DN standard deviation is only 0.45 and 0.77, respectively. The mean DN is zero in both cases. Since we are assuming an additive noise model with the Fourier filter, these noise images can be calculated by subtracting the corrected image from the original image.

- Apply the noise mask to the image spectrum as previously shown and calculate the inverse Fourier transform.

There are two places for manual intervention in this procedure, one for the width of the high-pass filter in step 1, which should be as small as possible while still removing most of the scene content, and the other for the threshold of step 2, which determines the extent of noise removal. An example for our image is shown in Fig. 7-14. The high-pass filter has a 1/e width of 0.05 cycles/pixel, i.e. one tenth of the highest frequency, and the threshold applied to the modified spectrum is simply midway between the minimum and maximum amplitude values. These parameters remove most of the noise in the image, but less than in Fig. 7-11.

Debanding

Banding refers to a striping-like noise that affects each multidetector scan of a system like TM. The noise is aggravated by bright scene elements, such as clouds, that cause detector saturation for part of the scan, and depends on the scan direction. Noise removal is complicated in geometrically-corrected imagery by the fact that relative detector calibration algorithms like those used for destriping are not valid if the data have been resampled in-track. The Fourier filter approach can be used, although the striping and banding components may

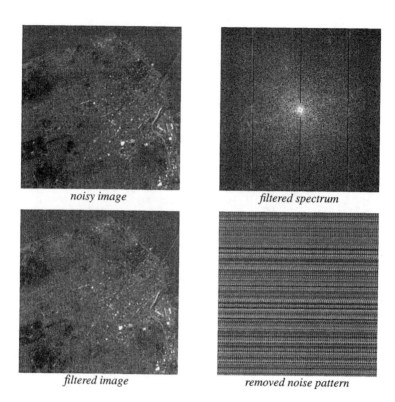

noisy image   filtered spectrum

filtered image   removed noise pattern

FIGURE 7-13. Application of the filter derived from the ocean area (Fig. 7-11) to another part of the same TM image. The Fourier spectrum is more complex, consistent with the detail in the image, but the noise characteristics are similar to those of the ocean area. The filtered image has no obvious artifacts due to the noise filter, even though the removed noise pattern is considerably different from that in Fig. 7-12. As explained in the text, this difference is in the amplitudes of the frequency components, not their frequencies. The noise DN standard deviation in this case is 0.69. Because this image has higher frequency content than the ocean segment, some image frequency components are also being removed here.

no longer align with single columns in the frequency domain if image rotation was included in the geometric corrections.

A simple, multistage convolution filter procedure has been described for this problem (Crippen, 1989). The process consists of three spatial filtering steps to isolate the noise from the image structure, followed by subtraction of the noise from the original image (Fig. 7-15). The 1 row-x-101 column low-pass filter (LPF) isolates the low-frequency signal and scan noise from the high-frequency

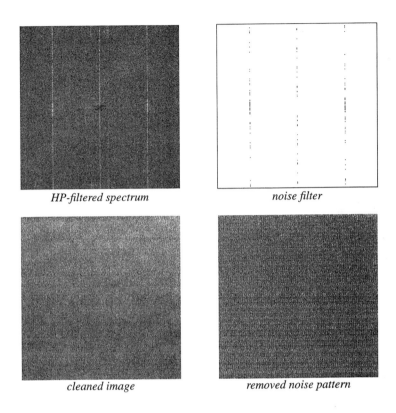

HP-filtered spectrum   *noise filter*

cleaned image   *removed noise pattern*

FIGURE 7-14. Automatic filter design for striping. The high-pass filtered amplitude spectrum of the ocean image is thresholded to yield the noise filter. The filtered image is in the lower left and the removed noise components in the lower right. The filtered image is comparable to those of Fig. 7-11, but the threshold parameter used here results in somewhat less noise removed.

image components. The 33 row-x-1 column high-pass filter (HPF) isolates the relatively high-frequency banding noise from the low-frequency signal. The use of 33 lines (rather than 32, the period of the TM forward/backward scans) arises from the adjustment of the original 30m TM pixel to 28.5m in the system-corrected product. The last filter, a 1 row-x-31 column LPF, suppresses diagonal artifacts that can be introduced by the previous filter. Finally, the isolated noise pattern is subtracted from the original image.

This debanding technique is illustrated with a TM band 4 image (Fig. 7-16). The three convolution filters are applied as prescribed (Fig. 7-17), with the exception of masking the water (low *DN*) pixels, and the resulting isolated noise

FIGURE 7-15. *The convolution filter algorithm proposed in (Crippen, 1989) to remove banding noise in system-corrected TM imagery. The series of filters produces an estimate of the noise pattern which is subtracted from the original image.*

pattern is subtracted from the original image (Fig. 7-18). The water masking step is essential to avoid edge artifacts at the boundary between land and water for this particular image, but appeared to be unnecessary in Crippen's examples. To process other bands, the same mask can be used; the NIR bands work well for creating the mask because of the low reflectance of water in the NIR. This technique is largely heuristic and cosmetic in nature, with no calibration basis for its derivation, unlike the *CDF*-based destriping algorithms.

noisy image

water mask

masked water area

padded with mean DN
(contrast stretched)

FIGURE 7-16. Landsat TM band 4 of San Francisco Bay from August 12, 1983, is shown at the top left. The water area exhibits TM banding, probably caused by a bright fog bank just off the left edge of the image (see Fig. 4-13 for another part of the same scene). A mask for water pixels is made by thresholding the image at a DN of 24. The zeros in the mask are then replaced by the mean DN of the water area to minimize edge artifacts from later processing.

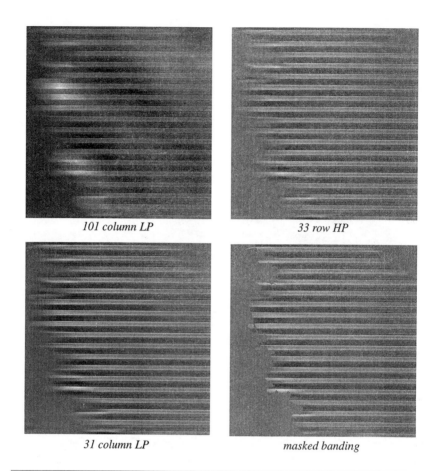

101 column LP   *33 row HP*

31 column LP   *masked banding*

FIGURE 7-17. Image sequence for the debanding process. Each filter is applied in sequence, as shown in Fig. 7-15. If the water mask was not used, the edge artifacts at the shoreline would appear in the final processed image.

7.3   Radiometric Calibration

There are several levels of radiometric calibration (Fig. 7-19). The first converts the sensor *DN*s to at-sensor radiances and requires sensor calibration information. The second is transformation of the at-sensor radiances to radiances at the earth's surface. This level is much more difficult to achieve, since it requires information about the atmospheric conditions at the time and location of the *cleaned image*   *masked water area*

FIGURE 7-18. *The final, debanded TM image. Most of the banding noise has been removed, and one can now see flow patterns in the Bay which were poorly depicted in the original data (compare to Fig. 7-16).*

image. That information can be in different forms, ranging from a simple categorization of the atmospheric condition as one of several "standard atmospheres," to estimates of certain parameters such as path radiance from the image data itself, to (ideally) coincident ground measurements. A promising approach appears to be the combination of image-based data with atmospheric modeling (Teillet and Fedosejevs, 1995). The final level of calibration to surface reflectance is achieved by correction for topographic slope and aspect, atmospheric path length variation due to topographic relief, particularly in mountainous terrain, solar spectral irradiance, solar path atmospheric transmittance, and down-scattered "skylight" radiance (see Chapter 2).

7.3.1 Sensor Calibration

There is some potential for confusion in the terminology of radiometric calibration. The sensor engineer uses Eq. (3 – 20) for calculations; here the sensor gain and offset have units of *DN*-per-unit radiance. This provides a "forward" calculation from predicted radiance to output sensor *DN*. The user of the data, however, receives *DN* values and wants to convert them to radiances in an "inverse" calculation. Therefore, the user wants calibration coefficients with units of radiance-per-*DN*. Unfortunately, these coefficients are also often referred to as "gain" and "offset" (EOSAT, 1993). To avoid any confusion here, we will call the user-oriented calibration coefficients, "cal_gain" and "cal_offset."

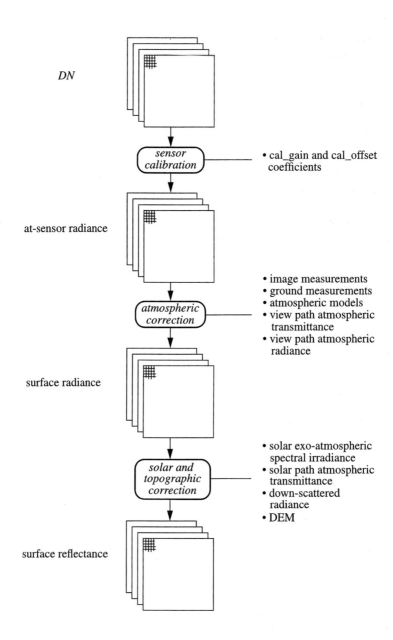

FIGURE 7-19. Data flow for calibration of remote sensing images to physical units.

The calibration gains and offsets for the Landsat-4 and -5 TM systems, measured preflight, are given in Table 7-3. These coefficients can be applied to TM pixel values in each band, DN_b, as follows,

$$\text{at-sensor:} \quad L_b^s = \text{cal\_gain}_b \cdot DN_b + \text{cal\_offset}_b \qquad (7-7)$$

to produce band-integrated radiance values, L_b^s. Although the sensor gain and offset are often assumed to be constant throughout the sensor's life, they can and usually do change over time. The Remote Sensing Group of the Optical Sciences Center, University of Arizona, has measured the gain response of the Landsat TM-5 since 1984 (Thome *et al.*, 1994). The gain shows a generally slow, but steady decrease over that time in most TM bands (Fig. 7-20). This degradation in sensor performance with time in orbit is usually attributed to material deposition on the sensor optics caused by outgassing from the system in vacuum. Similar degradations have been reported for the AVHRR system (Rao and Chen, 1994).

TABLE 7-3. *Pre-flight measurements of the TM calibration gain and offset coefficients. These are calculated using the procedure provided by EOSAT (EOSAT, 1993). Gain is given here in radiance/DN units and offset is radiance, with units W-m^{-2}-sr^{-1}-µm^{-1}.*

band	Landsat-4		Landsat-5	
	cal_gain	cal_offset	cal_gain	cal_offset
1	0.672	-3.361	0.642	-2.568
2	1.217	-6.085	1.274	-5.098
3	0.819	-4.917	0.979	-3.914
4	0.994	-9.936	0.925	-4.629
5	0.120	-0.7208	0.127	-0.763
6	0.0568	+1.252	0.0552	+1.238
7	0.0734	-0.367	0.0677	-0.0338

7.3.2 Atmospheric Correction

The modeling of Chapter 2 showed that the atmosphere plays a complex role in optical remote sensing. As shown there with MODTRAN simulations, the atmosphere dramatically alters the spectral nature of the radiation reaching the sensor. It is impossible to apply the complete model to every pixel in a typical remote-sensing image. Independent atmospheric data are almost never available for any given image, and certainly not over the full FOV. So, how are we to correct the imagery for atmospheric effects? A robust approach to the problem is to estimate

FIGURE 7-20. Time history of sensor gain for the Landsat-5 TM sensor in the non-thermal bands. Note all bands show a decrease, except for band 5. The value at zero days since launch is from the preflight laboratory calibration. The remainder are measured using a field calibration procedure, which includes atmospheric measurements and modeling, and reflectance measurements of a bright, spectrally and spatially uniform area. Data plotted from (Thome et al., 1994).

atmospheric parameters *from the image itself*. The parameters can then be refined with any available ancillary measurements and iterative runs of an atmospheric modeling program to achieve consistency between the atmospheric parameters and the image data. Any atmospheric correction that relies on physical modeling of the atmosphere, first requires correction of the sensor *DN*s to radiances at the sensor, i.e. sensor calibration, as discussed in Sect. 7.3.1.

This "bootstrap" approach has been the most common type of atmospheric correction since the early days of Landsat (Table 7-4). A usual simplification is to assume that the down-scattered atmospheric radiance, E_λ^d, in Eq. (2 – 10) is zero, so that, in band-integrated form,

$$at\text{-}sensor: \quad L_b^s(x, y) \cong L_b^{su}(x, y) + L_b^{sp} \ . \tag{7 – 8}$$

Assuming this equation is an equality and referring to Eq. (2 – 8), we can rewrite Eq. (7 – 8) as,

$$at\text{-}sensor: \quad L_b^s(x, y) = \tau_{vb} L_b(x, y) + L_b^{sp} \tag{7 – 9}$$

and solve for the surface radiance, L_b, in each band,

$$earth's\ surface: \quad L_b(x, y) = \frac{L_b^s(x, y) - L_b^{sp}}{\tau_{vb}} \tag{7 – 10}$$

In the simplest corrections for the visible wavelength region, the effort is focused on estimating the upwelling atmospheric path radiance, L_b^{sp}, with the view path transmittance assumed to be one, or at least a constant over wavelength. To a first approximation, this emphasis is not unreasonable, since the path radiance is the dominant atmospheric effect in the visible spectral region. The approach most commonly used requires identification of a "dark object" within the image, estimation of the signal level over that object, and subtraction of that level from every pixel in the image.[5] The rationale is that the only possible source for an observed signal from a dark object is the path radiance. The major weaknesses in this approach are in identifying a suitable object in a given scene, and once one is located, in the assumption of zero reflectance for that object (Teillet and Fedosejevs, 1995). Also, the DOS technique cannot correct for the path transmittance term in Eq. (7 – 10), which is especially important in TM bands 4, 5 and 7.

More recent efforts tend to use dark target radiance estimation, combined with atmospheric modeling. For example, Chavez has used a look-up table of atmospheric scattering dependencies (Table 7-5) to estimate the path radiance in

---

[5] Candidate dark objects are deep, clear lakes (Ahern *et al.*, 1977), shadows and asphalt paving (Chavez, 1989). This procedure is termed "dark object subtraction" (DOS).

TABLE 7-4. Examples of atmospheric calibration techniques for remote sensing images. Some of the techniques require the tasseled cap and NDVI spectral transforms described in Chapter 5.

sensor	approach	reference
MSS	band-to-band regression	(Potter and Mendolowitz, 1975)
MSS	all-band spectral covariance	(Switzer *et al.*, 1981)
airborne MSS	band-to-band regression	(Potter, 1984)
AVHRR	iterative estimation	(Singh and Cracknell, 1986)
MSS, TM	DOS with exponential scattering model	(Chavez, 1988)
TM	DOS with exponential scattering model downwelling atmospheric radiance measurements	(Chavez, 1989)
TM	pixel-by-pixel tasseled cap haze parameter	(Lavreau, 1991)
AVHRR	DOS NDVI AVHRR band 3	(Holben *et al.*, 1992)
airborne TMS, Landsat TM	ground and airborne solar measurements, atmospheric modeling code	(Wrigley *et al.*, 1992)
TM	comparison of ten DOS and atmospheric modeling code variations with field data	(Moran *et al.*, 1992)
TM	dark target, modeling code	(Teillet and Fedosejevs, 1995)
TM (all bands)	atmospheric modeling code, region histogram matching	(Richter, 1996a; Richter, 1996b)
TM	DOS with estimated atmospheric transmittance	(Chavez, 1996)

different bands (Chavez, 1989). The model used for a given image is determined by an estimated "haze" value from a blue or green waveband. Teillet and Fedosejevs have recommended the use of a dark target, with either a known or assumed reflectance, and detailed atmospheric modeling with the 5S code (Tanre *et al.*, 1990), but with no atmospheric measurments required, in a procedure that may be applied in a production environment (Teillet and Fedosejevs, 1995). They also make the point that significant errors can result from assuming the dark object has zero reflectance, even if the actual reflectance is only 0.01 or 0.02.

TABLE 7-5. *The discrete characterization of atmospheric conditions used in (Chavez, 1989).*

atmospheric condition	relative scattering model
very clear	$\lambda^{-4.0}$
clear	$\lambda^{-2.0}$
moderate	$\lambda^{-1.0}$
hazy	$\lambda^{-0.7}$
very hazy	$\lambda^{-0.5}$

7.3.3 Solar and Topographic Correction

To find the spectral reflectance of every pixel in an image requires more external information and further correction. Using Eq. (2 – 10) and solving Eq. (7 – 10) for reflectance in terms of the surface radiance, we obtain,

$$\rho_b(x, y) = \frac{\pi L_b(x, y)}{\tau_{sb} E_b^0 \cos[\theta(x, y)]} \qquad (7-11)$$

Thus, we need the surface radiance (as obtained in Sect. 7.3.2), solar path atmospheric transmittance, exo-atmospheric solar spectral irradiance and the incident angle (which depends on the solar angles and topography).[6] Again, in the visible, the atmospheric transmittance is reasonably predictable for different atmospheric conditions. Ideally, however, it should be measured at the time of image acquisition. The exo-atmospheric solar spectral irradiance is a well-known quantity and for flat terrain, the incident angle is readily obtained. For non-flat terrain,

---

[6] Remember that the atmospheric down-scattered radiance term was assumed to be zero earlier in this derivation.

calculation of the incident angle at each pixel requires a DEM at the same GSI as the sensor image.

We have assumed in all our derivations that the earth's surface was a Lambertian reflector, i.e. that it reflects equally in all directions. There is considerable evidence that this is not true for some types of surfaces, such as forests. A number of other analytical reflectance functions have been proposed, including the *Minnaert model*,

$$L_\lambda = L_n (\cos\theta)^{k(\lambda)} (\cos\phi)^{k(\lambda)-1} \qquad (7-12)$$

where ϕ is the radiation exitance angle to the surface normal vector, k is the empirical Minnaert constant, and L_n is the radiance if both θ and ϕ are zero. If k is one, Eq. $(7-12)$ reduces to a Lambertian model. The constant k is typically found empirically by regression of $\log(L_\lambda \cos\phi)$ against $\log(\cos\theta\cos\phi)$. The Minnaert model has been compared to the Lambertian model for Landsat MSS and TM imagery of mountainous, forested areas (Smith *et al.*, 1980; Colby, 1991; Itten and Meyer, 1993).

Image examples

An example of calibration of TM imagery for sensor response and atmospheric scattering is shown in Fig. 7-21. The pixel *DN* values are first calibrated to radiances using Table 7-3 and Eq. $(7-7)$. Band 1 becomes darker in the figure, reflecting the correction for its relatively high gain compared to bands 2 and 3. However, its relatively higher atmospheric scattering level remains. Then, an atmospheric scattering component is estimated from the lowest values within the Briones Reservoir and subtracted from every pixel. After this correction, band 1 becomes quite dark compared to bands 2 and 3. This final set of sensor- and path radiance-calibrated images displays the correct relative scene radiances (to the extent that the print density of the figure is a linear function of the pixel values).

Another example of the same calibration process is shown in the Cuprite, Nevada, TM image of Fig. 7-22. All six non-TIR bands are shown. Note how the calibration to scene radiance makes bands 5 and 7 appear dark. Their relatively high sensor gain has been removed, and the scene radiance still includes the solar irradiance factor, which falls off significantly in the SWIR. The latter can be removed by dividing each band by the corresponding TM band-integrated solar exo-atmospheric irradiance, which would leave the solar path atmospheric transmittance and topographic shading as the remaining external influences. Scene-based techniques can also be used, as described next for hyperspectral imagery.

Finally, two true-color composites of AVIRIS data are shown in Plate 7-1. These data are corrected first for path radiance by the dark object method, and then for incident solar irradiance and solar path atmospheric transmission by

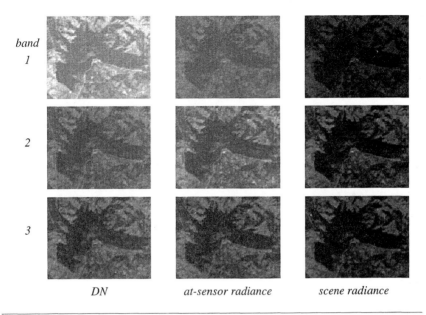

FIGURE 7-21. *Example calibration of TM bands 1, 2 and 3 using sensor calibration coefficients and DOS. All images are displayed with the same LUT to maintain their relative brightness and contrast. The first column shows the original imagery, the second column shows the effect of correction for the sensor pre-flight gain and bias coefficients, and the third column shows the effect of a subtractive correction for estimated atmospheric path radiance. The latter is determined from the darkest pixels within the Briones Reservoir. Atmospheric transmittance is assumed to be the same in each band.*

division by the spectral values of a bright target in each image. The latter step is a reasonable correction only if the bright target has a "white," or spectrally-neutral, spectrum.

7.3.4 Calibration and Normalization of Hyperspectral Imagery

There is special concern about atmospheric spectral transmittance and absorption characteristics and sensor calibration for hyperspectral imagery because of its substantially higher spectral resolution compared to broadband multispectral imagery. The specific reasons for this concern include the following:

FIGURE 7-22. Sensor DN-to-scene radiance calibration for a Landsat TM scene of Cuprite, Nevada, acquired on October 4, 1984. Again, atmospheric spectral transmittance is assumed constant across all bands.

- Hyperspectral sensor bands coinciding with narrow atmospheric absorption features or the edges of broader spectral features will be affected by the atmosphere differently than the neighboring bands.

- The band locations in imaging spectrometer systems are prone to small wavelength shifts under different operating conditions, particularly in airborne sensors.

- Many analysis algorithms for hyperspectral data require precise absorption band-depth measurements exactly at the wavelength of maximum absorption.

- From the computational standpoint alone, the calibration problem is much greater for hyperspectral systems.

A number of empirical techniques have been developed for calibration of hyperspectral data (Table 7-6). They produce relative or absolute calibrations in an empirical way, without the explicit use of atmospheric data and models; for this reason, they are more properly referred to as *normalization* techniques, rather than calibration techniques. The empirical line method adjusts the data for an arbitrary gain and offset, and therefore matches the linear model relating at-sensor radiance and reflectance. A detailed calibration of AVIRIS data using inversion of the 5S program's atmospheric calculations is described in (Zagolski and Gastellu-Etchegorry, 1995), and a comparison of 5S and the LOWTRAN-7 atmosphere modeling program for calibration of AVIRIS data is described in (Leprieur *et al.*, 1995). An advantage of imaging spectrometers over broadband sensors such as TM is that, by virtue of their high spectral resolution, it is possible to estimate atmospheric water vapor content from absorption bands and use that information to facilitate calibration of the imagery to reflectance (Gao and Goetz, 1990; Carrere and Conel, 1993; Gao *et al.*, 1993).

These different normalization procedures can be compared in terms of whether or not they compensate for the various external factors in remote-sensing radiometry (Table 7-7). On this basis, only the residual image approach appears to correct, at least in a heuristic fashion, for all major factors; the empirical line method is the next best technique, but requires extensive field measurements. Not all remote-sensing applications require full calibration to surface reflectance, however.

TABLE 7-6. Empirical normalization techniques for hyperspectral imagery that has previously been calibrated to at-sensor radiances.

technique	description	reference
residual image	scale each pixel's spectrum by a constant such that the value in a selected band equals the maximum value in that band for the entire scene subtract the average normalized radiance in each band over the entire scene from the normalized radiance in each band	(Marsh and McKeon, 1983)
continuum removal	generate a piecewise-linear or polynomial continuum across "peaks" of image spectrum and divide each pixel's spectrum by the continuum	(Clark and Roush, 1984)
internal average relative reflectance (IARR)	divide each pixel's spectrum by the average spectrum of the entire scene	(Kruse, 1988)
empirical line	band-by-band linear regression of pixel samples to field reflectance spectra for dark and bright targets	(Kruse *et al.*, 1990)
flat-field	divide each pixel's spectrum by the average spectrum of a spectrally-uniform, high reflectance area in the scene	(Rast *et al.*, 1991)

TABLE 7-7. Comparison of hyperspectral image normalization techniques in terms of their ability to compensate for various physical radiometric factors.

technique	view-path radiance	topography	solar irradiance	solar path atmospheric transmittance
residual images	✓	✓	✓	✓
continuum removal	X	X	✓	X
IARR	X	X	✓	✓
empirical line	✓	X	✓	✓
flat field	X	X	✓	✓

AVIRIS examples

An AVIRIS image of Cuprite, Nevada, will be used to illustrate the different normalization procedures. This area has been extensively measured, studied and mapped because of the presence of hydrothermally-altered minerals with little vegetative cover. These minerals have characteristic molecular absorption bands in the SWIR spectral region between 2 and 2.4μm (Chapter 1), and we will restrict our analysis to that spectral region.

A color composite of AVIRIS bands (at-sensor radiances) at 2.1, 2.2 and 2.3μm is shown in Plate 9-1. The colors are a clue that there is substantial minearological variation in this area, but the signatures are obscured by topographic shading and variation in surface brightness. Three sites are selected to represent the minerals alunite, buddingtonite and kaolinite, and three single pixels in different locations were selected as bright targets to use for the flat-field normalization. The identification of the mineral sites is aided by maps published in (Kruse *et al.*, 1990). Three-by-three pixel areas are averaged to reduce variability and the average spectral signature plotted in Fig. 7-23. Notice how these curves are largely dominated by the solar spectral irradiance. The atmospheric CO_2 absorption band at 2060nm is also apparent.

FIGURE 7-23. *AVIRIS at-sensor spectral radiances for the three mineral sites at Cuprite. The prominent absorption feature on the left is a CO_2 atmospheric band (compare to Fig. 2-4). The dips in the spectra between 2100 and 2200nm are the characteristic molecular abosrption bands of each mineral. The overall decrease in the radiance from 2000 to 2400nm is primarily due to the decrease in solar irradiance.*

The flat-field and IARR-normalized versions are calculated by dividing each band of the image by the bright target *DN* or the average *DN* of the band, respectively (Fig. 7-24). The atmospheric absorption features are nicely removed by either of these procedures, and the mineral absorption features are unaffected. They achieve this discrimination because the spectrum of every pixel in the scene is modified equally by the atmospheric transmittance, while the various mineral absorption features either tend to cancel each other in the full-scene averaging (IARR), or are not present to begin with in the correction function (flat-field).

These normalization techniques accomplish a significant correction of the hyperspectral radiance spectra, to the point where attempts at matching them to laboratory or field spectra may be successful. The data in Fig. 1-6 are replotted in Fig. 7-25 for comparison to the at-sensor radiances and IARR-corrected values. These two techniques produce similar results for this image, indicating that the average scene spectrum is nearly "white" (Fig. 7-26). Such a fortuitous situation is unlikely for more heterogeneous scenes, such as the one of Palo Alto (Plate 7-1).

The continuum removal procedure is somewhat different from the scene normalization techniques. In this case, the spectra of individual pixels are each adjusted by dividing them by the continuum. The modified spectra exhibit a flat background because of this operation, but also retain the atmospheric absorption features. In our case, the average spectra from each mineral site were normalized in this way (Fig. 7-27).

This completes our discussion of radiometric correction and calibration of remote-sensing imagery. It is probably one of the most logistically difficult tasks in remote sensing, and for that reason is too-often ignored or only partially achieved with image-based techniques. As was pointed out in Chapter 2, however, many types of remote-sensing analyses rely only on *relative* signature discrimination, and therefore do not require *absolute* radiometric calibration.

7.4   Distortion Correction

In addition to noise correction and radiometric calibration, remote-sensing imagery requires spatial distortion corrections to maximize its usefulness for information extraction. The distortion arises from scanner characteristics and their interaction with the orbital geometry and figure of the earth, as discussed in Chapter 3.

In correcting for distortion, we must essentially reposition pixels from their original locations in the data array into a specified reference grid. There are three components to the process:

- selection of suitable mathematical distortion model(s)

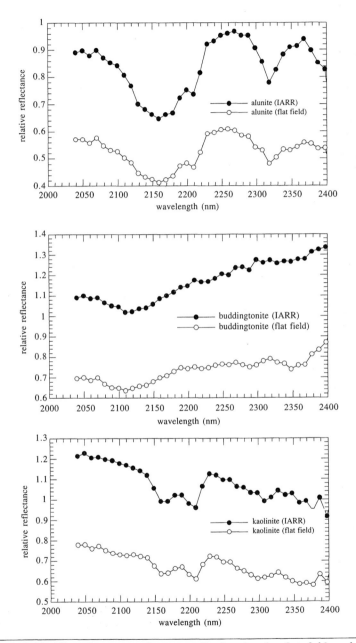

FIGURE 7-24. *Mineral spectra after normalization by the flat field and IARR techniques. Note the IARR result is greater than the flat field result, as expected.*

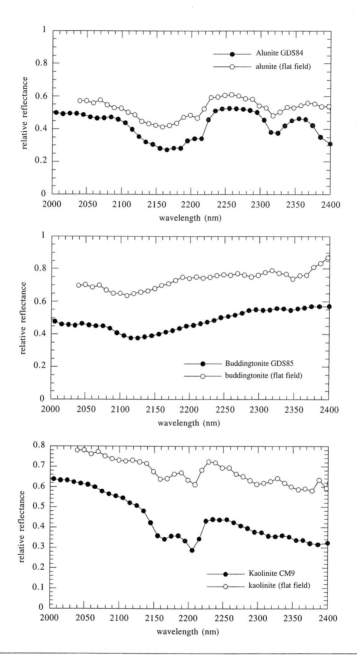

FIGURE 7-25. *Comparison of mineral reflectances and flat field-normalized relative reflectances from the 1990 AVIRIS image.*

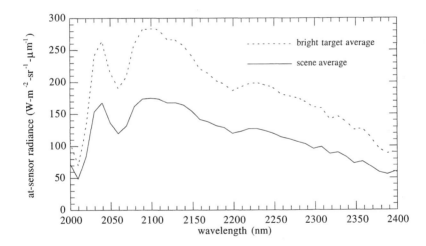

FIGURE 7-26. Comparison of the bright target and average scene spectra for the Cuprite scene. The full visible-to-SWIR spectrum is shown at the top and the SWIR region is expanded below.

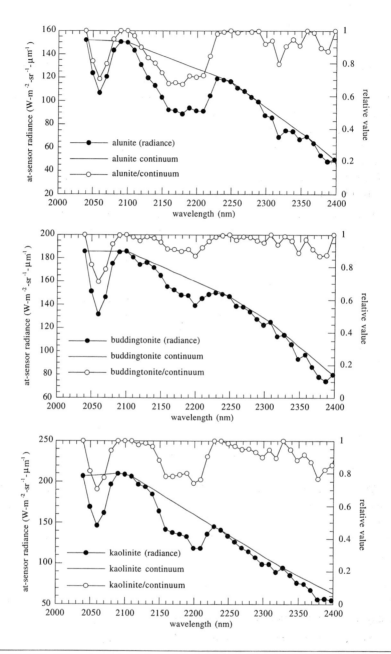

FIGURE 7-27. Mineral spectra adjusted by continuum removal. The continuum was defined manually as a piecewise-linear envelope enclosing the radiance spectra.

- coordinate transformation

- resampling (interpolation)

These three components are collectively known as *warping* (Wolberg, 1990). With the increasing complexity of remote-sensing systems, some having significant off-nadir viewing capability (e.g. SPOT and IRS-1C), the interest in multi-sensor image registration, and the increasing demands of earth scientists for precision, temporal resolution and repeatability in remote-sensing measurements, there is a continuing need for geometric processing algorithms that are more accurate, autonomous and efficient.

Various terms are used to describe geometric correction of imagery, and it is worthwhile defining them before proceeding:

- *Registration*: The alignment of one image to another image of the same area. Any two pixels at the same location in both images are then "in register" and represent two samples at the same point on the earth (Chapter 8).

- *Rectification*: The alignment of an image to a map so that the image is planimetric, just like the map (Jensen, 1996). Also known as *geo-referencing* (Swann *et al.*, 1988). An example rectification of Landsat TM data to map coordinates is shown in Fig. 7-28.

- *Geocoding*: A special case of rectification that includes scaling to a uniform, standard pixel *GSI*. The use of standard pixel sizes and coordinates permits convenient "layering" of images from different sensors and maps in a GIS.

- *Orthorectification*: Correction of the image, pixel-by-pixel, for topographic distortion. The result, in effect, is that every pixel appears to be viewing the earth from directly above, i.e. the image is in an orthographic projection.

Two approaches to rectification of satellite imagery are shown in Fig. 7-29. In Chapter 3, we described the pure satellite modeling approach, in which information on satellite position, attitude, orbit, and scan geometry are used with an Earth figure model to produce system-corrected products. This is the product that many users obtain from sources such as EOSAT or the USGS. Further correction to rectify the system-corrected imagery can be done with polynomial distortion functions and *Ground Control Points (GCPs)*. This second stage, polynomial correction, is the subject of this section. It results in satisfactory rectification in many applications, but has the following disadvantages:

- two resamplings are performed on the data, resulting in unnecessary degradation

original

rectified

FIGURE 7-28. Rectification of a Landsat TM band 4 image of Tucson, Arizona. The primary distortion in the original is due to the non-polar orbit; thus, rotation is the dominant correction (the major streets in Tucson are oriented north and south, and east and west). Note the corners of the rectified image are clipped by the image format; a larger output format would avoid this. Also, there are some areas in the rectified frame for which no input pixels exist in the original image. These pixels are filled with a uniform DN in the output.

- polynomials of any reasonable degree cannot correct for local topography
- a large number of GCPs are sometimes necessary for low residual error

It is possible to produce the scene-corrected, or precision level, product with a hybrid, single-stage correction and resampling (Westin, 1990), which presumably results in a superior product. Such correction, however, requires access to the raw imagery and detailed satellite ephemeris and attitude data, plus the necessary software algorithms.

Orthorectification requires a *Digital Elevation Model (DEM)*, since every pixel's location must be adjusted for topographic relief displacement. We will discuss this subject in the next chapter, where algorithms for automated registration of images and DEM extraction are also discussed.

7.4.1 Polynomial Distortion Models

The selection of an appropriate model (or models) is critical to the accuracy of any distortion correction. At least for satellite sensor or platform-induced distortions, there exist specific and accurate mathematical expressions for the distortion (Chapter 3). These types of models, however, require precise parameter input on sensor position, attitude and scan angle as a function of time.

Because of the difficulties in supplying the parameters for a precise sensor distortion model, data received by the user may contain residual geometric errors. Moreover, if the image is a lower level, system-corrected product (Chapter 1), it will not have undergone controlled rectification to a map projection. Therefore, it has become common practice in the user community to use a generic polynomial model for registering images to each other and to maps.

The polynomial model relates the global coordinates in the distorted image to those in the reference image or map,

$$x = \sum_{i=0}^{N} \sum_{j=0}^{N-i} a_{ij} x_{ref}^{i} y_{ref}^{j}$$

$$y = \sum_{i=0}^{N} \sum_{j=0}^{N-i} b_{ij} x_{ref}^{i} y_{ref}^{j}$$

$$(7-13)$$

There is no physical basis for the choice of polynomial functions in this task, unlike the models described in Chapter 3. Polynomials are widely used as an approximating function in all types of data analysis. The level of detail in the data that can be approximated depends directly on the order, N, of the polynomial.

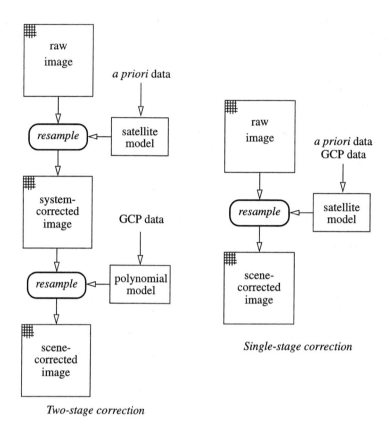

Two-stage correction

Single-stage correction

FIGURE 7-29. *Geometric processing data flow for the common two-stage process and a single-stage process for rectification. In the two-stage process, a generic polynomial model is used to remove residual distortions in the system-corrected imagery. The polynomial correction approach is widely used and available in all major software systems. A more sophisticated and efficient approach is to use physical models and include all corrections in a single resampling. (Adapted from (Westin, 1990).)*

If the input imagery has been processed accurately for systematic distortions, a linear polynomial may suffice for further correction. At worst, a quadratic polynomial (N equals 2) is sufficient for most problems in satellite remote sensing where the terrain relief is small and the FOV is not large,

$$x = a_{00} + a_{10}x_{ref} + a_{01}y_{ref} + a_{11}x_{ref}y_{ref} + a_{20}x_{ref}^2 + a_{02}y_{ref}^2$$

$$y = b_{00} + b_{10}x_{ref} + b_{01}y_{ref} + b_{11}x_{ref}y_{ref} + b_{20}x_{ref}^2 + b_{02}y_{ref}^2 .$$

$$(7 - 14)$$

The terms in Eq. $(7 - 14)$ can be interpreted as components in the total warp (Table 7-8). When used in combination, they can result in special warps, as shown in Fig. 7-30. The linear polynomial,

$$x = a_{00} + a_{10}x_{ref} + a_{01}y_{ref}$$
$$y = b_{00} + b_{10}x_{ref} + b_{01}y_{ref}$$

$(7 - 15)$

is known as an *affine* transformation (Wolf, 1983). It can simultaneously accommodate shift, scale and rotation. Moreover, Eq. $(7 - 15)$ can be written in a compact vector-matrix form,

$$\begin{bmatrix} x \\ y \end{bmatrix} = \begin{bmatrix} a_{10} & a_{01} \\ b_{10} & b_{01} \end{bmatrix} \begin{bmatrix} x_{ref} \\ y_{ref} \end{bmatrix} + \begin{bmatrix} a_{00} \\ b_{00} \end{bmatrix}$$

$(7 - 16)$

or, equivalently,

$$\boldsymbol{p} = \boldsymbol{T}\boldsymbol{p}_{ref} + \boldsymbol{T}_0.$$

$(7 - 17)$

TABLE 7-8. *Contributions of each coefficient in Eq. $(7 - 14)$ to the total warp.*

coefficient	warp component
a_{00}	shift in x
b_{00}	shift in y
a_{10}	scale in x
b_{01}	scale in y
a_{01}	shear in x
b_{10}	shear in y
a_{11}	y-dependent scale in x
b_{11}	x-dependent scale in y
a_{20}	nonlinear scale in x
b_{02}	nonlinear scale in y

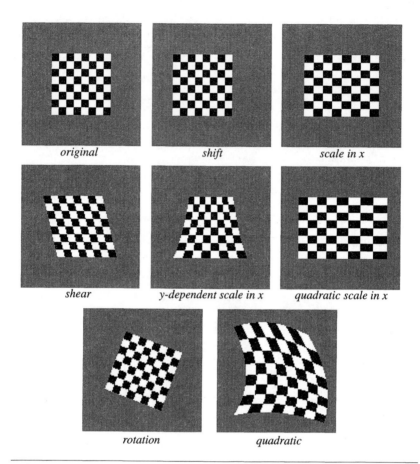

FIGURE 7-30. *Polynomial geometric warps. The upper two rows show 1-D warps in the x-direction. Nearest-neighbor resampling is used in these examples.*

The affine transformation can be used for approximate correction of satellite sensor distortions (Anuta, 1973; Steiner and Kirby, 1976; Richards, 1993). The matrix T takes on specific values for a particular sensor. For example, several affine transformations can be applied to correct scanner- and orbit-related distortions in Landsat MSS imagery and produce an approximately rectified product (Table 7-9). These individual matrices can be combined into a single matrix transformation,

$$T_{total} = T_1 T_2 T_3 \qquad (7-18)$$

for processing efficiency and to avoid multiple resamplings. This type of processing was common in the early years of the Landsat MSS because only the raw, uncorrected imagery was available. Much of the archived "legacy" imagery, dating to 1972, is in that raw format, so the need will remain for such simple system geometric correction techniques in the future.

TABLE 7-9. *Specific affine transformations for Landsat MSS data (Anuta, 1973; Richards, 1993). The various angles used are:*
    i : orbit inclination at the equator (degrees)
    ϕ: *geocentric latitude at frame center*
    θ: *orbit inclination at latitude* ϕ, $\theta = 90 - \text{acos}[\sin i / \cos \phi]$ *(degrees)*
    γ: *skew angle between the top and bottom of the image due to earth rotation*

distortion	transformation matrix
aspect ratio	$T_1 = \begin{bmatrix} 1 & 0 \\ 0 & 0.709 \end{bmatrix}$
earth rotation	$T_2 = \begin{bmatrix} 1 & -\alpha \\ 0 & 1 \end{bmatrix}$, $\alpha = \tan\gamma \cos\phi \cos\theta$
rotation to north	$T_3 = \begin{bmatrix} \cos\theta & -\sin\theta \\ \sin\theta & \cos\theta \end{bmatrix}$

Ground Control Points (GCPs)

The coefficients in the polynomial model Eq. (7 – 14) must be determined. The procedure is to define *Ground Control Points (GCPs)* to constrain the polynomial coefficients. GCPs should have the following characteristics:

- high contrast in all images of interest
- small feature size
- unchanging over time
- all are at the same elevation (unless topographic relief is being specifically addressed)

Examples include road intersections, corners of agricultural fields, small islands and river features. It is advisable to confirm that the GCPs have not changed over the time interval between the images being registered. This is particularly important for GCPs that rely on water levels for their location. Finding corresponding GCPs in an image and a map is generally more difficult than in two images because the map is an abstraction of the area's features and is often considerably older than the image.

Visual location of GCPs is the most widely-used paradigm in Landsat and SPOT remote-sensing applications. Wherever available, man-made features prove the most reliable for GCPs. The primary drawbacks are the difficulty in finding satisfactory control points that are well-distributed across the image area and the intensive manual labor involved. It can also be difficult, if not impossible, to locate GCPs in low resolution imagery such as from the AVHRR. That is one reason why there has been considerable interest in accurate orbit and platform modeling for the AVHRR, as described in Chapter 3.

To see how GCPs are used to find the polynomial coefficients, suppose we have located M pairs of GCPs in the distorted image and reference (image or map) coordinate systems. Assuming the global polynomial distortion model, we can write, for *each* pair m of GCPs, a polynomial equation of degree N in each variable, x and y,[7]

$$
\begin{aligned}
x_m &= a_{00} + a_{10}x_{refm} + a_{01}y_{refm} + a_{11}x_{refm}y_{refm} + a_{20}x_{refm}^2 + a_{02}y_{refm}^2 \\
y_m &= b_{00} + b_{10}x_{refm} + b_{01}y_{refm} + b_{11}x_{refm}y_{refm} + b_{20}x_{refm}^2 + b_{02}y_{refm}^2,
\end{aligned} \quad (7-19)
$$

leading to M pairs of equations. This set of equations can be written in vector-matrix form for the x coordinates of the image as,

$$
\begin{bmatrix} x_1 \\ x_2 \\ \vdots \\ x_M \end{bmatrix} = \begin{bmatrix} 1 & x_{ref1} & y_{ref1} & x_{ref1}y_{ref1} & x_{ref1}^2 & y_{ref1}^2 \\ 1 & x_{ref2} & y_{ref2} & x_{ref2}y_{ref2} & x_{ref2}^2 & y_{ref2}^2 \\ \vdots & \vdots & \vdots & \vdots & \vdots & \vdots \\ 1 & x_{refM} & y_{refM} & x_{refM}y_{refM} & x_{refM}^2 & y_{refM}^2 \end{bmatrix} \begin{bmatrix} a_{00} \\ a_{10} \\ a_{01} \\ a_{11} \\ a_{20} \\ a_{02} \end{bmatrix} \quad (7-20)
$$

or,

$$
X = WA \quad\quad\quad (7-21)
$$

and similarly for the y coordinates of the image,

$$
Y = WB. \quad\quad\quad (7-22)
$$

---

[7] N is two for a quadratic polynomial.

Now, if M equals the number of polynomial coefficients K,[8] we have an exact solution by inverting the M x M matrix W,

$$M = K: \quad \begin{aligned} A &= W^{-1}X \\ B &= W^{-1}Y, \end{aligned} \qquad (7-23)$$

and the error in the polynomial fit will be exactly zero at every GCP. It is usually desirable to have more GCPs because some may be in error. In this case we write,

$$M \geq K: \quad \begin{aligned} X &= WA + \varepsilon_X \\ Y &= WB + \varepsilon_Y \end{aligned} \qquad (7-24)$$

where the added terms represent anticipated error vectors in the GCP locations. But when M is greater than K, the M x K matrix W cannot be inverted. In that case, it is possible to calculate a so-called *pseudoinverse* solution (Wolberg, 1990),

$$M \geq K: \quad \begin{aligned} \hat{A} &= (W^T W)^{-1} W^T X \\ \hat{B} &= (W^T W)^{-1} W^T Y. \end{aligned} \qquad (7-25)$$

which is equivalent to a least-squares solution for A and B that minimizes the total squared error in the polynomial fit to the GCPs,

$$\begin{aligned} min[\varepsilon_X^T \varepsilon_X] &= (X - W\hat{A})^T (X - W\hat{A}) \\ min[\varepsilon_Y^T \varepsilon_Y] &= (Y - W\hat{B})^T (Y - W\hat{B}). \end{aligned} \qquad (7-26)$$

The numerical stability of the solution for W can be improved by using matrix decomposition techniques instead of the pseudoinverse solution (Wolberg, 1990).

It is important to realize that the goodness-of-fit exhibited by the polynomial is *only* that. It does *not* indicate the validity of the polynomial as a model for the physical distortion. The fit may be quite good at the GCPs (zero error, in fact, if M equals K), while simultaneously there may be large error at other points. Although not always done in practice, any use of GCPs for distortion correction should include a GCP subset for control of the polynomial, i.e. determination of its coefficients, and another, distinct GCP subset for evaluation of the residual errors at other points. The latter subset may be called just *Ground Points (GPs)*, since they are not used for control of the model. Such a procedure is analogous to the use of independent training and testing sites in multispectral classification (Chapter 9).

---

[8] $K = (N + 1)(N + 2)/2$

An example of GCP selection and polynomial fitting can be generated from the aerial photograph and map in Fig. 1-2. A set of six GCPs and four test GPs are shown on the photograph and the scanned map in Fig. 7-31. The direct mapping of image points to map points is shown in Fig. 7-32. The dominate distortion between the two seems to be a rotation (the image needs to be rotated clockwise), and the GPs appear to be consistent with the GCPs. The GCP pairs were used to find the polynomial distortion function for K values of three (affine) to six (quadratic). The RMS deviation at the GCPs and GPs for each polynomial is plotted in Fig. 7-33. As the number of terms in the polynomial is increased, the RMS deviation at the GCPs decreases, eventually to zero when K equals M. At the same time the polynomial is providing a better fit to the GCPs, the RMS deviation at the GPs increases. This implies that the higher-order polynomial is not necessarily a better model for whatever actual distortion exists between the image and map.

The selection of GCPs for a registration or rectification task is typically an iterative process in which an initial set of GCPs are used to find the polynomial. The GCP coordinates in the reference coordinate system are subjected to the polynomial transformation, and the goodness-of-fit to the actual GCP locations examined in the image coordinate system. If the transformed GCP coordinates for any GCP are found to exceed a specified deviation from the image GCP, that point is assumed to be in error and either remeasured or dropped from the list of candidate GCPs. An example is shown in Fig. 7-34, where initially six GCPs were selected. After a trial fit with a bilinear polynomial, one point was considered to be in error since it deviated from the polynomial approximation by more than a pixel. After dropping that point from the analysis, the polynomial fit was repeated and the remaining GCPs showed smaller deviations from the polynomial model. The process is stopped when the total error at the GCPs falls within an acceptable level.

Care should be exercised in application of global polynomials to uncorrected images with a large FOV. For example, cross-track panoramic distortion in whiskbroom sensor images is caused by uniform time sampling during the scan sweep (Chapter 3). The image scale (e.g. in m/pixel) off-nadir is related to that at nadir by Eq. (3 – 30). This equation is graphed in Fig. 7-35, along with the quadratic polynomial fit to Eq. (3 – 30) over ±50° (approximately the *FOV* of the AVHRR). An error of up to 15% at the extremes of the scan will occur if a global quadratic polynomial is used to approximate the actual distortion function. The error can be reduced to less than 5% with a quartic polynomial. Additional error may result when only a few GCPs are used to calculate the fit, as would be the case in practice. A quadratic polynomial model for panoramic distortion is quite good, however, over the much smaller ±5° *FOV* of the Landsat MSS or TM.

As a final example of the use and limitations of global polynomial distortion mappings, we will try to rectify the image of Fig. 3-20. A rectangular grid was

FIGURE 7-31. GCP location for rectifying the aerial photograph (top) to the scanned map (bottom). Six GCPs are selected for control (black crosses with white center) and four GPs for testing (white crosses with black center). The contrast of the image and map are purposely made low to emphasize the GCPs and GPs in this figure.

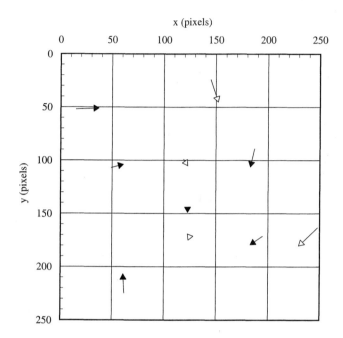

FIGURE 7-32. *Direct mapping of the image GCPs (black arrowheads) and GPs (white arrowheads) to those in the map. The coordinate system is line- and sample-based and has the origin at the upper left. Note the clockwise rotation, with also some scale differential around the borders indicated by the outer arrows pointing inward to the center.*

created approximating the corrected geometry of the fields (a map was not available), six GCPs selected in the image, and affine and quadratic warps performed. The results clearly show that higher frequency distortions are present (Fig. 7-36). It is possible that a fifth, or higher-order polynomial may improve the situation, but not eliminate all distortion. Even if platform attitude data were available, it would have to be sampled very frequently (say every 10 or 20 image lines) to account for the rapid changes evident in the image. This example also serves to emphasize the importance of stabilized platforms in airborne remote sensing.

The global polynomial model can be modified to better match distortion over large areas, or rapidly changing distortions such as those caused by topographic relief or platform attitude, by subdividing the image into small, contiguous sections. The distortion within each section is then modeled by a separate polynomial, and the results pieced together. For example, the coordinate transformation within a quadrilateral defined by four control points may be modeled by a

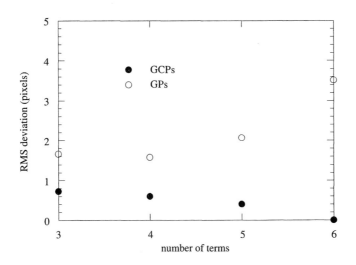

FIGURE 7-33. RMS deviations between the GCP and GP locations as predicted by the fitted-polynomial and their actual locations, for different numbers of terms in the polynomial. Note the deviation at the GPs is larger than that at the GCPs by a pixel or more.

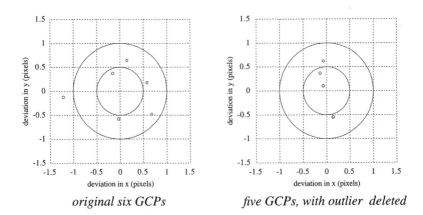

original six GCPs   *five GCPs, with outlier deleted*

FIGURE 7-34. Refinement of GCPs. The plot on the left shows that one GCP deviates from the affine polynomial model by more than one pixel in the x direction. The RMS deviation for all GCPs is 0.76 pixels. After the offending point is removed from the GCP list, and the polynomial is fit to the remaining five GCPs, the plot on the right results, and the RMS deviation is reduced to 0.49 pixels.

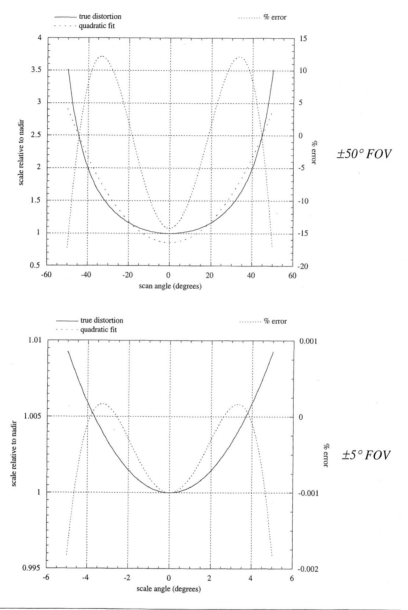

FIGURE 7-35. Comparison of actual scanner panoramic distortion and a polynomial model over the FOVs of the AVHRR and Landsat sensors.

*affine
correction*

*quadratic
correction*

FIGURE 7-36. *Attempted global rectifications of the airborne ASAS image of Fig. 3-20
using six GCPs and a synthesized orthogonal grid. The affine transformation accounts
for most of the global distortion, and the quadratic transformation adds little
improvement. To rectify this image with global polynomials would require more GCPs
and a higher-order polynomial or a piecewise polynomial correction, as described in
the text.*

power series polynomial with the first four terms of Eq. (7 – 14). By defining a
net of contiguous quadrilaterals across the image, each with a different polynomial transformation, a complex transformation can be accomplished by piecewise approximation (Fig. 7-37). A dense net of control points will generally yield
higher accuracy, but a large number of GCPs, distributed across the image, may
be difficult to achieve in some images. Also, a four term polynomial is discontinuous across the boundary between two quadrilaterals, so ground features that
cross the boundary may be discontinuous after processing. The use of contiguous
triangles between three control points, with a linear polynomial transformation in

each triangle, eliminates the boundary discontinuities.[9] The global distortion is then modeled by a piecewise set of planes, much like a faceted surface.

The technique of piecewise distortion modeling was originally used by the Jet Propulsion Laboratory (Castleman, 1996) to correct severely-distorted planetary spacecraft images. Applications to airborne scanner imagery are presented in (Craig and Green, 1987) and (Devereux *et al.*, 1990), and application to SPOT imagery is reported in (Chen and Lee, 1992).

7.4.2 Coordinate Transformation

Once the coordinate transformation f is found, it is used to define a mapping from the reference frame coordinates (x_{ref}, y_{ref}) to the distorted image frame coordinates (x, y),

$$(x, y) = f(x_{ref}, y_{ref}) .$$

$(7 - 27)$

The transformation of Eq. $(7 - 27)$ is implemented by stepping through the integer coordinates (x_{ref}, y_{ref}) one-by-one, and calculating the transformed (x, y) values. Since the (x, y) coordinates will not be integer values in general, a new pixel must be estimated between existing pixels by an interpolation process known as *resampling*. The corrected output image is then created by "filling" each location (x_{ref}, y_{ref}) of a new, initially empty array, with the pixel calculated at (x, y) in the distorted image, as depicted in Fig. 7-38. This procedure, which may seem backwards at first, is preferable to a mapping from (x, y) to (x_{ref}, y_{ref}) because it avoids "overlapping" pixels and "holes" that have no assigned pixels in the output image.

Map projections

Maps are the spatial framework for analysis of remote-sensing images. However, every map has spatial distortion since it is a projection of the 3-D spherical earth onto a 2-D plane. A wide variety of projection schemes have been developed by trading off one type of distortion for another (Gilbert, 1974; Snyder, 1987). For example, the relative areas of objects on the ground may be preserved at the expense of increased distortion in distances or shapes. The mathematical definitions of several common map projections are given in Table 7-10.

In general, the smaller the area of interest, the less important are the differences among map projections. Global datasets are being assembled as mosaics of many remote sensing images to support the increased interest in long-term

---

[9] However, the *direction* of features can still change across boundaries.

piecewise quadrilateral warping

piecewise triangle warping

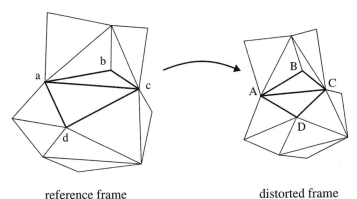

reference frame   distorted frame

FIGURE 7-37. *Piecewise polynomial mapping. At the top, a net of quadrilaterals is defined by groups of four GCPs in both coordinate systems. The mapping function within each quadrilateral is a four-term polynomial. At the bottom, the same GCPs are used to define a net of triangles; the mapping function within each triangle is a three term, affine polynomial. The coefficients for the polynomial within each polygon, for example triangles a-b-c and a-c-d, will in general be different. The triangular network can be generated by Delaunay triangulation (Devereux et al., 1990; Chen and Lee, 1992), which is the dual of Dirichlet tessellation (also known as the Voronoi diagram) used in pattern recognition (Jain and Dubes, 1988; Schürmann, 1996) and in modeling of spatial statistics (Ripley, 1981).*

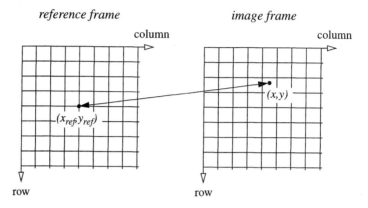

FIGURE 7-38. *The two-way relationship between the reference and distorted coordinate systems, expressed as rows and columns in the digital data. The arrow from (x_{ref}, y_{ref}) to (x,y) indicates a pointer, given by Eq. (7 – 27), from an integer coordinate in the reference frame to a (generally) non-integer coordinate in the image frame. The arrow from (x,y) to (x_{ref}, y_{ref}) indicates the transfer of an estimated pixel value, resampled at the non-integer location (x,y) in the original, distorted image, to the output location (x_{ref}, y_{ref}) in the reference frame.*

changes of the earth's environment. Global coverage in a single-map coordinate system demands special consideration (Steinwand, 1994; Steinwand *et al.*, 1995).

7.4.3 Resampling

The interpolation of new pixels between existing pixels is required to implement the geometric transformation as previously explained. The geometry of resampling is illustrated in Fig. 7-39, where a new pixel is to be resampled at the location *R*. Resampling can be thought of as convolution of the distorted image with a moving window function, as in spatial filtering. The resampled output values are calculated, however, *between* original pixels. Therefore, the resampling weighting function must be defined as a continuous function, rather than the discrete array used in convolution filtering.

The fastest scheme for calculating a resampled pixel is *nearest-neighbor* assignment (sometimes called *zero-order interpolation*). For the value of each new pixel at (x_{ref}, y_{ref}) in the output image, the value of the original pixel nearest

TABLE 7-10. Projection plane equations for several common map projections (Moik, 1980). The latitude of a point on the Earth is φ and its longitude is λ. The projected map coordinates, x and y, are called "easting" and "northing," respectively. R is the equatorial radius of the Earth and ε is the Earth's eccentricity (Chapter 3). The subscripted values of latitude and longitude pertain to the definition of a particular projection.

projection	x	y
polar stereo-graphic	$2R\tan\left(\dfrac{\pi}{2}-\varphi\right)\sin\lambda$	$2R\tan\left(\dfrac{\pi}{2}-\varphi\right)\cos\lambda$
Mercator	$R\lambda$	$R\lambda\ln\left[\tan\left(\dfrac{\pi}{4}+\dfrac{\varphi}{2}\right)\left(\dfrac{1-\varepsilon\sin\varphi}{1+\varepsilon\sin\varphi}\right)^{\varepsilon/2}\right]$
oblique Mercator	$R\operatorname{atan}\left[\dfrac{\cos\varphi\sin(\lambda-\lambda_p)}{\sin\varphi\cos\varphi_p-\beta\sin\varphi_p}\right]$ $\beta = \cos\varphi\cos(\lambda-\lambda_p)$	$\dfrac{R}{2}\ln\left[\dfrac{1+\alpha+\beta\cos\varphi_p}{1-\alpha+\beta\cos\varphi_p}\right]$ $\alpha = \sin\varphi\sin\varphi_p$
transverse Mercator	$R\operatorname{atan}\left[\cos\varphi\sin(\lambda-\lambda_p)\right]$	$\dfrac{R}{2}\ln\left[\dfrac{1+\beta}{1-\beta}\right]$
Lambert normal conic	$\rho\sin\theta$	$\rho_0-\rho\cos\theta$ $\rho = \dfrac{R\cos\varphi_1}{\sin\varphi_0}\left[\dfrac{\tan\left(\dfrac{\pi}{4}-\dfrac{\varphi}{2}\right)}{\tan\left(\dfrac{\pi}{4}-\dfrac{\varphi_1}{2}\right)}\right]^{\sin\varphi_0}$ $\theta = \lambda\sin\varphi_0$ $\sin\varphi_0 = \dfrac{\ln\left(\dfrac{\cos\varphi_1}{\cos\varphi_2}\right)}{\ln\left[\dfrac{\tan\left(\dfrac{\pi}{4}-\dfrac{\varphi_1}{2}\right)}{\tan\left(\dfrac{\pi}{4}-\dfrac{\varphi_2}{2}\right)}\right]}$

to (x,y) is selected. In Fig. 7-39, this would result in pixel C at the output location. Because of this "round-off" property, geometric discontinuities on the order of plus or minus one-half pixel are introduced into the processed image. This effect is often negligible in displayed images, but may be important for subsequent numerical analyses. The significant advantage of nearest-neighbor assignment over other algorithms is that no calculations are required to derive the output pixel value, once the location (x,y) has been calculated, which must be

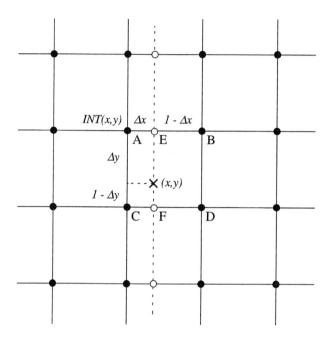

FIGURE 7-39. *Geometry for resampling a new pixel at (x,y). Pixels in the distorted image (solid circles) are located at integer row and column coordinates in the image frame (Fig. 7-38). The coordinates of the local origin at pixel A are the integer parts and the offsets ($\Delta x, \Delta y$) are the fractional parts of (x,y) as calculated in Eq. (7 – 27). For the two-step, separable implementation of bilinear resampling, the pixels E and F are first calculated by interpolation between A and B, and C and D, respectively. Then the DN at (x,y) is estimated by interpolation between E and F. If cubic resampling is used, the four intermediate pixels (open circles) are interpolated from the surrounding four pixels in each row, and then used to estimate the DN at (x,y).*

done in any case. Nearest-neighbor assignment is equivalent to convolving the input image with a uniform spatial weighting function that is one sample interval wide, as shown in Fig. 7-40.

A smoother interpolated image is obtained with *bilinear* (first-order) resampling. This algorithm uses the four input pixels surrounding the point (x,y) to estimate the output pixel. Bilinear resampling is usually implemented by first convolving the input image along its rows, creating new, resampled columns of pixels, and then along the new columns, with a triangle weighting function in

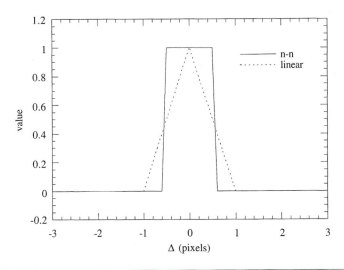

FIGURE 7-40. Nearest-neighbor and linear resampling spatial-weighting functions. The distance, Δ, is measured from the location, (x,y), of the resampled pixel (Fig. 7-39).

both directions (Fig. 7-40). The total equation for the example in Fig. 7-39 would be,

$$DN = [\Delta x DN_B + (1 - \Delta x)DN_A](1 - \Delta y)$$
$$+ [\Delta x DN_D + (1 - \Delta x)DN_C]\Delta y \qquad (7-28)$$

The difference between the nearest-neighbor and bilinear algorithms may be appreciated by the comparison in Fig. 7-41. The image has been enlarged digitally by the two methods to illustrate the continuous nature of bilinear resampling versus the discrete nature of nearest-neighbor resampling. The difference between the two resampling functions is especially apparent in a surface plot of the resampled image (Fig. 7-42). Bilinear resampling produces a smoother surface, at the expense of being considerably slower than nearest-neighbor resampling, because Eq. (7 – 27) must be calculated for every output pixel.

The smoothing incurred with bilinear interpolation may be avoided with *cubic* (second-order) interpolation, at the expense of more computation. The cubic interpolating function is a piecewise cubic polynomial that approximates the theoretically ideal interpolation function for imagery, the *sinc* function (Pratt, 1991).[10] The sinc function is not used for image interpolation because a large pixel neighborhood is required for accurate results. The cubic interpolator yields results approximating those from a sinc function and requires only a 4-by-4 neighborhood in the input image. The cubic resampling function is actually a *parametric cubic convolution (PCC)* family of functions, defined by a single parameter α,

$$
\begin{aligned}
r(\Delta;\alpha) &= (\alpha + 2)|\Delta|^3 - (\alpha + 3)|\Delta|^2 + 1 & |\Delta| \le 1 \\
&= \alpha(|\Delta|^3 - 5|\Delta|^2 + 8|\Delta| - 4) & 1 \le |\Delta| \le 2 & \qquad (7-29) \\
&= 0 & |\Delta| \ge 2
\end{aligned}
$$

where Δ is the local offset in either x or y, as appropriate (Park and Schowengerdt, 1983).

A drawback of cubic resampling is that it produces *DN* overshoot on either side of sharp edges. The magnitude of the overshoot is directly proportional to the magnitude of α. Although this characteristic contributes to the visual sharpness of the processed image, it is undesirable for further numerical analyses where radiometric accuracy is of prime importance. It has been shown by (Keys, 1981) and (Park and Schowengerdt, 1983) that -0.5 is an optimal value for α (rather than -1, as is commonly used), and the value of -0.5 was implemented early in the production of TM data (Fischel, 1984). The two interpolation functions are compared to the sinc function in Fig. 7-43.

The effect of the nearest-neighbor, linear, and PCC resampling functions on Landsat TM imagery can be seen in Fig. 7-44. Difference images between the nearest-neighbor and bilinear and between the bilinear and PCC results are shown in Fig. 7-45. PCC generally does not yield a pronounced visual improvement over bilinear interpolation and requires more computation time. For applications in which the image sharpness is not critical, nearest-neighbor resampling may produce a satisfactory product with a large savings in computer processing time compared to the other interpolation functions.

Nearest-neighbor resampling does not introduce new pixel *DN* vectors into the image statistical distribution, whereas bilinear and bicubic resampling can create new vectors. This is sometimes used as an argument for using only nearest-neighbor resampling if the image will later be classified by statistical techniques (Chapter 9). The effect of resampling on the quality of TM system-corrected imagery, relative to uncorrected imagery, has been reported in (Wrigley *et al.*, 1984). A comparison of bilinear and bicubic resampling on subsequent co-occurrence texture calculations (see Chapter 4) is presented in (Roy and Dikshit, 1994), on multispectral classification in (Dikshit and Roy, 1996), and on the NDVI calculated from resampled AVHRR imagery in (Khan *et al.*, 1995).

---

[10] Defined as $sinc(x) = \dfrac{\sin(\pi x)}{\pi x}$.

FIGURE 7-41. Image magnification using nearest-neighbor and bilinear resampling. Although the bilinear resampling is smooth, there remains a jagged "stairstep" pattern along the curved road at the top. This is aliasing that originated in the digitization of the photograph.

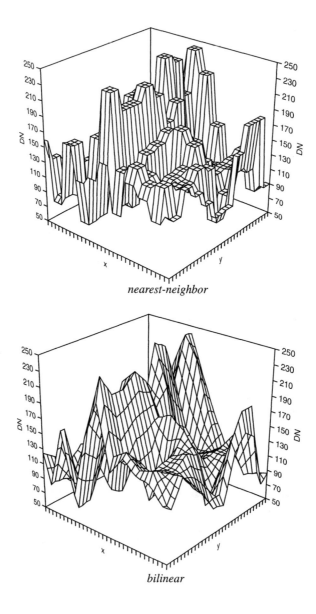

FIGURE 7-42. Surface plots of DN(x,y) from the 3X-zoomed images in Fig. 7-41 for
nearest-neighbor resampling and bilinear resampling.

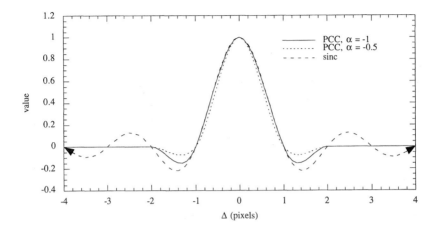

FIGURE 7-43. *Parametric cubic convolution and sinc resampling spatial-weighting functions. Note the sinc function continues on beyond the graph and never goes completely to zero, whereas the PCC functions are bounded between ±2 pixels.*

7.5   Summary

Remote-sensing images require various types of systematic corrections for noise and geometric distortion. They also require radiometric calibration for accurate comparisons between dates and sensors. The important aspects of these operations are:

- Noise correction is necessary only if the noise will affect information extraction. It is best done before any resampling of the imagery.

- Local noise is removed with a variety of spatial convolution filters, while global periodic noise is removed with Fourier filters. Uncorrelated noise in multispectral images may be removed by spectral decorrelation techniques, such as the *PCT*.

- Rectification of system-corrected products to maps can be implemented over limited areas with moderate topographic relief by global polynomial functions.

nearest-neighbor   *bilinear*

PCC (α = –0.5)   *PCC (α = –1.0)*

FIGURE 7-44. A portion of the rectified TM image of Fig. 7-28, as obtained with four different resampling functions. Nearest-neighbor resampling results in a blockiness, particularly for diagonal features. Bilinear resampling is continuous, while PCC results in a slightly sharper image. The visual difference between PCC with α equal to –0.5 and –1.0 is small; a value of –1.0 produces slightly more edge sharpening and overshoot.

- Resampling for geometric correction alters the local spatial and global radiometric properties of the image. Nearest-neighbor resampling does not introduce any new pixel spectral vectors into the data, bilinear resampling will average spectral vectors within their original *DN* range, and cubic resampling introduces new spectral vectors outside the original *DN* range.

nearest-neighbor – bilinear   *bilinear – PCC*

FIGURE 7-45. Images of the differences between nearest-neighbor and bilinear resampling and bilinear and PCC with $\alpha = -0.5$. *Negative differences are dark and positive differences are light. These difference maps illustrate the blockiness of nearest-neighbor relative to bilinear, and the fact that PCC has a high-boost filtering effect relative to bilinear. The difference image on the right actually has some resemblance to a high-pass filtered image, which is expected from a transfer function analysis of resampling (Park and Schowengerdt, 1983).*

In the next chapter, we look at the spatial registration of images and some techniques for image fusion. Resampling and relative radiometric calibration play important roles in both topics.

7.6 Exercises

7–1. Specify a mathematical form for the frequency domain Gaussian notch-filter for:

- in-track striping in pushbroom imagery
- electronic interference noise at 45° to the cross-track direction
- cross-track periodic noise with a period of four pixels

7–2. Show in detail how Eq. (7 – 12) and linear regression are used to find the Minnaert constant k.

7–3. Explain how each hyperspectral normalization algorithm in Table 7-7 achieves the indicated corrective capabilities.

7–4. Show that convolution of a row of pixels with the triangle weighting function of Fig. 7-40 and with calculation of output values between the original pixels, i.e. resampling, is equivalent to linear interpolation.

7–5. Estimate the *DN* of the resampled pixel at ✕ using bilinear resampling:

CHAPTER 8

Image Registration and Fusion

8.1 Introduction

Spatial registration of multidate or multisensor images is required for many applications in remote sensing, such as the construction of image mosaics, DEM generation from stereo pairs, orthorectification, and multitemporal analysis. Registration is the process which makes the pixels in two images precisely coincide to the same points on the ground. Although the images are then in *relative* registration, the *absolute* ground coordinates of the pixels may be unknown. Registration is easier to achieve with imagery that has been previously corrected for sensor and orbit distortions. Once registered, the images can be combined, or fused, in a way that improves information extraction. Examples of fusion include extraction of DEMs from stereo image pairs and compositing of images with different spatial and spectral resolutions.

8.2   What Is Registration?

Overlapping coverage from multiple images is obtained in several ways:

- the same satellite sensor on revisits, either in the same orbit (Table 1-4) or by pointing off-nadir in different orbits
- neighboring orbits of the same satellite sensor, which are typically several days apart
- different satellite sensors
- satellite and airborne sensors

These possibilities are illustrated in Fig. 8-1. It is difficult, but not impossible, to obtain coverage on the same date from multiple sensors. In all of these cases, the images are initially unregistered.

To register such images, we need to establish a coordinate transformation that relates the pixel coordinates (row and column) of one image to those of the other, or relates both to a common reference. This can be done with a few GCPs and a suitable distortion model, as described in Chapter 7. Such an approach is often sufficient for satellite imagery of areas with little to moderate topographic relief, but the manual identification and measurement of GCPs is tedious, particularly in a production environment. Moreover, topographic variations in aerial or high resolution satellite imagery are usually at spatial frequencies that are too high to be accurately modeled by global polynomials. We therefore need ways to increase the GCP density so that lower-order models, such as piecewise polynomials, may be used. Solutions to these problems are the subject of the first part of this chapter.

8.3   Automated GCP Location

Automated GCP location in two images consists of two steps. The first *extracts spatial features* from each image. Then, the features are paired by *correspondance matching*. Different types of spatial features have been used, including points, lines and regions (Table 8-1). The success of the process depends, in part, on the similarity of the features in the two images. Temporal, view angle and sensor differences all adversely affect correspondance matching of GCP features.

8.3.1 Area Correlation

Manual determination of single-pixel GCPs was described in Chapter 7. The local spatial context of the GCP is visually and mentally incorporated in that

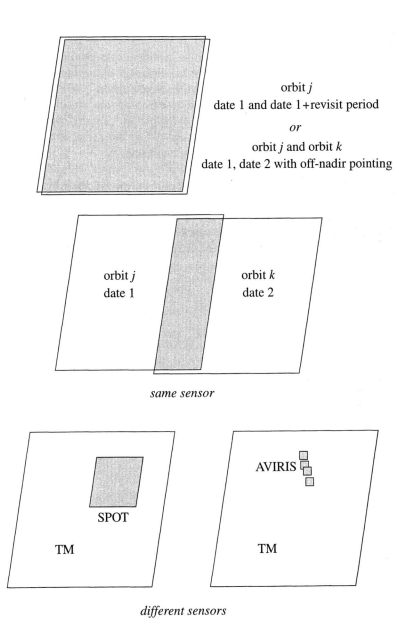

FIGURE 8-1. *Ways to obtain multitemporal data of the same area from a single satellite sensor or two different sensors. The ground coverage of the images is shown.*

process. Similarly, small areas in each image ("chips" or "patches") can serve as the spatial features for automated registration. It is not necessary to precisely specify the location of corresponding chips in both images, because spatial cross-correlation will be used to determine the required shift for registration. Thus, the most tedious part of manual GCP selection is automated. The chips should be small enough that a simple shift is sufficient for registration of each pair of chips, i.e. we do not want internal distortions within a chip.[1] The differences between the shifts determined from the different chips define any global rotation, skew or other misalignments between the full images (Chapter 7).

To calculate the cross-correlation, an *N*-by-*N* "target" chip *T* (sometimes called a "template") is selected in the reference image and an *M*-by-*M* "search" chip *S*, with *M* greater than *N*, is selected in the distorted image. The cross-correlation between the two areas (sometimes called "template matching") is calculated by sliding the target chip over the central *L*-by-*L* region of the search area (Fig. 8-2), multiplying the two arrays pixel-by-pixel, and summing the result,

$$\sum_{m=1}^{N} \sum_{n=1}^{N} T_{mn} S_{i+m, j+n} \qquad (8-1)$$

The target and search areas do not have to be square; the only requirement is that the search area be larger than the target area. Subpixel precision can be achieved by interpolating the *L*-by-*L* correlation surface to estimate the point of maximum correlation, which presumably indicates the shift needed to register the two chips.

To prevent false correlation peaks arising from changes in the image DN over the search area, the cross-correlation is usually normalized in the following way to give the *normalized cross-correlation* (Hall, 1979; Rosenfeld and Kak, 1982; Pratt, 1991; Gonzalez and Woods, 1992),

$$R_{ij} = \left[\sum_{m=1}^{N} \sum_{n=1}^{N} T_{mn} S_{i+m, j+n} \right] / K_1 K_2 , \qquad (8-2)$$

where

$$K_1 = \left[\sum_{m=1}^{N} \sum_{n=1}^{N} T_{mn}^2 \right]^{1/2} , \qquad (8-3)$$

---

[1] However, this is unavoidable in the presence of high terrain relief. Any internal distortions in one image chip relative to its mate in the other image can be thought of as adding to the dissimilarity between the two features.

TABLE 8-1. *Some examples of image registration work. Reviews of automated registration techniques are provided in (Brown, 1992) and (Fonseca and Manjunath, 1996).*

image types	features	reference
airborne MSS band/band Apollo 9 SO65 photo band/band	area correlation using the FFT	(Anuta, 1970)
simulated	area correlation	(Pratt, 1974)
MSS/airborne	edges, shapes	(Henderson *et al.*, 1985)
HCMM day/night MSS/TMS	regions	(Goshtasby *et al.*, 1986)
airborne scanner	points	(Craig and Green, 1987)
TM/MSS MSS/TMS	points	(Goshtasby, 1988; Goshtasby, 1993)
TM/TM	points	(Ton and Jain, 1989)
TM/SPOT	regions	(Ventura *et al.*, 1990)
TM/TM	area correlation	(Scambos *et al.*, 1992)
SPOT/SPOT MSS/MSS MSS/SPOT TM/SPOT	wavelet transform	(Djamdji *et al.*, 1993b)
balloon-borne photos	wavelet transform	(Zheng and Chellappa, 1993)
TM/SPOT	regions	(Flusser and Suk, 1994)
airborne/airborne TM/SPOT TM/Seasat	contours	(Li *et al.*, 1995)
airborne/airborne	points	(Liang and Heipke, 1996)
airborne/airborne	FFT phase correlation	(Reddy and Chatterji, 1996)

and

$$K_2 = \left[\sum_{m=1}^{N} \sum_{n=1}^{N} S^2_{i+m, j+n} \right]^{1/2} . \qquad (8-4)$$

Since K_1 is the same for all shifts, it can be ignored in locating the relative maximum correlation. This normalization will remove any local gain differences in the two images to improve their similarity. In the spatial calculation of

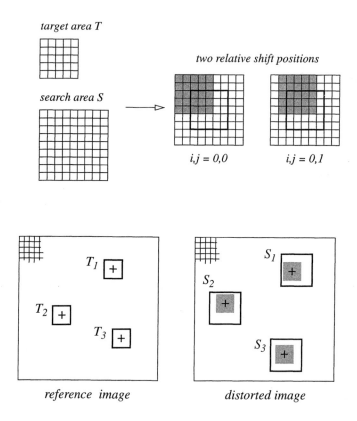

FIGURE 8-2. *Area correlation for image registration. A 5 x 5 pixel target area, T, and a 9 x 9 search area, S, are shown at the top. The DN arrays are correlated by "sliding" the target patch over the search area and calculating Eq. (8 – 2) at each possible shift position (only the first two are shown; the bold square shows the L x L possible shift locations)). In this example, N and L are each five. At the bottom, we have a distorted image which is to be registered to the reference image. Three patches are used to find the distorted image coordinates at which maximum cross-correlation is obtained, possibly at fractional row and column values. The crosses in the distorted and reference images then represent the same ground point which is, in effect, a GCP. With these three GCPs, a global affine transformation can be performed.*

Eq. (8 – 2), the coordinate system is "local," i.e. (m,n) is defined relative to the same pixel in each of the patches. One must keep track of the "global" coordinates, of course, in the original image frames.

Sometimes the mean DN within each chip is subtracted from the data before multiplication, yielding the *cross-correlation coefficient*,

$$r_{ij} = \frac{\displaystyle\sum_{m=1}^{N}\sum_{n=1}^{N}(T_{mn}-\mu_T)(S_{i+m,\,j+n}-\mu_S)}{\left[\displaystyle\sum_{m=1}^{N}\sum_{n=1}^{N}(T_{mn}-\mu_T)^2\right]^{1/2}\left[\displaystyle\sum_{m=1}^{N}\sum_{n=1}^{N}(S_{i+m,\,j+n}-\mu_S)^2\right]^{1/2}} \quad . \qquad (8-5)$$

The benefit gained is that any local additive bias differences between the two images are removed, further increasing their similarity. An example is shown in Fig. 8-3. The correlation is successful even though the two images, eight months apart, are not radiometrically matched prior to the operation.

The spatial cross-correlation requires on the order of N^2L^2 operations. The numerator in Eq. (8 – 2) can be computed via the *Fast Fourier Transform (FFT)*, which may be more efficient than direct spatial correlation (Anuta, 1970). Techniques that significantly increase the speed of spatial domain correlations have been described in (Barnea and Silverman, 1972) and applied to Landsat images in (Bernstein, 1976). These *Sequential Similarity Detection Algorithms (SSDAs)* use a small number of randomly-located pixels within the target and search areas to quickly find the approximate point of registration, followed by full calculations for shifts in the vicinity of the estimate for precise registration. The spatial similarity metric used in this case is the sum of the absolute difference, pixel-by-pixel, for different shifts,

$$D_{ij} = \sum_{m=1}^{N}\sum_{n=1}^{N}\left|T_{mn}-S_{i+m,\,j+n}\right| . \qquad (8-6)$$

If the cumulative value of D_{ij} exceeds a specified threshold for a small number of pixels, it is assumed that the current shift is not a candidate for the point of registration, and the algorithm moves to the next candidate location.

Finally, we note that the correspondance matching aspect of registration is intrinsically solved by area correlation. The specification of areas T and S implies that they contain the same ground area. With other spatial features, correspondance matching must be explicitly addressed (Ton and Jain, 1989; Ventura *et al.*, 1990; Liang and Heipke, 1996).

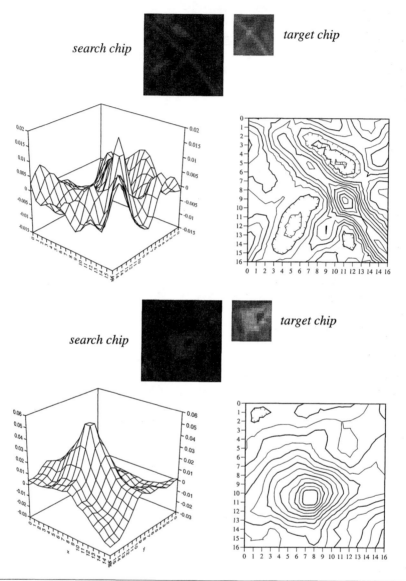

FIGURE 8-3. Example cross-correlation coefficient surfaces obtained from multitemporal TM image chips of San Jose, California, acquired on December 31, 1982 (left) and August 12, 1983 (right). Note how the correlation surface shape mimics that of the target, particularly for the highway intersection. The cross-correlation peaks can be estimated to a fractional pixel shift by contour interpolation, as shown, or by fitting a 2-D function such as a bivariate polynomial.

Relation to spatial statistics

In Chapter 4, we described several measures of spatial similarity, such as the covariance and semivariogram functions. While not defined in exactly the same way, the correlation coefficient of Eq. (8 − 5) is clearly another measure of similarity, in this case between two different but similar images. A useful way to think of area correlation is that if the two images are identical, the correlation becomes a spatial statistic on a single image and behaves similarly, with a peak for zero lag (shift) and a decrease as the lag increases. This is the limiting and optimal case for correlation between two images; in reality, they always differ to some degree. The more dissimilar the two images, the lower the correlation, and for quite different images, *there is no guarantee of a peak in the correlation at the point of registration.* Particularly difficult examples are registration of optical and radar images (Hall, 1979), or multitemporal images from different seasons. It is for these reasons that derived spatial features based on scene regions, or the boundaries between regions, are of interest for registration.

8.3.2 Other Spatial Features for Registration

Registration of two images can be viewed as a scale-space problem ranging from global registration (where we make a coarse estimate of the transformation, encompassing global rotation, scale difference and offset) to a local registration (where we refine the global model to accomodate high spatial frequency distortions, such as topographic parallax). The scale-space transform techniques discussed in Chapter 6 are therefore applicable to this problem. Zero-crossings of operators such as the Laplacian-of-Gaussian (LoG) filter have been used as spatial features in autonomous registration (Greenfield, 1991; Schenk *et al.*, 1991).

A wavelet transform technique for automatically locating numerous GCPs in two images has been developed and applied to SPOT registration with MSS and TM (Djamdji *et al.*, 1993b). The features used are local DN maxima in the thresholded high frequency components (see Chapter 6), which occur at major image features such as rivers, topographic ridges or valleys, or roads. Typically, over 100 GCPs are automatically generated by the algorithm, depending on the scene content. The correspondance matching of GCPs in the two images is then achieved (in an unspecified way in the previous reference) and a global polynomial defined using the techniques of Chapter 7. A registration accuracy of about 0.6 pixel (of the image with the larger *GSI*) is claimed.

Another scale-space approach to registration of images with similar *GSI*s operates directly in the spatial domain. A resolution pyramid is created for each image, and coarse disparity estimates are formed at the highest levels using area correlation. These estimates are propagated by scaling to the next lower level to constrain the correlation search area. The estimated disparity is refined as the analysis propagates down the pyramid. This approach is described in detail next.

8.4   Orthorectification

As discussed in Chapter 3, ground objects at different elevations exhibit displacements (disparities) in their image positions. Using the principles of photogrammetry, these disparities can be used to estimate the elevation differences between points. This information can then be used to make the image *orthographic*, with every point in its correct location relative to other points, regardless of elevation. It is as if all perspective distortion is removed, and every point on the ground is viewed from directly above.

Correction of remote-sensing imagery for terrain displacement requires a *Digital Elevation Model (DEM)*, consisting of a spatial grid of elevation values. Traditionally, DEMs have been created with analytical stereo plotters that can digitize elevation profiles extracted manually from stereo aerial photograph pairs. These DEMs tend to be at a lower resolution than the aerial photography. In the last few years, practical techniques have been developed for precise local registration of digital image stereo pairs. Such techniques can construct a high resolution DEM at approximately the same *GSI* as the imagery.

8.4.1   Low-Resolution DEM

The U. S. Geological Survey has a large database of DEMs within the United States at *GSI*s of 240m and 30m. The 30m DEMs are used to rectify aerial photographs and produce *digital orthophotoquads* corresponding to existing 7.5 minute topographic map quadrangles as part of the National Spatial Data Infrastructure program. We will follow the description of the image processing aspects of this process presented in (Hood *et al.*, 1989).

The overall process is depicted in Fig. 8-4. For aerial photography, the camera model consists of an *internal orientation model* that relates the scanned image coordinates to the camera reference frame given by the fiducial marks, and an *external orientation model* that describes the 3-D perspective geometry of a frame camera. If the camera and scanner geometry are good, the internal model can be simply an affine transformation (Chapter 7). The external model is governed by the *collinearity equations* (Wolf, 1983),

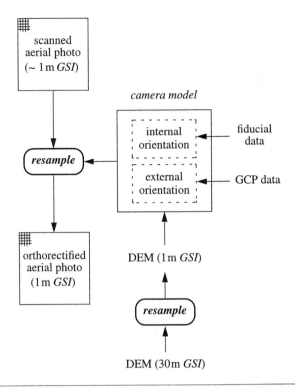

FIGURE 8-4. The process used to create a digital orthorectified photograph (commonly called an "orthophoto") from aerial photography and a pre-existing low-resolution DEM. The target GSI of the final product is 1m, so the DEM must be resampled to estimate elevations on that grid. All coordinate transformations are calculated in the camera model and only one resampling of the image is necessary.

$$x = -f \cdot \frac{m_{11}(X_p - X_0) + m_{12}(Y_p - Y_0) + m_{13}(Z_p - Z_0)}{m_{31}(X_p - X_0) + m_{32}(Y_p - Y_0) + m_{33}(Z_p - Z_0)}$$

$$y = -f \cdot \frac{m_{21}(X_p - X_0) + m_{22}(Y_p - Y_0) + m_{23}(Z_p - Z_0)}{m_{31}(X_p - X_0) + m_{32}(Y_p - Y_0) + m_{33}(Z_p - Z_0)}$$

$$(8-7)$$

where (x,y) are the photograph coordinates in the fiducial system, (X_p, Y_p, Z_p) and (X_0, Y_0, Z_0) are the coordinates of a ground point p and the camera station, respectively, m_{ij} are the elements of the transformation matrix (which contain information about the tilt of the camera relative to the terrain), and f is the camera focal length. The same equations apply to pushbroom imaging, with the x (in-track) coordinate set to zero and a time dependence introduced in the parameters on the right side of Eq. (8 – 7) (Westin, 1990).

In practice, the coefficients m_{ij} are found by a small set (at least three) of GCPs with known (X_p, Y_p, Z_p) and (x, y). The original low-resolution DEM is resampled, typically with bilinear interpolation (Hood et al., 1989), to the desired high-resolution grid, and each DEM point is transformed by the collinearity and internal orientation equations to obtain the corresponding scanned photograph coordinate. Since that coordinate is not likely to be at an existing scanned image pixel, a new pixel has to be resampled between the scanned image pixels.

To illustrate, a scanned aerial frame is shown in Fig. 8-5. The *GSI* of this image is about 1.2m. The 30m DEM, previously calculated from other aerial stereo photography, is shown in Fig. 8-6, along with a shaded relief image with approximately the same solar angles as those for Fig. 8-5. The orthorectified aerial photograph appears in Fig. 8-7. The rectification achieved can be seen by examining the two crossed roads in the upper center. In the original photograph they are distorted because of the perspective distortion caused by terrain relief; after rectification they are straight as in the map (Fig. 8-8).

8.4.2 High-Resolution DEM

If a high-resolution DEM with the desired final *GSI* is available, the previous process can be used, with the intermediate DEM resampling step eliminated. However, there are few DEMs at the spatial resolutions of aerial or high-resolution (5m or less *GSI*) satellite imagery, and for many areas of the world, no DEMs exist. Using area correlation as described at the beginning of this chapter, it is possible to calculate a DEM from a stereo image pair, and with a *GSI* that is about the same as that of the imagery. The basic equation describing the dependence of image disparity on elevation differences was derived in Chapter 3 and is repeated here,

$$\Delta Z \cong \Delta p \frac{H^2}{fB} = \Delta p \times \frac{H}{f} \times \frac{H}{B}. \qquad (8 - 8)$$

The relative elevation difference of two image points is proportional to their relative disparity, Δp. If we perform area correlations over many candidate GCPs, we can then calculate a spatially-dense DEM. To increase the density of GCPs and reduce the computation necessary at full resolution, an image pyramid is used, as described next. Stereo images can be acquired with satellite sensors in two ways: overlap in images from adjacent orbits,[2] or images taken at different

---

[2] This *sidelap* amounts to about 30 percent of the cross-track *GFOV* for TM images at 35° latitude (Ehlers and Welch, 1987).

FIGURE 8-5. *The full frame as scanned from a National Aerial Photography Program (NAPP) aerial photograph of Harrisburg, Pennsylvania, acquired on April 8, 1993. The eight marks around the border are the camera fiducial marks, exposed onto every frame and used to orient the internal camera model that relates the scan coordinates to the camera coordinates. This is necessary because the scan may not be exactly orthogonal to the camera frame. (The image and DEM data in this example are courtesy of George Lee and Brian Bennett of the U.S. Geological Survey, Western Mapping Center.)*

FIGURE 8-6. The Harrisburg NE quarter quad DEM (left) and a shaded relief representation (right). As we have seen before, the shaded relief image is similar to the remote-sensing image (compare to Fig. 8-5).

view angles by a pointable sensor such as SPOT. A brief survey of research in DEM extraction from remote-sensing imagery is given in Table 8-2.

Hierarchical warp stereo

The local neighborhood correlation of two images at one location tells us nothing about the global distortion between them. To measure that, we would have to do neighborhood correlations at several locations across the image. If the distortion is complex with abrupt changes, as might be caused by topographic relief, then the local correlation would have to be done at every pixel of the two images. Although that is possible, we like to avoid such intensive computations. Furthermore, the SNR of the correlation calculation can be improved by using larger window areas, but that would only further increase the computational burden.

We will describe the *Hierarchical Warp Stereo (HWS)* approach as a detailed example of DEM calculation from digital stereo image pairs (Quam, 1987; Filiberti *et al.*, 1994). The process is depicted in Fig. 8-9. The "warp image" is to be registered to the "reference image." A Gaussian pyramid, as described in Chapter 6, is first created for each image (*REDUCE*). The low-resolution layers

FIGURE 8-7. The digital orthophoto (left) and the corresponding portion of the quadrangle map (right). Note the precise alignment, independent of elevation. The topographic map image was scanned and resampled to a 2.4m GSI from the original 1:24,000 scale paper map (USGS, 1995).

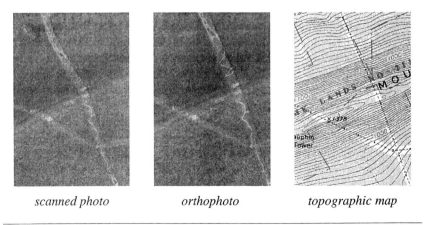

scanned photo	*orthophoto*	*topographic map*

FIGURE 8-8. Evidence of the orthographic projection achieved by the digital orthophoto process. The "roads" are access right-of-ways to power lines.

TABLE 8-2. Example approaches to determining elevation from stereo remote-sensing imagery. The elevation accuracy achieved is generally on the order of the sensor GSI.

sensor	technique	elevation accuracy (m)	reference
aerial photos	patch correlation	–	(Panton, 1978)
SIR-B	HWS on a massively parallel processor	–	(Ramapriyan *et al.*, 1986)
TM	area correlation	42 RMS	(Ehlers and Welch, 1987)
SPOT	film images on analytical stereoplotter	7.5 RMS	(Konecny *et al.*, 1987)
SPOT	film images on analytical stereoplotter	7 maximum	(Rodriguez *et al.*, 1988)
SPOT	edge matching, area correlation	12-17 RMS	(Brockelbank and Tam, 1991)
SPOT	edge point matching and correlation	26-30 RMS	(Tateishi and Akutsu, 1992)
IRS-1A LISS-II	area correlation	34-38 RMS	(Rao *et al.*, 1996)

are used to find coarse disparities which are refined at the next level (*REFINE*). Correlations are localized to the most likely region by transforming the coordinates of the warp image using the current disparity estimates (*warp*). The local area correlations are then performed for every pixel's neighborhood as described earlier (*match*), and the calculated disparities are added to the input disparities, which are propagated by scaling to the next, higher resolution layer (*EXPAND*). A scheme needs to be included to fill areas in the disparity map ("holes") where the spatial correlation is too low to be useful. That occurs in image areas that have little spatial detail or contrast. Holes can be filled by an interpolation routine that uses the nearest valid disparity values.

To demonstrate the HWS algorithm, the stereo pair of aerial photographs in Fig. 8-10 will be used. Ground control is not available for the raw disparity map extraction, and the focal length and altitude of the camera are not known. A target area of 13 x 13 pixels and a search area of 29 x 29 pixels were used, allowing for a maximum disparity of ±8 pixels (Filiberti *et al.*, 1994). The calculated disparity map appears similar to a DEM derived by interpolation of a contour map, but there is a low-to-high trend, or tilt, in the disparity map (Fig. 8-11).[3] The trend can be removed and the disparities calibrated to elevation values by fitting a

---

[3]   A camera bubble level imaged in the border of each aerial frame also indicated that there was relative airplane pitch between the two frames in the flight direction.

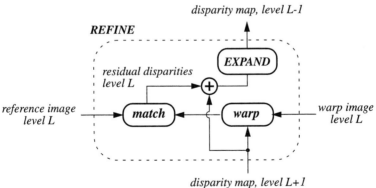

FIGURE 8-9. The HWS algorithm. At each level of the pyramid, the REFINE operation improves the resolution and accuracy of the disparity map. The match operation is local area correlation, centered on every pixel (after (Quam, 1987)).

planar function through three corners of the two maps and adjusting the disparity data to match the topographic data at those three points. Such a correction procedure is not generally recommended; the internal and external camera orientation process described for orthorectification should be followed when fiducial and GCP data are available. We have, in effect, achieved approximate orientation *after* DEM generation by referencing disparities to the existing topographic map.

Since we already have a topographic contour map in this case, why go to the effort of processing the scanned aerial photographs with the HWS algorithm? The answer is that with a correlation process, such as HWS, the DEM is derived directly from the digital imagery, with little manual intervention compared to the analytical stereo plotter approach. The derived DEM is also at a higher ground resolution, as evidenced in Fig. 8-12. Finally, generation of the dense, gridded DEM from the topographic contour map requires digitizing and interpolation between the contours, which further degrades the already low-resolution topographic data.

8.5   Multi-image Fusion

Remote-sensor systems are designed within often competing constraints, among the most important being the trade off between spatial resolution and signal-to-noise ratio (SNR). Since multispectral, and to a greater extent hyperspectral, sensors have reduced spectral bandwidths compared to panchromatic sensors, they typically have a larger IFOV in order to collect more photons and maintain image SNR. Modern sensors such as SPOT and future Landsats have a set of multispectral bands and a co-registered higher spatial resolution panchromatic band. With appropriate algorithms it is possible to combine these data and produce imagery with the best characteristics of both, namely high spatial and high spectral resolution. This process is known variously as multisensor *merging*, *fusion*, or *sharpening* (of the lower resolution image).

The low- and high-resolution images must be geometrically registered prior to fusion. The higher resolution image is used as the reference to which the lower resolution image is registered. Therefore, the lower resolution image is up-sampled to match the *GSI* of the higher resolution image. The GCP approach (Chapter 7) or automated registration approaches described in this chapter can be used for this step. The dominant relative distortion is the *GSI* difference, but shift and rotation, and possibly higher-order distortions, are also normally present.

Effective multisensor image fusion also requires radiometric correlation between the two images, i.e. they must have some reasonable degree of similarity. Artifacts in fused images arise from poor spectral correlation. As a familiar example, suppose we have a VNIR multispectral sensor (e.g. TM) and a higher reference image   HWS disparity map

warp image   interpolated contour map

FIGURE 8-10. *Stereo pair of aerial photos acquired at 12:20P.M. on November 13, 1975, over Cuprite, Nevada. The flight direction is vertical, so the primary parallax is also vertical. However, the photographs are uncontrolled with respect to ground coordinates and in fact are both tilted in the flight direction relative to the terrain horizontal plane. The "noise" around the border of the disparity map is outside the region of overlap between the two photographs. A gridded DEM obtained by interpolation of 40-foot elevation contours on a 1:24,000 topographic map is also shown. (Aerial photography courtesy of Mike Abrams, Jet Propulsion Laboratory, and photo scanning courtesy U.S. Geological Survey, Western Mapping Center; HWS results and interpolated contour map produced by Daniel Filiberti, University of Arizona.)*

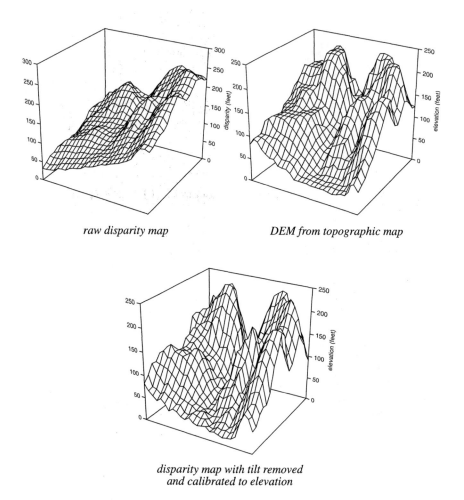

raw disparity map   *DEM from topographic map*

disparity map with tilt removed
and calibrated to elevation

FIGURE 8-11. The raw disparity map shows the effect of uncorrected camera tilt relative to the topographic map DEM. The z-coordinates at three corners of both maps were used to make a planar adjustment to the HWS-derived DEM, producing the corrected map at the bottom. The structure in the HWS-derived DEM now clearly matches that in the topographic map. The plots are for a 300 x 300 pixel area in the lower right of the full DEM (Fig. 8-10) and are subsampled by a factor of about 15 to produce uncluttered surface plots.

interpolated topographic map   *HWS-derived DEM*

FIGURE 8-12. *Subsets of the interpolated topographic map and HWS-derived DEM, corrected for tilt and z-scale as described in the text (compare to Fig. 8-10). A contouring algorithm was run on both DEMs. The greater detail in the HWS-derived DEM is obvious. That does not mean, of course, that it is necessarily more accurate; it should be evaluated against a dataset with higher resolution than the interpolated contour map to evaluate absolute accuracy.*

resolution panchromatic band (e.g. SPOT). Since the panchromatic band's sensitivity does not extend into the near IR, images with vegetation will show good correlation between the visible bands and panchromatic band, and poor correlation between the near IR band and panchromatic band. Thus, false color IR composites of fused imagery will tend to have artifacts, particularly near vegetation-soil boundaries, where the original image contrast reverses between the visible and near IR bands. Fusion techniques can also be applied to diverse sensor imagery, for example satellite multispectral imagery and aerial photography or SAR. However, the same cautions on radiometric correlation apply.

Many different sharpening algorithms have been demonstrated (Table 8-3). They generally fall into two categories, spatial domain and spectral space techniques (component replacement). We will illustrate both in the following sections.

8.5.1 Spatial Domain Fusion

The idea in this case is to somehow transfer the high frequency content of the higher resolution image to the lower resolution image. One example of this approach was presented in (Schowengerdt, 1980). Bands 1 and 4 of the Landsat MSS were degraded by spatial averaging to 240m and 400m and recombined with the original band 2 by the following pixel-by-pixel operation,

$$R_{ijk} = M_{ijk} + HPH_{ij} \qquad (8-9)$$

where R_{ijk} is the reconstructed (fused) image in band k, M_{ijk} is the lower resolution multispectral image in band k and HPH_{ij} is a high-pass version of H_{ij}, the higher-resolution image (in this case band 2). Note Eq. $(8-9)$ is like the high-boost filter (HBF) of Eq. $(6-5)$, except that the original component M_{ijk} and the high-pass component HPH_{ij} come from two different images. The success of the reconstruction in Eq. $(8-9)$ is thus expected to depend on the radiometric correlation between the high frequency components of the two images. Although the color composite of R_{ij1}, R_{ij4} and M_{ij2} was visually sharper than the composite of the degraded bands 1 and 4 with the original band 2, there were artifacts wherever band 2 was negatively correlated with these bands, in particular near vegetation/soil boundaries. A heuristic, spatially-variable, multiplicative weighting function, K_{ijk}, was included to improve the fusion in these regions,

$$R_{ijk} = M_{ijk} + K_{ijk}HPH_{ij}. \qquad (8-10)$$

TABLE 8-3. *Multisensor and multispectral image fusion experiments.*

low resolution image	high resolution image	*GSI* ratio	technique	reference
MSS	airborne SAR	–	FCC	(Daily *et al.*, 1979)
MSS	Seasat SAR	3:1	HFM	(Wong and Orth, 1980)
simulated from MSS-1 and -4	MSS-2	5:1, 3:1	HBF	(Schowengerdt, 1980)
MSS HCMM	RBV MSS	2.7:1 7.5:1	HSI	(Haydn *et al.*, 1982)
simulated SPOT-XS	simulated SPOT-P	2:1	HFM HFM, LUT	(Cliche *et al.*, 1985) (Price, 1987)
SPOT-XS	SPOT-P	2:1	HFM IHS radiometric	(Pradines, 1986) (Carper *et al.*, 1990) (Pellemans *et al.*, 1993)
TM	aerial photography	7:1	image addition	(Chavez, 1986)
TM	airborne SAR	2.5:1	HSI	(Harris *et al.*, 1990)
TM-6	TM-1 – 5, 7	4:1	HBF LUT	(Tom *et al.*, 1985) (Moran, 1990)
TM, SPOT-XS	SPOT-P	3:1, 2:1	HSI	(Ehlers, 1991)
TM	SIR-B SAR	2.3:1	HSI	(Welch and Ehlers, 1988)
TM	SPOT-P	3:1	HSI HBF, HSI, PCT HSI, PCT HFM	(Welch and Ehlers, 1987) (Chavez *et al.*, 1991) (Shettigara, 1992) (Munechika *et al.*, 1993)
AVIRIS	aerial photography	6:1	HFM	(Filiberti *et al.*, 1994)
TM	SPOT-P	3:1	wavelet transform	(Yocky, 1996)
SPOT-XS	SPOT-P	2:1	wavelet transform	(Garguet-Duport *et al.*, 1996)

High frequency modulation

In the *High Frequency Modulation (HFM)* algorithm, the high-resolution panchromatic image, H, is multiplied by each band of the multispectral, low-resolution image, M, and normalized by a low-pass filtered version of H, LPH, to estimate an enhanced multispectral image in band k,

$$R_{ijk} = M_{ijk}H_{ij}/LPH_{ij} \qquad (8-11)$$

Thus, the algorithm assumes that the enhanced (sharpened) multispectral image in band k is simply proportional to the corresponding high-resolution image at each pixel. The constant of proportionality is a *spatially-variable gain factor*, K_{ijk},

$$K_{ijk} = M_{ijk}/LPH_{ij}, \qquad (8-12)$$

making Eq. (8 – 11),

$$R_{ijk} = K_{ijk}H_{ij}. \qquad (8-13)$$

Using the two-component spatial frequency image model introduced at the beginning of Chapter 6,

$$H_{ij} = LPH_{ij} + HPH_{ij}, \qquad (8-14)$$

we can rewrite Eq. (8 – 11) as,

$$\begin{aligned} R_{ijk} &= M_{ijk}(LPH_{ij} + HPH_{ij})/LPH_{ij} \\ &= M_{ijk}(1 + HPH_{ij}/LPH_{ij}) \qquad (8-15) \\ &= M_{ijk} + K_{ijk}HPH_{ij} \end{aligned}$$

which is the same form as Eq. (8 – 10), with a specific spatial weighting function, K_{ijk}, given by Eq. (8 – 12). Thus, the HFM technique is equivalent to a high-boost filter of M, with the high frequency components coming from the higher resolution image H and weighted by K.

Filter design

Low- and high-pass filters must be applied to the high-resolution image to implement spatial domain fusion. These filters cannot be arbitrary because they define the radiometric normalization between the two images. For example, in Eq. (8 – 12), it would seem reasonable to match the LPF to the effective PSF of the multispectral image relative to that of the high-resolution panchromatic image. This intuitive relationship can be derived mathematically, but is beyond our present scope (Schowengerdt and Filiberti, 1994). Essentially, using the net relative PSF between the low- and high-resolution images maintains the local radiance from the original to the reconstructed images. The net PSF consists of three components:

- the *GIFOV* difference, e.g. 3 x 3 when merging SPOT panchromatic and TM multispectral images
- the remaining sensor PSF difference (see Chapter 3)
- the resampling function used to register the two images (see Chapter 7)

These three components are convolved with each other to obtain the net PSF, as explained in Chapter 3. Examples are presented later illustrating the improvement obtained with the net PSF compared to a box filter based only on the *GIFOV* difference.

8.5.2 Spectral Domain Fusion

As discussed in Chapter 5, we can transform a multispectral image into a new space in which one image represents the correlated component, for example, PC_1 in the principal components (*PC*) space or intensity in a perceptual color space. Spectral domain fusion techniques replace this component with the higher resolution image and transform the result back to the image space. The *PC* transform can be used for any number of bands (Fig. 8-13), but color transforms can only be applied to three band images (Fig. 8-14).

It is important in both approaches that the original and replacement components are radiometrically correlated. It often is not sufficient to linearly scale the minimum and maximum DNs if the two image histograms have different shapes; a reference stretch (Chapter 5) is more useful in that case and will partially compensate for weak correlation between the original component and the high-resolution replacement.

8.5.3 Scale-Space Fusion

Fusion of multiresolution imagery seems to be a natural application of the scale-space concepts discussed in Chapter 6. In particular, the wavelet transform has been used to fuse SPOT multispectral and panchromatic imagery (Garguet-Duport *et al.*, 1996) and Landsat TM and SPOT panchromatic imagery (Yocky, 1996). In combining the SPOT multispectral and panchromatic images, the pan image is first reference stretched three times, each time to match one of the multispectral band histograms. The first level of the wavelet transform (since the *GSI*s differ by a factor of two) is computed for each of these modified pan images. A synthetic level 1 multispectral wavelet decomposition is constructed

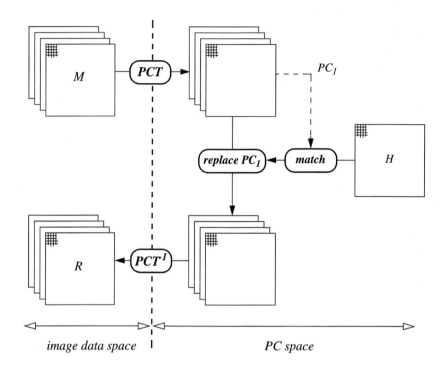

FIGURE 8-13. *Image fusion using the PCT and replacement of the first principal component. The multispectral image M has been previously registered to the high-resolution image H.*

using each original multispectral image (20m *GSI*) and the three high-frequency components (L_xH_y, H_xL_y, H_xH_y; see Chapter 6) from the corresponding modified pan image. These components are also at 20m *GSI* because of the down-sampling in the wavelet transform. The inverse wavelet transform is then calculated, resulting in three reconstructed multispectral images at 10m *GSI*. A similar approach was used to fuse SPOT panchromatic and TM multispectral images, except that a five level wavelet pyramid was constructed for both images (after they were registered and resampled to the same *GSI*) and all the SPOT data at the highest level was replaced by all the TM data at the same level (Yocky, 1996). Both of these wavelet-based techniques are similar to HFM, with the wavelet algorithm used to perform the necessary scale resampling.

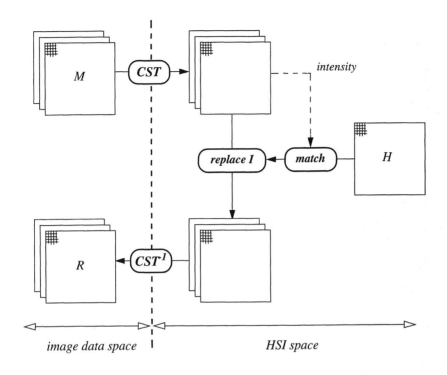

FIGURE 8-14. *Image fusion using a color space transform (CST) and replacement of*
the intensity component. The multispectral image M has been previously registered to
the high resolution image H.

8.5.4 Image Fusion Examples

A TM image of San Francisco acquired on December 31, 1982, and a SPOT
panchromatic (P) image acquired on June 11, 1993, will be used to illustrate the
fusion process. GCPs were located in the full frame and the TM image was
registered to the SPOT image (which is geocoded) using bilinear resampling. The
affine polynomial fit to the GCPs was good to about one-third TM pixel, or one
SPOT pixel.

These two images, in addition to being 11 years apart, are six months apart
seasonally. The length and direction of shadows are therefore at nearly the
extremes expected for seasonal multitemporal images, which will decrease the
radiometric correlation between them. In Plate 8-1, the results of the PCT, HSI
and HFM algorithms are shown. In the spectral domain replacement algorithms,
the SPOT-P image was reference stretched to match either the PC_1 or intensity

component, as appropriate. In the first example of downtown San Francisco, the surface materials are primarily concrete, asphalt and metal, with few vegetated areas. We would expect such materials to have relatively uniform spectral response across the bands 2, 3 and 4 used in the composite, and therefore that the SPOT-P image would be correlated to each band (Fig. 8-15). The fusion results are visually good for all the algorithms, particularly in view of the shadowing differences. For comparison, the results for the HFM algorithm with a 3 x 3 box LPF and with a full PSF applied to the SPOT image, as described earlier, are shown. The latter results in substantial improvement.

In the second example area containing Golden Gate Park, there is about fifty percent vegetation and significant shadowing from hills in the TM image (Plate 8-2). In this case, the SPOT-P image is significantly less correlated to the TM bands, and to the PC_1 and intensity components (Fig. 8-16). Even with the reference stretch preprocessing, the resulting composites are much less satisfactory than for the city area. The reconstructed color in the spectral domain algorithms is poor because of the lack of correlation between the TM and SPOT data. The HFM approach appears to be less sensitive to the lack of correlation than the color space-based algorithms; in a comparison of the related HBF technique to the PC and intensity replacement techniques, the same result was noted in (Chavez *et al.*, 1991).

In summary, the quality of fused multisensor images will be improved by attention to the following points:

- To reduce scene-related factors, the images should be as close to the same date and time as possible, and the area should have little topographic relief.

- To reduce sensor-related factors, the spectral band of the high resolution image should be as similar as possible to that of the replaced low resolution component.

- To reduce residual radiometric artifacts, the high resolution image should be globally contrast matched to the replaced component. This can be accomplished by relative radiometric calibration of the two sensors or by a reference contrast stretch.

These factors are less important when the fused images are from regions of the spectrum with different remote-sensing phenomenologies (Chapter 2). Then there is no *a priori* reason to assume radiometric correlation between the images. For example, fusion of low-resolution thermal imagery and higher resolution, multispectral visible imagery can provide useful interpretation clues (Haydn *et al.*, 1982). A CST fusion approach, with the thermal image replacing the hue component of a three-band multispectral image, is particularly effective. Another example is the use of SAR imagery as a replacement for the intensity component of 3 bands of optical imagery.

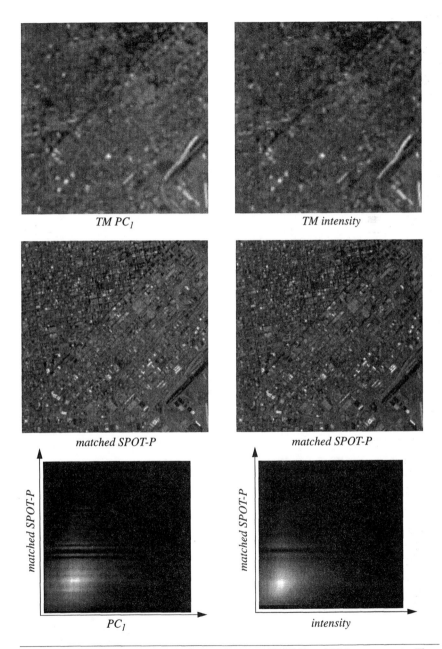

FIGURE 8-15. *City segment correlations between SPOT-P and TM components. The SPOT image is © CNES/SPOTImage, 1995 (SPOTImage, 1995).*

TM PC₁   *TM intensity*

matched SPOT-P   *matched SPOT-P*

FIGURE 8-16. *Park segment correlations between SPOT-P and TM components. The SPOT image is © CNES/SPOTImage, 1995 (SPOTImage, 1995).*

8.6   Summary

Image-to-image registration is required for many types of remote-sensing analyses. Tools for automating and improving the accuracy of the registration process beyond manually-selected GCPs were described in this chapter. Registered images can then be fused for information extraction. The major contributions of this chapter are:

- Pixel-by-pixel registration requires spatial feature extraction and matching in both images. Scale-space data structures are an efficient way to include large and small scale distortions.

- High-resolution elevation maps can be derived by pixel-by-pixel registration of stereo images.

- The quality of multiresolution image fusion depends strongly on the radiometric correlation between the fused images. Weak correlation results in poor fusion.

In the next chapter, the production of thematic maps from remote-sensing images is discussed. The mathematical and algorithmic tools needed for this process are altogether different than those described so far, but the quality of the resulting maps depends on many of the factors described in preceding chapters.

8.7   Exercises

8–1. Why is it impossible to obtain stereo coverage from the bottom portion of one Landsat TM scene and the upper portion of the next scene, even though they overlap?

8–2. The cross-correlation coefficient of Eq. $(8 - 5)$ is commonly used to register image patches. Which environmental and calibration factors in remote-sensing imagery are removed by this normalization?

8–3. Derive a mathematical expression in image space for PCT fusion of a low-resolution multispectral image and a high-resolution panchromatic image. For simplicity, assume they are registered and have the same *GSI*, even though the *GIFOV*s are different.

8–4. Generate data flow diagrams, similar to Fig. 8-13 and Fig. 8-14, for the two wavelet-based fusion techniques described in Sect. 8.5.3.

CHAPTER 9

Thematic Classification

9.1 Introduction

A *thematic map* shows the spatial distribution of identifiable earth surface features; it provides an *informational description* over a given area, rather than a data description. Image classification is the process used to produce thematic maps from imagery. The themes can range, for example, from categories such as soil, vegetation and surface water in a general description of a rural area, to different types of soil, vegetation and water depth or clarity for a more detailed description. It is implied in the construction of a thematic map from remote-sensing imagery that the categories selected for the map are distinguishable in the image data. As described in the previous chapters, a number of factors can cause confusion among spectral signatures, including topography, shadowing, atmospheric variability, sensor calibration changes, and class mixing within the *GIFOV*. Although some of these effects can be modeled, some cannot (with any reasonable amount of effort), and so they must be treated simply as statistical variability. In this chapter, we will look at classification algorithms, and in particular, how their performance depends on the physical and data models discussed in Chapters 2, 3 and 4.

9.2   The Classification Process

Traditionally, thematic classification of an image involves several steps, as shown in Fig. 9-1:

- *feature extraction*–Transform the multispectral image by a spatial or spectral transform to a feature image. Examples are selection of a subset of bands, a PCT to reduce the data dimensionality, or a spatial smoothing filter. This step is optional, i.e. the multispectral image can be used directly, if desired.

- *training*–Extract the pixels to be used for training the classifier to recognize certain categories, or *classes*. Determine the discriminant functions in the feature space which allow assignment of a class label to each pixel. This step is either supervised by the analyst or unsupervised.

- *labeling*–Apply the discriminant functions to the entire feature image and label all pixels. If the training was supervised, the labels are associated with the discriminant functions; if it was unsupervised, the analyst must now supervise the labeling. The output map consists of one label for each pixel.

The end result is to convert the numerical image data into descriptive labels that categorize different surface materials or conditions. A large reduction in data quantity takes place; the multispectral image, consisting of several to hundreds of bands with at least 8 bits/pixel/band, has been reduced to a map consisting of perhaps as few as a dozen or so category labels.[1] The map can therefore be stored with binary encoding as a single band file with less than 8 bits/pixel. By virtue of the labeling process, however, we have presumably converted the data into something that has informational value.

Classification is sometimes used as a compression tool for efficient data transmission (Fig. 9-2). The look-up table from pixel vectors to labels is called a *codebook*, which is used to encode the data at the transmitter and decode the data at the receiver. If the codebook consists of a table linking each unique pixel feature vector to a unique label, then the process is *lossless*; the labels received at the receiver can be decoded exactly into the corresponding pixel vectors. To the extent that different pixel vectors are mapped to the same label, the process is *lossy*. The performance of the system is then judged by how well it preserves the original numerical content of the image. Thus, from this perspective, the classifier produces an *approximation* to the original image, which naturally leads to

---

[1] Some complex and detailed mapping projects may approach as many as 30 or 40 categories for a single dataset, but this is relatively uncommon.

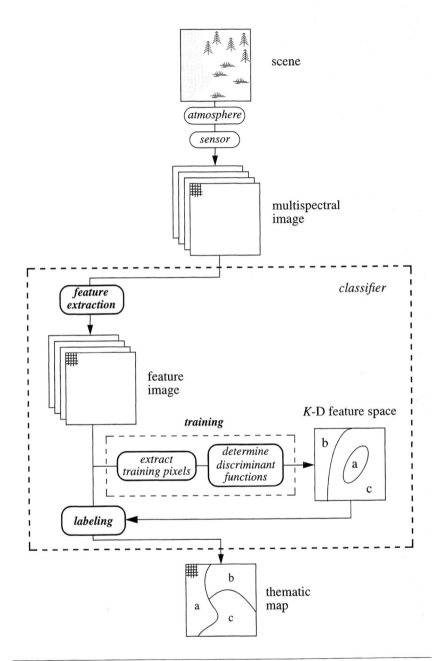

FIGURE 9-1. The data flow in a classification process.

evaluating the classifier on the basis of the accuracy of that approximation. We will use this approach to compare classification results in many of the examples in this chapter. It provides an objective, numerical evaluation criterion which avoids difficulties with specification of appropriate test sites for estimating the *label accuracy* of the thematic map. The latter approach is widely used in classification applications, however, because analysts are interested in the accuracy of the map *per se*.

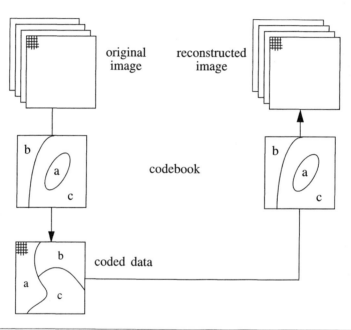

FIGURE 9-2. *Classification as a data compression technique. The training step of Fig. 9-1 has been previously performed to create the codebook. The left column is the encoding stage that takes place at the source transmitter, and the right column is the decoding stage that takes place at the receiver. The decoded image may or may not be a perfect reconstruction, as discussed in the text.*

9.2.1 The Importance of Image Scale and Resolution

In traditional photointerpretive mapping from aerial photographs, the *geometric scale* of the photographs, defined as the ratio of flight altitude to camera focal length, is of primary importance. The spatial resolution in the focal plane of a film camera is determined by the lens and film optical spread functions and the noise characteristics of the film. These factors result in an *effective* IFOV for the system that is independent of altitude. Therefore, the level of discernable ground detail in the photography is inversely related to the image scale. A 1:24,000 scale aerial photograph reveals more surface detail than a 1:48,000 scale photograph taken with the same camera system at twice the altitude. Mapmakers have come to associate the type of information that can be reliably extracted from aerial photographs with the photographic scale (Avery and Berlin, 1992).

With digital satellite imagery, the situation is essentially the same, except that scale is not an explicit characteristic of the data. The relevant measures in this case are the *GSI* and *GIFOV*, for example, 30 meters per pixel for TM imagery or 10 meters per pixel for SPOT panchromatic imagery. Clearly, the degree and type of information that can be extracted from digital imagery depends on the *GSI* and *GIFOV* (see Chapter 4). For example, a *GSI* and *GIFOV* of one meter per pixel expected from the new generation of commercial satellite systems may allow identification of larger road vehicles *by type*, a task which is completely out of the question in TM imagery.[2] Similarly, it is possible to map some features of urban areas with TM data, but not with AVHRR data.

A widely-used, hierarchical land cover/land use classification system, reproduced in (Table 9-1), was published in (Anderson *et al.*, 1976). It consists of two general levels, which are easily expanded to higher levels, as shown for the residential class in Table 9-2. The higher levels III (or even IV, if sufficient information is available) can be adapted to particular applications or regions, within the framework of the fixed-category structure in level I and II, as defined by the U. S. Geological Survey (Avery and Berlin, 1992).

The *GIFOV* required to map particular types of objects, for example, roads (Benjamin and Gaydos, 1990) and general temperate land cover (Markham and Townshend, 1981) has been evaluated. It is generally accepted that the Anderson Level I categories can be reliably mapped using Landsat MSS imagery, and Level II categories with TM and SPOT multispectral imagery (Lillesand and Kiefer, 1987). Level III requires SPOT panchromatic or higher resolution imagery.

9.2.2 The Notion of Similarity

Similarity between pixels, or groups of pixels, is a fundamental concept behind many image-processing algorithms. In classification, for example, we want to label areas on the ground that have similar physical characteristics. We do this by grouping data with similar characteristics, namely *class signatures*. A major question is, therefore: how well do the class data signatures in the image correspond to the class physical characteristics that actually distinguish one cate-

---

[2] The military community has long used a guideline of three to five "resolution elements" across the linear dimension of an object as a threshold for visual target identification; for example, distinguishing a tank from a truck.

TABLE 9-1. A land-use and land-cover classification hierarchy (Anderson et al., 1976).

Level I	Level II
1 urban/built-up land	11 residential 12 commercial and services 13 industrial 14 transportation, communication, and utilities 15 industrial and commercial complexes 16 mixed urban/built-up land 17 other urban/built-up land
2 agricultural land	21 cropland and pasture 22 orchards, groves, vineyards, nurseries and ornamental horticultural areas 23 confined feeding operations 24 other agricultural land
3 rangeland	31 herbaceous rangeland 32 shrub and brush rangeland 23 mixed rangeland
4 forest land	41 deciduous forest land 42 evergreen forest land 43 mixed forest land
5 water	51 streams and canals 52 lakes 53 reservoirs 54 bays and estuaries
6 wetland	61 forested wetland 62 non-forested wetland
7 barren land	71 dry salt flats 72 beaches 73 sandy areas other than beaches 74 bare exposed rock 75 strip mines, quarries, and gravel pits 76 transitional areas 77 mixed barren land
8 tundra	81 shrub and brush tundra 82 herbaceous tundra 83 bare ground tundra 84 wet tundra 85 mixed tundra
9 perennial snow or ice	91 perennial snowfields 92 glaciers

TABLE 9-2. *An example expansion of the Anderson classification scheme (Anderson et al., 1976).*

Level II	Level III
11 residential	111 single family units 112 multifamily units 113 group quarters 114 residential hotels 115 mobile home parks 116 transient lodgings 117 other

gory from another? An example of the type of semantic problem that can arise is in the mapping of urban land-*use* classes, such as "urban residential" or "light industrial." They typically are composed of several land-*cover* types (such as vegetation, pavement, and different types of roof coverings), each with a different spectral signature. Within an area of particular land-use, therefore, several spectral classes occur, resulting in a heterogeneous spectral signature whose characteristics depend on the proportions of each of the component land-cover types, which in turn changes from pixel-to-pixel. Thus, for land-use mapping, we must look for more complex relationships between the physical measurements, the pixel-by-pixel multispectral image, and the map classes of interest. Sometimes, remote-sensing data is insufficient for the task, and additional, so-called *ancillary* information is required.

From the discussions and examples of previous chapters, it is clear that the spectral signature of a given surface material is not characterized by a single, deterministic spectral vector, but rather a *distribution* of vectors. In Fig. 9-3, we show two possible cases for the relative distributions of training data from three classes. *To a large extent, our ability to perform an accurate classification of a given multispectral image is determined by the extent of overlap between class signatures.* An optimum compromise can be achieved with the maximum-likelihood, or Bayes, classifier which minimizes the total error in the classification if our estimate of the underlying probability distributions is correct. Another approach accepts the fact that class signatures overlap and expresses that as likelihoods of membership in each class, i.e. a *fuzzy* classification, as explained next.

9.2.3 Hard Versus Soft Classification

The notion of a thematic map, as traditionally used in geography, geology and other earth science disciplines, presumes that every spot on the ground can be

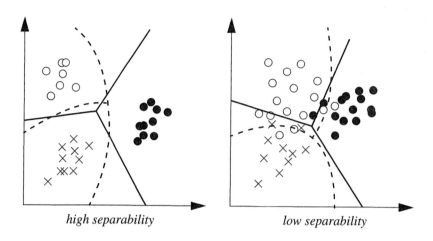

high separability   *low separability*

FIGURE 9-3. *Two possible situations for training data in feature space and candidate decision boundaries. If the training classes are highly separable, there are many potential decision region partitions that can separate the classes without error; either the solid or dashed lines would do equally well. If the training data from different classes overlap, then a smooth decision boundary is impossible without misclassifications. The exact form of the decision boundary is then critical to the classification error.*

---

labeled as belonging to one, and only one, category. Such discrete categorization is convenient and appealing in its simplicity, but is not a particularly accurate protrayal of real landscapes, and, in fact, is inconsistent with the high resolution, numerical nature of remote-sensing data.

Remote sensing provides, in effect, a continuous labeling fuction with as many as 2^{QK} possible categories for data with Q bits per pixel per band and K bands. When this nearly continuous measurement space is compressed by classification into relatively few, discrete labels, we ignore a large amount of the information contained in the data in order to obtain the relative simplification of the thematic map.[3]

Most classification algorithms produce a "likelihood" function for the assignment of a class label to each pixel. A hard classification is produced by selecting that class label with the greatest likelihood of being correct (the *Winner-Take-All (WTA)* algorithm, as it is termed in the artificial neural network literature). The feature space decision boundaries for a *hard* classification are well-defined (Fig. 9-4). If the likelihood values are retained, however, allowing for multiple

---

[3] The actual number of *distinguishable* data categories is somewhat reduced by sensor noise.

labels at each pixel, a *soft* classification is obtained. The feature space decision boundaries for a soft classification may be thought of as ill-defined, or *fuzzy*. What do the likelihood values represent? The most obvious interpretation is that they represent the relative proportions of each category within the spatially- and spectrally-integrated multispectral vector of the pixel. We will introduce classification as a "hard" algorithm initially, because that is the traditional view. Later in this chapter, "soft" decision algorithms are addressed.

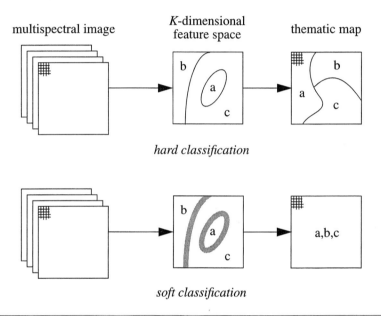

FIGURE 9-4. *One way to view the difference between hard and soft classification. In both cases pixel spectral vectors from the multispectral image are passed through the decision space for labeling. In a hard classification the decision is "winner-take-all," with only one label being permitted at each pixel. In a soft classification, the decision is multivalued, with the possibility of more than one label per pixel. Each label has an associated likelihood of being correct. These likelihoods can be interpreted in a number of ways, one of which is that they indicate the proportion of each category within the pixel, à la the mixing model discussed in Sect. 9.8.1.*

9.3 Feature Extraction

The multispectral image data can be used directly in a classification, but it contains all the various external influences, such as atmospheric scattering and topographic relief, described in earlier chapters. Also, the data are often highly correlated between spectral bands, resulting in inefficient analysis. Furthermore, image-derived features, such as measures of spatial structure, may provide more useful information for classification. Thus, it is prudent to consider various pre-classification transformations to extract the greatest amount of information from the original image. Most of the techniques described in Chapters 5, 6 and 7 are useful in that regard. As pointed out in a number of cases, some features tend to suppress undesired variability in remote-sensing signatures, e.g. spectral ratios can suppress topographic variation. In general, it is wise to use such features because they allow the classifier to better distinguish spectral classes. Separability analysis, discussed later, provides a way to determine the most effective subset of a given set of features.

9.4   *Training the Classifier*

In order to classify an image into categories of interest, the classification algorithm needs to be trained to distinguish those categories. Representative category samples, known as *prototypes*, *exemplars*, or simply *training samples*, are used for this purpose. After the classifier is trained to "recognize" the different categories represented by the training samples, the "rules" that were developed during training are used to label all pixels in the image as one (hard classification) or more (soft classification) of the training categories.

If we are only interested in classifying a single scene relative to training signatures derived from pixels within that scene, atmospheric factors are not particularly critical, as long as they are consistent over the whole scene. They then influence the training and unknown pixel spectral vectors equally and do not affect their relative positions in spectral space. If the atmosphere varies significantly across the area of interest (for example, due to haze or smoke), then spatially-dependent correction is needed, as demonstrated in (Richter, 1996a). If training data from one image are to be used to classify a second image, then either both images should be corrected for atmospheric effects (Chapter 7), or the second should be normalized to the first (Chapter 5).

The training of a classification algorithm can be either *supervised*, in which case the prototype pixel samples are already labeled by virtue of ground truth, existing maps, or photointerpretation, or *unsupervised*, in which the prototype pixels are not labeled, but have been determined to have distinguishing intrinsic data characteristics.[4]

9.4.1 Supervised Training

The themes of interest are determined by the application. In geology, one may wish to map different mineral types. In forestry, the themes could be different tree species, or perhaps healthy and diseased trees. Agricultural classes would normally include different crop types, fallow fields and degrees of soil moisture. For supervised training, the analyst must select representative pixels for each of the categories. If the multispectral image contains sufficiently distinct visual cues, it may be possible to find suitable training areas by visual examination. Frequently, however, one must resort to additional sources of information, such as field data or existing maps, to find representative areas for each class. The process of finding and verifying training areas can therefore be rather labor intensive.

With supervised training, it is important that the training area be a homogeneous sample of the respective class, but at the same time include the range of variability for the class. Thus, more than one training area per class is often used. If there is considerable within-class variability, the selection of training sites can be laborious, and it is impossible to be entirely certain that a comprehensive set of training samples for each class has been specified. Example training sites for three easily distinguished classes are shown in Plate 9-1.

In many cases it is impossible to obtain homogeneous sites. A common problem is sparse vegetation, which complicates attempts to map both vegetation and soils. One technique for improving training data under these conditions is to "clean" the sites of outlying pixels (in feature space) before developing the final class signatures (Maxwell, 1976). This can be done by classifying the training pixels according to their given signatures. Some training pixels will likely be misclassified, or at least have a low likelihood of being in the correct class. Those pixels are then excluded from the training set and the class signatures are recalculated from the remaining pixels. Another approach excludes those pixels within the training sites that do not meet certain spatial and spectral homogeneity criteria (Arai, 1992).

In certain applications, the nature of the classes makes manual training site delineation difficult, for example, an "asphalt" class consisting of narrow roads, or a "water" class for a river. In these cases, a semi-automated delineation is useful. Efficient tools have been developed that, starting from a seed pixel specified by the analyst, use *region-growing* algorithms to define an irregular site

---

[4]   Ground "truth" refers to any knowledge of an area which is, for all practical purposes, certain. It implies that field visits have been made to specific sites and that the surface materials at those sites have been accurately mapped and measured. It is sometimes used, less appropriately, to describe knowledge from other sources, such as aerial photographs.

(Buchheim and Lillesand, 1989). The concepts behind region-growing algorithms are discussed in Sect. 9.7.

Separability analysis

Once the training pixels are selected, the *features* to be used by the classifier should be specified. These may be all or a subset of the original multispectral bands, or derived features, such as the principal components (Chapter 5). Since supervised training is driven by the desired map themes, and not by characteristics of the data itself, there is no assurance that the classes will actually be distinguishable from one another. A *separability* analysis can be performed on the training data to estimate the expected error in the classification for various feature combinations (Swain and Davis, 1978). The results may suggest that some of the initial features be dropped before classification of the full image.

Finding an appropriate definition for interclass separability is not trivial. Simple measures of the separation of the means, such as the *city block*, *Euclidean* and *angular* distances would at first seem to be good candidates (Table 9-3). These simple distance measures are shown in Fig. 9-5 for the distance between two vectors in 3-D. They do not account, however, for overlap of class distributions due to their variances and are therefore not particularly good as measures of separability. The *normalized city block* measure is better, in that it is proportional to the separation of the class means and inversely proportional to their standard deviations. If the means are equal, however, it will be zero regardless of the class variances, which does not make sense for a statistical classifier based on probabilities. For this reason, probability-based measures have also been defined.

The *Mahalanobis* separability measure is a multivariate generalization of the Euclidean measure for normal distributions. It is always zero if the class means are equal. The *divergence* and *Bhattacharyya distance* measures avoid this problem. Divergence equals zero only if the class means *and* covariance matrices are equal. A problem remains with both, however, in that they increase without bound for large class separations, and do not asymptotically converge to one, as does the probability of correct classification. A *transformed divergence*, based on the ratio of probabilities for classes *a* and *b*, does exhibit this behavior. The *Jeffries-Matusita distance* depends on the difference between the probability functions for *a* and *b* and is similar to the probability of correct classification for large class separations, but requires more computation than does transformed divergence (Swain and Davis, 1978).

Separability can also be used to determine the combination of features that is best, on average, at distinguishing among the given classes. As a tool for feature selection, a measure of separability is typically computed for all possible pairs of classes[5] and for all combinations of q features, out of K total features (Jensen, 1996). The average separability over all class pairs is computed, and the subset of

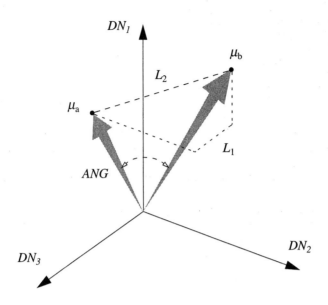

FIGURE 9-5. *The L_1, L_2, and ANG distance measures depicted for two vectors in 3-D.*
Note that ANG, the arc-cosine of the normalized inner (dot) product of the two vectors,
is independent of the length of either vector. This property makes it useful for
classification of data that is not corrected for topographic shading (refer to Fig. 4-38).

features that produces the highest average separability is found. One may then
use that subset for classification and save computation time in the classification
stage. It is important to match the separability measure to the classifier for the
feature subset analysis. For example, one should use Euclidean distance if the
nearest-mean algorithm will be used for classification, or transformed
divergence, if the maximum-likelihood algorithm will be used.

9.4.2 Unsupervised Training

For unsupervised training, the analyst employs a computer algorithm that locates
concentrations of feature vectors within a heterogeneous sample of pixels. These
so-called *clusters* are then assumed to represent classes in the image and are used
to calculate class signatures. They remain to be identified (labeled), however, and

---

[5] There are $M(M-1)/2$ possible pairs out of M classes. In general, for R classes out
of a total of M classes, there are (ignoring order) $M!/[R!(M-R)!]$ possible
combinations.

TABLE 9-3. Distance measures between two distributions in feature space. The city block, Euclidean, and angular measures ignore the covariances of the distributions. The normalized city block and MH measures are extensions that include covariance information for each class. The last five measures assume normal class distributions for a and b. All of these distance measures are scalars. Derivations of the normal distribution-based distance measures can be found in many books on statistical pattern recognition, including (Duda and Hart, 1973; Swain and Davis, 1978; Richards, 1993)

name	formula
city block	$L_1 = \|\mu_a - \mu_b\| = \sum\limits_{k=1}^{K} \|m_{ak} - m_{bk}\|$
Euclidean	$L_2 = \|\mu_a - \mu_b\| = [(\mu_a - \mu_b)^T (\mu_a - \mu_b)]^{1/2}$ $= \left[\sum\limits_{k=1}^{K} (m_{ak} - m_{bk})^2 \right]^{1/2}$
angular	$ANG = acos\left(\dfrac{\mu_a^T \mu_b}{\|\mu_a\| \|\mu_b\|} \right)$
normalized city block	$NL_1 = \sum\limits_{k=1}^{K} \dfrac{\|m_{ak} - m_{bk}\|}{(\sqrt{c_{ak}} + \sqrt{c_{bk}})/2}$
Mahalanobis	$MH = \left[(\mu_a - \mu_b)^T \left(\dfrac{C_a + C_b}{2} \right)^{-1} (\mu_a - \mu_b) \right]^{1/2}$
divergence	$D = \dfrac{1}{2} tr\left[(C_a - C_b)(C_b^{-1} - C_a^{-1}) \right]$ $+ \dfrac{1}{2} tr\left[(C_a^{-1} + C_b^{-1})(\mu_a - \mu_b)(\mu_a - \mu_b)^T \right]$
transformed divergence	$D^t = 2[1 - e^{-D/8}]$
Bhattacharyya	$B = \dfrac{1}{8} MH + \dfrac{1}{2} \ln\left[\dfrac{\|(C_a + C_b)/2\|}{(\|C_a\|\|C_b\|)^{1/2}} \right]$
Jeffries-Matusita	$JM = \left[2(1 - e^{-B}) \right]^{1/2}$

may or may not correspond to classes of interest to the analyst. Supervised and unsupervised training thus complement each other; in the former, the analyst imposes knowledge on the analysis to constrain the classes and their characteristics, and in the latter, an algorithm determines the inherent structure of the data, unconstrained by external knowledge.

The term "cluster" is misleading in most cases. It implies that there is a distinct grouping of pixel vectors in a localized region of the multidimensional data space. This is seldom the case. In order to create a comprehensive set of classes that span the full data space, a large training sample must be used. Due to subpixel class mixing (Sect. 9.8), sensor noise, topographic shading and other factors, the data distribution is typically diffuse (see simulated scattergrams in Chapter 4 and real scattergrams in Chapter 5). Competitive algorithms, such as *K*-means described next, find an optimal partitioning of the data distribution into the requested number of subdivisions. The final mean vectors resulting from the clustering will be at the centroids of each subdivision.

In defining image areas for unsupervised training, the analyst does not need to be concerned with the homogeneity of the sites. Heterogeneous sites can be purposely chosen to insure that all classes of interest and their respective within-class variabilities are included. Even the full image (perhaps subsampled to reduce computation time) may be used in the clustering algorithm for a "wall-to-wall" description. The assignment of identifying labels to each cluster may be done by the analyst after training or after classification of the full image. Because unsupervised training does not require any information about the area, beyond that in the image itself, it can be useful to delineate relatively homogeneous areas as potential supervised training sites.

K-means clustering algorithm

One of the more common clustering methods is the *K*-means algorithm (Duda and Hart, 1973). In Fig. 9-6 we show the application of the *K*-means algorithm to a two-dimensional set of idealized data. In the first step of the algorithm, an initial mean vector ("seed" or "attractor") is arbitrarily specified for each of *K* clusters.[6] Each pixel of the training set is then assigned to the class whose mean vector is closest to the pixel vector, thus forming the first set of decision boundaries.[7] A new set of cluster mean vectors is then calculated from this classification, and the pixels are reassigned accordingly. In each iteration, the *K* means will tend to gravitate towards concentrations of data within their currently-assigned region of feature space. The iterations are continued until there is no significant

---

[6]   Not to be confused with *K* data dimensions. The use of *K* for the number of clusters is common convention.

[7]   This is the *nearest-mean algorithm* discussed in Sect. 9.6.4.

change in pixel assignments from one iteration to the next. The criterion for ending the iterative process can be defined in terms of the *net mean migration* from one iteration to the next. Specifically, this is the magnitude change of the mean vectors from iteration $i - 1$ to iteration i, summed over all K clusters,

$$\Delta\mu(i) = \sum_{k=1}^{K} |\mu_k(i) - \mu_{k-1}(i-1)| \quad . \tag{9-1}$$

The final, stable result is not sensitive to the initial specification of seed vectors, but more iterations may be required for convergence if the final vectors are not close to the seed vectors. A typical iteration sequence for the net migration of the cluster means is shown in Fig. 9-7. The final cluster mean vectors may be used to classify the entire image with a minimum-distance classifier in one additional pass, or the covariance matrices of the clusters may be calculated and used with the mean vectors in a maximum-likelihood classification.

The number of ways to determine clusters in data has been limited only by the ingenuity of researchers in defining clustering criteria. Clustering of image data is discussed in a number of books (Duda and Hart, 1973; Jain and Dubes, 1988; Fukunaga, 1990), and Fortran computer programs can be found in (Anderberg, 1973; Hartigan, 1975). All commonly-used algorithms employ iterative calculations to find decision boundaries for the given data and optimization criteria. The *ISODATA* algorithm (Ball and Hall, 1967) is a common modification of the *K*-means algorithm and includes merging of clusters if their separation is below a threshold, and splitting of a single cluster into two clusters, if it becomes too large. Unfortunately, as the number of capabilities grows, so does the list of parameters that must be set by the user!

Clustering examples

Unfortunately, the idealized data distribution of Fig. 9-6 is not particularly realistic. Most of the time, image data does not exist in such distinct clusters, but rather in a large "blob," which may contain only hints of internal concentrations. We will use bands 3 and 4 of a Landsat TM image to illustrate the process of clustering on actual image data. The area is near Marana, Arizona, and contains irrigated agricultural fields, most of which are fallow. Surrounding the agricultural area is desert, with a mixture of soils and thin vegetation cover (Plate 9-1). The image scattergram reveals that the two bands are strongly correlated because of the large areas of bare soil, ranging from dark to light. The few fields that have a growing crop appear in the upper left of the band 4 versus band 3 scattergram (see Chapter 4 for a discussion). Because the area of growing crop is relatively small, a clustering algorithm will be dominated by the soil pixels.

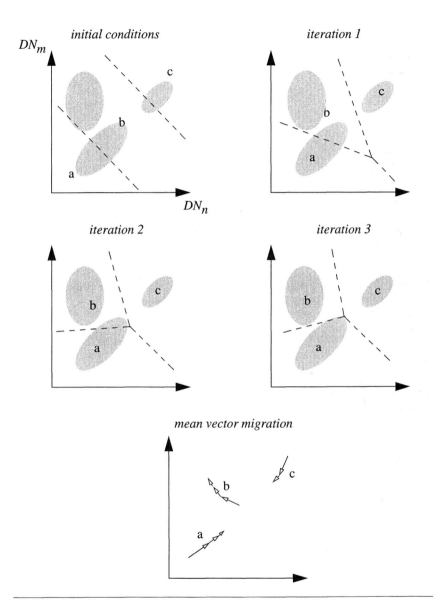

FIGURE 9-6. *An idealized data distribution during three iterations of the K-means clustering algorithm with the nearest-mean decision criterion. The data is depicted as three distinct clusters, with the current, estimated mean vector for each class located at a, b and c. The initial "seed" locations are equidistant along the feature space diagonal. In the bottom figure, the movement of the estimated cluster means is shown.*

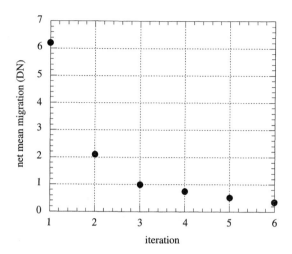

FIGURE 9-7. Typical behavior of the net mean migration from one iteration to the next in the K-means algorithm. The plot is for real image data.

In Fig. 9-8, we see the clustering results of a K-means algorithm (using the Euclidean distance measure) for different values of K. Note that the crop fields do not appear in a unique cluster until six clusters are used. This is also seen in Fig. 9-9, where the final cluster means are overlayed on the scattergram. The initial cluster seeds were distributed along the first principal component axis of the data. As predicted, the cluster mean locations are dominated by the large population of soil pixels, and not until six clusters have been requested, do we see the appearance of a unique cluster for the crop pixels. This cluster remains and shows some movement towards the vegetation part of the spectral space when seven clusters are used.

One may view cluster analysis as an attempt to "fit" the data with relatively few, distinct signatures (the cluster means).[8] With only two clusters, the fit is quite poor, but improves with each additional cluster. In Fig. 9-10, the residual magnitude error at every pixel is shown for each case in Fig. 9-8. We see that the overall fit steadily improves as the number of clusters increases, but that the largest error (in the active crop fields) remains high until six clusters are used, when the error drops dramatically. This dependence is clearly apparent if the residual error is plotted against the number of clusters (Fig. 9-11).

---

[8] This is the basis of a data compression technique, *vector quantization* (Pratt, 1991).

FIGURE 9-8. Final cluster maps for different numbers of clusters.

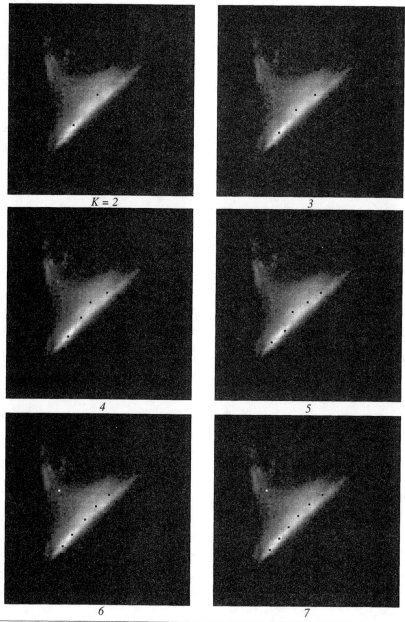

FIGURE 9-9. Band 4 versus band 3 scattergrams for Fig. 9-8 with the final cluster means. The new cluster center that first appears for K equal to 6 represents vegetation.

FIGURE 9-10. *Residual magnitude error maps between the original image and the approximation given by the cluster mean DNs. Note the greatly reduced error in the crop fields when six or more clusters are used.*

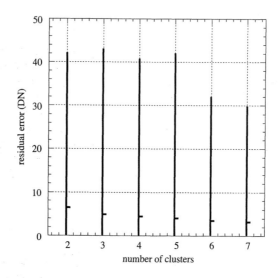

FIGURE 9-11. The residual error in approximating the two-band image with cluster mean DNs, versus the number of clusters used. The range from minimum to maximum error is shown, with the crossbar indicating the average error. Note the average error steadily decreases as the number of clusters increases, but the maximum error does not until six clusters are used, where it drops 25%.

9.4.3 Hybrid Supervised/Unsupervised Training

Because supervised training does not necessarily result in class signatures that are numerically separable in feature space, and unsupervised training does not necessarily result in classes that are meaningful to the analyst, a combined approach has the potential to satisfy both requirements. Unsupervised training is first performed on the data and an unlabeled cluster map of the training area is produced using the obtained clusters. Typically, a large number of clusters, say 50 or more, is used to insure adequate data representation. The analyst then evaluates the map with field survey data, aerial photographs and other reference data and assigns class labels to the clusters. Normally, some clusters must be subdivided or combined to make this correspondence; the analyst might also choose to modify the class labels based on the clustering results. The labeled cluster data can then be used in a final supervised classification, or the labeled cluster map can be simply accepted as the final map.

9.5   Nonparametric Classification

Classification algorithms may be grouped into one of two types: *parametric* or *nonparametric*. Parametric algorithms assume a particular class statistical distribution, commonly the normal distribution, and require estimates of the distribution parameters, such as the mean vector and covariance matrix, for classification. Nonparametric algorithms make no assumptions about the probability distribution and are often considered *robust* because they may work well for a wide variety of class distributions, *as long as the class signatures are reasonably distinct*. Of course, parametric algorithms should yield good results under the same conditions, even if the assumed class distribution is invalid. See the earlier discussion about Fig. 9-3.

9.5.1 Level-Slice Classifier

This classifier, also known as a *box classifier*, is perhaps the simplest of all classification methods. A set of K-dimensional boxes, centered at the estimated class mean vectors, are placed in K-dimensional feature space. If an unlabeled pixel vector lies within one of the boxes, it is assigned that class label (Fig. 9-12). By its nature, the level-slice algorithm also creates an "unlabeled" class, consisting of all pixel vectors that do not fall within any of the designated boxes. This characteristic can be used to prevent overestimation of class populations caused by outliers.

Specification of the box limits is typically in terms of the data extent in each dimension, for example ± one standard deviation about the mean. Since the boxes are aligned with the axes of the data, the actual labeling can be achieved with appropriately specified hardware LUTs in an interactive manner (Chapter 1). A complication occurs if a pixel vector falls within two or more boxes (the boxes can overlap unless that is explicitly prevented). A decision on a pixel's label must then be made with another algorithm, such as the nearest-mean.[9]

---

[9]   We have a bit of semantic difficulty here. If the location of the boxes of the level-slice classifier are dependent on the training-class mean vectors, and the box sizes are dependent on the class standard deviations, then we in fact have a *parametric* classifier! If the boxes are not so constrained, but are designed interactively by examination of spectral space scattergrams, as implemented in some software systems, then we have a *nonparametric* classifier. In either case, the addition of the nearest-mean classifier to resolve conflicts is a parametric modification.

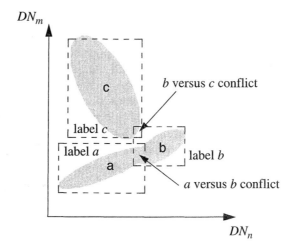

FIGURE 9-12. Level-slice decision boundaries for three classes in two dimensions. Pixel labeling within the conflict regions must be handled with an alternate decision, for example by the nearest-mean classifier.

Given that class DN distributions are not typically aligned with the *K*-D data axes because of correlation between spectral bands, the level-slice classifier does not fit remote-sensing data particularly well. A modification, the *parallelepiped* classifier, allows hyperdimensional boxes that align with the data clusters and whose faces are parallelograms rather than rectangles.

9.5.2 Histogram Estimation Classifier

A classifier based on *K*-D histogram estimation was proposed in (Skidmore and Turner, 1988). The basic classifier consists of three steps:

- Supervised training pixels are used to construct the feature-space histogram *of each defined class* in *K*-D. The histograms are each normalized properly by the respective total number of pixels in each training class, as in Eq. $(4 - 1)$, accounting for different training set sizes in each class.

- Every spectral vector "cell" in *K*-D is checked to find the class with the most histogram counts, and that class label is assigned to the cell. A LUT that maps spectral vector to class label is thus created.

- Unlabeled pixels are classified by the LUT.

The primary benefit of this classifier is that it is fast in the classification stage because of the LUT. However, a serious drawback is caused by the sparseness of the estimated class histograms. Training pixel class populations are, by definition, small samples of the overall class population. Therefore, it is unlikely that every class spectral vector will be represented in the training data, leading to empty cells in the K-D space. Ways to reduce this problem include rebinning the histogram cells to a coarser *DN* resolution (Skidmore and Turner, 1988) and filling of empty cells by convolution filtering in the K-D space (Dymond, 1993). The underlying histogram distribution can also be approximated by *Parzen estimation*, consisting of superposition of component "kernel" distributions (James, 1985; Fukunaga, 1990).

9.5.3 Nearest-Neighbors Classifier

A number of nonparametric classification schemes assign labels to unknown pixels according to the labels of neighboring training vectors in feature space. They include:

- *nearest-neighbor*–assign the same label as that of the nearest training pixel
- *k nearest-neighbors*–assign label according to the majority label of *k* nearest-neighbor training pixels
- *distance-weighted k nearest-neighbors*–assign weights to the labels of *k* nearest-neighbor training pixels, inversely proportional to their Euclidean distance from the unknown pixel, and assign the label with the highest aggregate weight

These and similar algorithms are described and evaluated in (Hardin, 1994). They tend to be slow because distances from each unknown pixel to many of the training pixels must be calculated. Modifications to speed the classification are presented in (Hardin and Thomson, 1992). If the training class signatures are compact and well separated, the nearest-neighbor algorithms will produce results similar to those of the parametric nearest-mean algorithm. For discussions of the relationship between the nearest-neighbor classifier and the maximum-likelihood classifier, see (James, 1985) and (Fukunaga, 1990).

9.5.4 Artificial Neural Network (ANN) Classifier

The ANN algorithm is a recently popular nonparametric approach to classification. It differs significantly from the level-slice and histogram estimation algorithms, in that the decision boundaries are not fixed by a deterministic rule applied to the prototype training signatures, but are determined in an iterative fashion by minimizing an error criterion on the labeling of the training data. In that sense, ANNs are similar to clustering algorithms. Examples of the use of ANNs for remote-sensing image classification are listed in Table 9-4. An excellent presentation of neural network classification and its relationship to traditional statistical classification is given in (Schürmann, 1996).

A basic network is shown in Fig. 9-13.[10] This network has three layers; the middle (or "hidden") layer and the output layer contain *processing elements* at each node. The input layer nodes, on the other hand, are simply an interface to the input data and do not do any processing. The input patterns are the features used for classification. In the simplest case, they are the multispectral vectors of the training pixels, one band per node. Other features, such as a spatial neighborhood of pixels or multitemporal spectral vectors, can also be used (Paola and Schowengerdt, 1995a).

Within each processing node, we have a summation and transformation (Fig. 9-14). At each hidden layer node, j, the following operation is performed on the input pattern, p_i, producing the output, h_j,

$$\text{hidden layer:} \quad \begin{aligned} S_j &= \sum_i w_{ji} p_i \\ h_j &= f(S_j) \end{aligned} \qquad (9-2)$$

which is directed to each output layer node, k, where the output, o_k, is calculated,

$$\text{output layer:} \quad \begin{aligned} S_k &= \sum_j w_{kj} h_j \\ o_k &= f(S_k). \end{aligned} \qquad (9-3)$$

The most widely-used transformation function is the *sigmoid function*,

$$f(S) = \frac{1}{1 + e^{-S}}, \qquad (9-4)$$

shown in Fig. 9-15. Other functions, for example a hard threshold with no gradient, may be used, but it seems that the end result of the network is not particularly sensitive to the specification of f. The shape of f can affect the convergence rate during training, however.

---

[10] There are actually *many* ANN variants. We will discuss only a basic network architecture.

TABLE 9-4. Example applications of ANNs in remote-sensing image classification. Note the wide range of sensor and feature data that have been used. A survey and analysis of papers published prior to 1994 can be found in (Paola and Schowengerdt, 1995a).

data	reference
AVHRR, SMMR	(Key *et al.*, 1989)
MSS, DEM	(Benediktsson *et al.*, 1990a)
TM	(Ritter and Hepner, 1990)
multidate SPOT	(Kanellopoulos *et al.*, 1991)
TM	(Liu and Xiao, 1991)
TM	(Bischof *et al.*, 1992)
TM	(Heermann and Khazenie, 1992)
SPOT	(Kanellopoulos *et al.*, 1992)
SPOT, texture	(Civco, 1993)
SPOT, texture	(Dreyer, 1993)
TM	(Salu and Tilton, 1993)
polarimetric SAR	(Hara *et al.*, 1994)
TM	(Yoshida and Omatu, 1994)
TM, texture	(Augusteijn *et al.*, 1995)
TM ratios	(Baraldi and Parmiggiani, 1995)
AVIRIS	(Benediktsson *et al.*, 1995)
SPOT	(Chen *et al.*, 1995)
airborne SAR	(Foody, 1995; Foody *et al.*, 1995)
polarimetric multifrequency SAR	(Hara *et al.*, 1995)
aerial photography	(Kepuska and Mason, 1995)
TM	(Paola and Schowengerdt, 1995b)
airborne TM, airborne SAR	(Serpico and Roli, 1995)
AVHRR	(Yhann and Simpson, 1995)
ASAS, multiangle	(Abuelgasim *et al.*, 1996)

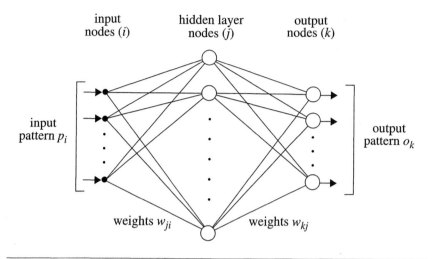

FIGURE 9-13. The traditional structure of a three-layer ANN.

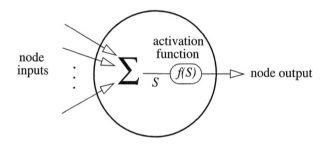

FIGURE 9-14. The components of a processing element.

Back-propagation algorithm

The discrimination capability of an ANN is contained in its weights. During training, they are iteratively adjusted towards a configuration that allows the network to distinguish the prototype patterns of interest. The back-propagation algorithm minimizes the squared error over all patterns at the output of the network and was the first successful approach to training the network of Fig. 9-13 (Rumelhart *et al.*, 1986; Lippmann, 1987). It belongs to the family of iterative, gradient descent algorithms and consists of the following steps:

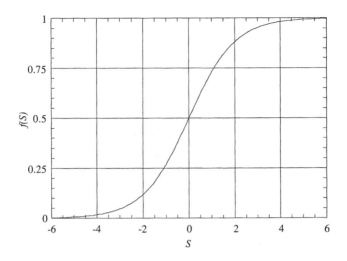

FIGURE 9-15. The sigmoid activation function.

1. Select training pixels for each class and specify the desired output vector d_k for class k, typically $d_m = 0.9$ ($m = k$) and $d_m = 0.1$ ($m \neq k$). These are the target values that the ANN attempts to produce.

2. Initialize weights as random numbers between 0 and 1 (typically small values near zero are used).

3. Set the frequency for weight updating to one of the following:

    • after every training pixel (sequential)
    • after all training pixels in each class
    • after all training pixels in all classes (batch)

   Batch training is commonly used as it minimizes the frequency of weight updates; the remainder of the algorithm below assumes batch training.

4. Propagate the training data forward through the net, one pixel at a time.

5. After each training pixel is propagated forward through the network, calculate the output o and accumulate the total error relative to the desired output d (the 1/2 factor is a mathematical convenience),

$$\frac{\|\varepsilon\|^2}{2} = \frac{1}{2} \sum_{p=1}^{P} \sum_{k} (d_k - o_k)^2 \qquad (9-5)$$

Repeat over all P training patterns (pixels) for batch training.

6. After all training pixels have been used, adjust the weight w_{kj} by,

$$\Delta w_{kj} = LR \frac{\partial \varepsilon}{\partial w_{kj}}$$

$$= LR \sum_{p=1}^{P} (d_k - o_k) \frac{d}{dS} f(S) \bigg|_{S_k} h_j \qquad (9-6)$$

where *LR* is a *Learning Rate* parameter used to control the speed of convergence.

7. Adjust the weight w_{ji} by,

$$\Delta w_{ji} = LR \sum_{p=1}^{P} \left\{ \frac{d}{dS} f(S) \bigg|_{S_j} \sum_{k} \left[(d_k - o_k) \frac{d}{dS} f(S) \bigg|_{S_k} w_{kj} \right] p_i \right\} \qquad (9-7)$$

8. Repeat steps 4 through 7 until ε < threshold.

From Eq. (9 – 6) and Eq. (9 – 7), we see the origin of the term "back-propagation." The errors in the output vector, *o*, are propagated backwards through the network links to adjust their weights (Richards, 1993).

The convergence properties of this algorithm can be problematic. First of all, the back-propagation algorithm is notoriously slow. Other, faster ways to obtain convergence of the weights have been developed, and efficient preprocessors have been proposed (Gyer, 1992). Secondly, the network may not reach a "globally" optimal solution that minimizes the output error, but become stuck in a "local" minimum in the error surface. A fraction of the previous iteration's Δw can be added to the current iteration's Δw to accelerate the convergence out of local minima and towards a global minimum; the added fraction is called a *momentum parameter* (Pao, 1989). Finally, because the weights are initially random, the network *does not converge to the same result in repeated runs*. It is estimated that this stochastic property may contribute 5% or more to the residual uncertainty at the end of convergence (Paola and Schowengerdt, 1997).

As noted, there are a wide assortment of ANN architectures. A survey of their use in remote-sensing classification reveals few guidelines for particular architectures or for specifying the number of hidden layers or nodes (Paola and Schowengerdt, 1995a). For a fully-connected three-layer network as in Fig. 9-13, with H hidden layer nodes, K bands and L classes, the number of free parameters is (Paola and Schowengerdt, 1995b),

$$N_{ANN} = \text{number of weights}$$
$$= H(K + L) \,.$$

$(9-8)$

This represents the *degrees-of-freedom* in the ANN. Similarly, a maximum-likelihood parametric classifier requires mean vector and covariance matrix values,

$$N_{ML} = \text{number of mean values} + \text{number of unique covariance values}$$

$$= LK + L\left(\frac{K^2 + K}{2}\right)$$

$(9-9)$

$$= LK\frac{(K+3)}{2} \,.$$

These are the degrees-of-freedom in a maximum-likelihood classifier. If we want to compare the two classifiers, a reasonable way to constrain the ANN is to make the number of weights equal to the number of parameters required by a maximum-likelihood classifier, and solve for the number of hidden layer nodes,

$$H = \frac{LK(K+3)}{2(K+L)} \,.$$

$(9-10)$

This depends weakly on the number of classes, but strongly on the number of bands (Fig. 9-16). For a hyperspectral dataset of 200 bands with 10 classes, 1000 hidden layer nodes are needed to match the number of parameters in a Gaussian maximum-likelihood classifier. For TM non-TIR data, 20 nodes are sufficient, even for as many as 20 classes. The back-propagation training time is dependent on the number of hidden layer nodes, so it is desirable to use only as many nodes as needed for satisfactory classification performance. The previous analysis provides only a guideline; it does indicate that the ANNs in some published examples may be larger than needed for the problem at hand.

9.5.5 Nonparametric Classification Examples

Supervised classifications of the Marana image will be used to illustrate the differences among the nonparametric classifiers. Again, for clarity, we will use only bands 3 and 4, which serve to distinguish soil and vegetation. Three well-discriminated classes and training sites are defined by visual examination of the CIR composite (Plate 9-1). In Fig. 9-17, we see the result of the level-slice classifier with a threshold of $\pm 4\sigma$ in each band. A large portion of the image is not classified with this threshold. A larger threshold could be used to include more

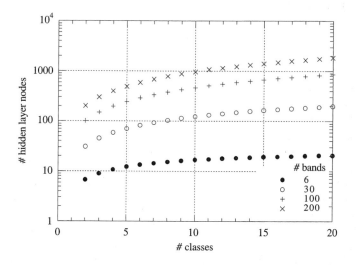

FIGURE 9-16. The number of hidden layer nodes required in a 3-layer ANN to match the degrees-of-freedom in a maximum-likelihood classifier.

thematic map   DN feature space

FIGURE 9-17. Level-slice classification results in the image space and the decision regions in the feature space. The class means are shown as dots in the feature space, which ranges [0,80] in both axes.

image pixels, but they would be less similar to the training set and the map error would be greater.

The ANN classifier's training stage is iterative—it converges to a stable solution by small incremental changes in the system weights, which in turn determine the decision boundaries. The same image and training data were used for ANN classification with a 2-3-3 network, i.e. two input nodes (two spectral bands), three hidden layer nodes, and three output nodes (three classes). The network was trained in a "batch" mode. The mean-squared and maximum errors at the output nodes as a function of the training cycle number, are shown in Fig. 9-18. There is little improvement in the training pixel classification after about 2000 cycles. The learning rate and momentum parameters change rapidly early in the process and become relatively stable once the network starts to converge at about 500 cycles.

The decision boundaries and hard classification maps are shown in Fig. 9-19 for three points during the iterative training process. At cycle 250, the crop class does not even appear in the decision space or classification map. This is evidently because of its small population relative to the dark and light soil classes. The ANN output node for "crop" has a nonzero value, but is less than that for the other two nodes at this point in the training. If the network is trained to 5000 cycles, we see that the decision boundaries change, but there is little change in the final hard map, because most of the feature space change takes place where there are few image spectral vectors (Fig. 9-20 and Fig. 9-42). Interestingly, though, the narrow, unpaved roads ("light soil") between the crop fields appear to be better mapped after this level of training. The mean-squared error at the output nodes drops from 0.115 at 1000 cycles to 0.074 at 5000 cycles.

The output node values for each class are shown in Fig. 9-21 as soft classification maps. Higher *GL*s in the maps indicate larger output node values in the ANN. We can see additional evidence here that the crop class does not become competitive with the other two classes until somewhere between cycles 500 and 750. The ANN output values can be related to the spatial-spectral mixing of surface components (Foody, 1996; Schowengerdt, 1997).

Finally, we note that the feature-space decision boundaries for ANNs are typically more adapted to the data distribution than the boundaries produced by parametric classifiers, such as Gaussian maximum-likelihood. An example for a 12 class urban land use/land cover classification of TM imagery is shown in Plate 9-1 (Paola and Schowengerdt, 1995b). We will comment more on this point later, after a review of parametric classification.

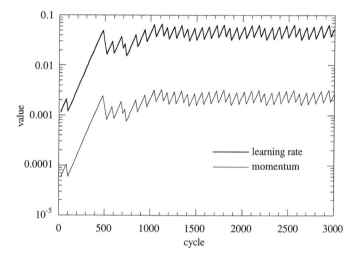

FIGURE 9-18. The behavior of the output-node errors and learning rate and momentum parameters during training of the network. The network does not start to converge to a stable solution until about cycle 500. At this point, the learning rate and momentum do not need to be increased further; instead they are periodically adjusted every five cycles to maintain the convergence of the network. The overall shape of these curves is typical, but will vary in detail from one dataset to another.

FIGURE 9-19. Hard maps and decision boundaries at three stages of the back-propagation algorithm. (The processing and results in this section were provided by Justin Paola, Oasis Research Center.)

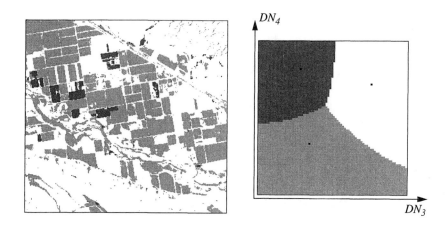

FIGURE 9-20. Final ANN hard classification map and decision boundaries at 5000 cycles.

9.6   Parametric Classification

Parametric algorithms rely on assumptions about the form of the probability distribution for each class. The most notable example is *maximum-likelihood*, which explictly uses a probability model to determine the decision boundaries. The necessary parameters for the model are estimated from training data. A useful property of parametric classifiers, that is not possible with nonparametric classifiers, is theoretical estimation of classifier error from the assumed distributions.

9.6.1 Estimation of Model Parameters

If a sufficiently large number of representative training pixels are available in each class, we can calculate the class histograms and use them as approximations to the continuous probability density functions of an infinite sample of data (Chapter 4). These *class-conditional* probability density functions, $p(f|i)$, have unit area and describe the probability of a pixel having a feature vector f given that the pixel is in class i.

Each probability density function (histogram) may be weighted by the *a priori* probability, $p(i)$, that class i occurs in the image area of interest. These scaled probability functions, $p(f|i)p(i)$, do not have unit area. In remote sensing, the

FIGURE 9-21. *Soft classification maps at four stages of the back-propagation training process. The greylevel in each map is proportional to the class output-node value and all maps are scaled by the same LUT, so direct brightness comparisions are possible. Notice how the output-node values for all three classes increase as more training cycles are used.*

a priori probabilities may be estimated from external information sources such as ground surveys, existing maps or historical data. For example, suppose the goal is to determine the proportion of crop types planted during a particular season from classification of a Landsat image. We might reasonably set the *a priori* probabilities to historical estimates of the percentage of each crop type in the area. In most cases, however, reliable *a priori* probabilities are difficult to obtain and are assumed equal for all classes. A discussion of the use of *a priori* probabilities in remote sensing is presented in (Strahler, 1980) and a simple example is given in Sect. 9.6.5.

To make a classification decision for a pixel, we need to know the *a posteriori* probabilities that the pixel belongs to each of the training classes *i*, given that the pixel has the feature vector *f*. This probability, $p(i|f)$, may be calculated with *Bayes' Rule*,

$$p(i|f) = \frac{p(f|i)p(i)}{p(f)} \qquad (9-11)$$

where

$$p(f) = \sum_i p(f|i)p(i) . \qquad (9-12)$$

A *decision rule* may now be formed with the *a posteriori* probabilities of Eq. (9 – 11). It is intuitively satisfying to assign the pixel to a particular class if its *a posteriori* probability is greater than that for all other classes; this is the rule for the *maximum-likelihood* classifier. Since $p(f)$ is the same for all classes per Eq. (9 – 12), it can be ignored in a comparison of *a posteriori* probabilities, and we can write the *Bayes decision rule* as,

If *p(i|f) > p(j|f), for all j≠i*, assign pixel to class *i*

where *p(i|f)* is found from Eq. (9 – 11), the training data distributions *p(f|i)*, and the *a priori* probabilities *p(i)*. In the very unlikely situation that the two *a posteriori* probabilities are exactly equal, a decision cannot be made from the class probabilities. A tie-breaking process then must be employed, such as using the classification of a neighboring, previously classified pixel or randomly choosing either class. It can be shown that the Bayes decision rule minimizes the average probability of error over the entire classified data set, if all the classes have normal (Gaussian) probability density functions (Duda and Hart, 1973).

9.6.2 Discriminant Functions

The Bayes decision rule may be restated as,

If $D_i(f) \geq D_j(f)$, *for all j≠i*, assign pixel to class i

where the *discrimant function* for class i, $D_i(f)$, is given by,

$$D_i(f) = p(i|f)p(f) = p(f|i)p(i). \tag{9-13}$$

Setting D equal to the *a posteriori* probabilities results in a Bayes optimal classification, but is not the only choice that produces the same result. The decision boundary does not change under any monotonic transformation, for example,

$$D_i(f) = Ap(i|f)p(f) + B = Ap(f|i)p(i) + B \tag{9-14}$$

or,

$$D_i(f) = \ln[p(i|f)p(f)] = \ln[p(f|i)p(i)] \tag{9-15}$$

are both valid discriminant functions. The latter transform is advantageous under a normal distribution assumption, as we will see next.

9.6.3 The Normal Distribution Model

To apply the parametric classification approach described previously, we must decide on an appropriate probability model. Now, for optical remote sensing,[11] it is almost universal that the *normal* (or Gaussian) distribution is chosen for the following reasons:

- It provides a mathematically tractable, analytical solution to the decision boundaries
- Image samples selected for supervised training often exhibit normal-like distributions[12]

Although normal distributions are widely assumed, it is occasionally useful to remember that all of the theory and equations to this point *do not require* that assumption, and may be applied to data with *any* probability distribution. The

---

[11] The normal distribution is not appropriate for SAR image statistics.

[12] The reader should be aware that there is a paucity of quantitative data in the literature to support this assumption. Although an arbitrary data distribution can be modeled as a *sum* of Gaussian kernel distributions by Parzen histogram estimation, such an approach is rare in remote sensing analyses. In any case, the *net* distribution would not be Gaussian.

normal distribution does lead, however, to some nice mathematical results, as described in the following.

The transformation of Eq. (9 – 15) is particularly useful if the class probability distributions are normal. A K-D, maximum-likelihood discriminant function for class i is then, from Eq. (4 – 18) and Eq. (9 – 15),

$$D_i(f) = \ln[p(i)] - \frac{1}{2}\left\{ K\ln[2\pi] + \ln|C_i| + (f - \mu_i)^T C_i^{-1}(f - \mu_i) \right\} \quad . (9-16)$$

Only the last term must be calculated at each pixel.[13] For any two classes a and b, the decision boundary is found by setting $D_a(f)$ equal to $D_b(f)$ and solving for f. This is equivalent to setting,

$$\ln[p(a|f)p(f)] = \ln[p(b|f)p(f)] \tag{9-17}$$

or,

$$p(a|f) = p(b|f) \ . \tag{9-18}$$

The value of f on the decision boundary is that for which the *a posteriori* distributions are equal. To one side of the boundary the decision favors class a and to the other side the decision favors class b (Fig. 9-22). The logarithmic discriminant functions of Eq. (9 – 16) are quadratic functions of f; in two dimensions, they intersect in quadratic decision boundaries, as depicted in a general way in Fig. 9-23. A probability threshold can be used to exclude outliers, i.e. pixels with a low probability of belonging to any class, as shown later with image examples.

The total probability of classification error is given by the area under the overlapping portions of the *a posteriori* probability functions. The total probability of error is the sum of the probabilities that an incorrect decision was made on either side of the class partition. It is easy to see that the Bayes optimal partition minimizes this error because a shift of the partition to the right or left will include a larger area from one class or the other, thereby increasing the total error.

It is instructive at this point to note again the role of the *a priori* probabilities. From Eq. (9 – 18) we see that the decision boundary will move to the left if $p(b)$ is greater than $p(a)$ and to the right if $p(a)$ is greater than $p(b)$. Even if reasonable estimates of the *a priori* probabilities are available, we may choose to bias them heavily if the significance of an error in one class is much greater than for others. For example, consider a hypothetical project to locate all occurrences of a rare class. The actual *a priori* probability of that class would be very low, but we could assign an artificially high *a priori* probability to insure that no occurrences

---

[13] The quadratic term is, in fact, the Mahalanobis distance between the feature vector f and the distribution of class i; see Table 9-3.

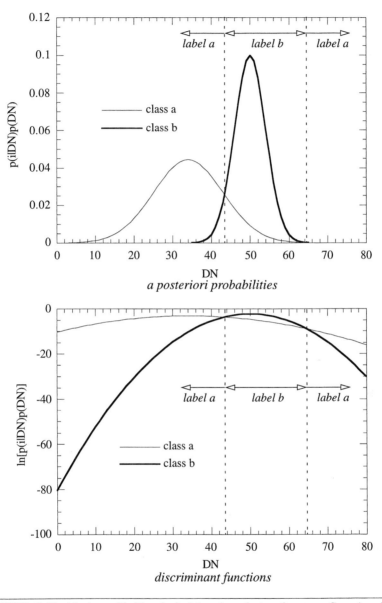

FIGURE 9-22. Maximum-likelihood decision boundaries for two Gaussian DN distributions in one dimension. The class statistics are: $\mu_a = 34$, $\sigma_a = 9$, $\mu_b = 50$, $\sigma_b = 4$. Note the decision boundary on the right is not visible in the upper graph because of the ordinate scale, but becomes clear in the discriminant function graph below.

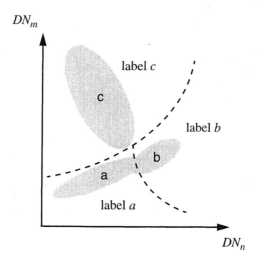

FIGURE 9-23. Maximum-likelihood decision boundaries for three classes in two dimensions, with Gaussian distributions for each class. The boundaries are quadratics in 2-D. Their dependence on the covariance matrices of the individual classes is shown in (Duda and Hart, 1973).

are missed. The penalty would be many pixels incorrectly classified as the rare class. These "false alarms" would then have to be removed by site visits or by referencing other data, such as aerial photography.

9.6.4 The Nearest-Mean Classifier

If we assume that the covariance matrices of L classes are equal, i.e.

$$C_i = C_0 \qquad (9-19)$$

and that the *a priori* probabilities are equal,

$$p(i) = 1/L , \qquad (9-20)$$

then the discriminant function of Eq. $(9-16)$ becomes,

$$D_i(f) = A - \frac{1}{2}(f - \mu_i)^T C_0^{-1}(f - \mu_i) \qquad (9-21)$$

where the constant A is given by,

$$A = \ln[1/L] - \frac{1}{2}\left\{ \ln[2\pi] + \ln[|C_0|] \right\} \qquad (9-22)$$

and may be ignored in a comparison of D among different classes. The right-hand term in Eq. $(9-21)$ depends only on the feature vector f and the class mean μ_i, since C_0 is assumed to be common to all classes. The expansion of this term results in a quadratic equation in f, whose quadratic term is independent of the class i (Duda and Hart, 1973), and may therefore be combined with the constant A for comparison among different classes. Thus Eq. $(9-21)$ reduces to a form that is linear in f, meaning that the decision boundaries are hyperplanes (linear functions in two dimensions), in contrast to the hyperquadric functions of Eq. $(9-16)$ in the general covariance case.

If the covariance matrices are further constrained to be diagonal, i.e. the features are uncorrelated, and to have equal variance along each feature axis, i.e.

$$C_0 = \begin{bmatrix} c_0 & \cdots & 0 \\ \vdots & & \vdots \\ 0 & \cdots & c_0 \end{bmatrix} \qquad (9-23)$$

then

$$D_i(f) = A - \frac{(f-\mu_i)^T(f-\mu_i)}{2c_0} \quad . \qquad (9-24)$$

The quantity $(f-\mu_i)^T(f-\mu_i)$ is the square of the L_2 distance between the vectors f and μ_i (Table 9-3). Since A and c_0 are the same for all classes, Eq. $(9-24)$ represents the discriminant functions for the *nearest-mean*, or *minimum-distance classifier*. $D_i(f)$ will be largest (because of the minus sign) for that class i with minimum L_2, i.e. the class with the nearest mean.

The following points can be made about the nearest-mean algorithm:

- It ignores the covariance matrices of the prototype classes.

- The L_1-distance function can be used for computational efficiency, but the mathematical relationship to the maximum-likelihood algorithm would no longer apply. The nearest-mean decision boundaries for L_1 are piecewise linear and approximate the linear boundaries obtained with L_2 (Schowengerdt, 1983).

An idealized depiction of the nearest-mean decision boundaries is shown in Fig. 9-24. The nearest-mean algorithm does not create an "unlabeled" class automatically, unless an upper limit is applied to the distance from the mean. Then, pixels whose distance from the nearest class mean is greater than that threshold are assigned to the "unlabeled" class.

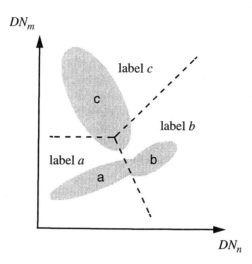

FIGURE 9-24. *Nearest-mean decision boundaries for three classes in two dimensions, using the L_2-distance measure. Distance thresholds could be implemented either as in the level-slice algorithm, or as circular boundaries, centered at each class mean, at a specified radius.*

9.6.5 Parametric Classification Examples

The hard classification map and decision boundaries for the L_2 nearest-mean classifier are shown for the Marana TM image and training data used earlier in Fig. 9-25. No outlier threshold is implemented, so every pixel is assigned a label, unlike the level-slice classifier.

A maximum-likelihood classification of the same data is shown in Fig. 9-26. Note the quadratic decision boundaries, characteristic of the Gaussian distribution assumption. A small threshold results in a large number of excluded pixels and shows the elliptical nature of the distribution model for each class. This high sensitivity to the threshold indicates that, for these data, the training class distributions are compact in the feature space.

To illustrate the modeling of class distributions with Gaussian probability functions, we choose a relatively simple, single-band example. In Fig. 9-27, an MSS band 4 image of Lake Anna, Virginia, is shown with three training sites, two for water and one for vegetation. Initially, we will use only one site for water (the larger one in the lower right) and model both classes with Gaussian distributions having the training site means and standard deviations (Fig. 9-28). Two classifi-

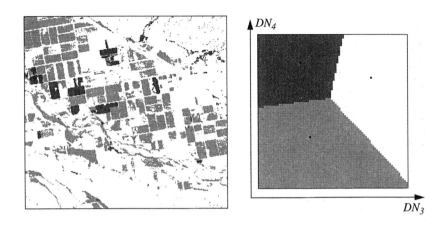

FIGURE 9-25. Nearest-mean classification results in image and feature space.

cations are shown in Fig. 9-29, nearest-mean and maximum-liklihood with equal *a priori* probabilities. The nearest-mean classification map appears more accurate in that the lake is mapped in its entirety, while the maximum-likelihood classification map does not correctly classify the upper arms of the lake.

Some insight is gained into why the classifiers are producing significantly different results by looking at the total image histogram and Gaussian models for each class (Fig. 9-28). The data is bimodal, with high separability between the two classes. The nearest-mean decision boundary is halfway between the two class means, while the maximum-likelihood boundary is much further to the left because of the small standard deviation of the water class. Also, we can now see the role of the *a priori* probabilities, which are assumed equal for the two classes. Because the water occupies a much smaller portion of the image than vegetation, its histogram mode is smaller than that for vegetation. However, that is not reflected in the *a priori* probabilities, leading to a poor model for the data. Estimating the proportion of water to be about 5% and vegetation about 95%, the *a posteriori* probabilities match the histogram much better (Fig. 9-30). Unfortunately, this does not significantly increase the area classified as water, because the decision boundary moves to the left by only a small amount.

The problem with the water class is that its variability within the scene is not sufficiently represented by the single training class. The width of the Gaussian distribution model for that one site is less than the total image histogram indicates. Therefore, we include the second water site in one of the arms of the lake, and "pool" the statistics for the two sites by averaging their DN means and

$T = 0\%$

$T = 1\%$

FIGURE 9-26. Maximum-likelihood classification results in image and feature space. Note the disjoint decision regions for the class "light soil" if no probability threshold is used. A threshold limits the distance from the mean. The threshold is specified as the percentage excluded volume under the Gaussian model distributions of each class. The fact that a low threshold of 1% results in such a large population of unlabeled pixels implies that the distributions for each class are highly localized near their means.

adding their DN variances. The resulting Gaussian model now has a larger variance and fits the data histogram much better (Fig. 9-31). A second maximum-likelihood classification produces an improved mapping of water   (Fig. 9-32).

FIGURE 9-27. Lake Anna Landsat MSS band 4 image with three training sites. The site for vegetation is the large one in the lower left, the initial site for water is in the lower right, and the second site for water is the small one in the far left arm of the lake. These sites were selected by visual interpretation of a color infrared composite of bands 4, 3 and 2.

The nearest-mean classifier performs well in either case, because it uses only the information about the class means, and not their variances.

A comparison of parametric maximum-likelihood decision boundaries to nonparametric ANN decision boundaries for a 12-class urban land use/land cover TM classification is shown in Plate 9-1. The ANN boundaries adapt to the data during the training stage, and are not constrained by any parametric assumption. In this particular experiment, the ANN and maximum-likelihood classification maps differed at 35% of the pixels, with the ANN map being superior for certain classes (Paola and Schowengerdt, 1995b).

Unfortunately, we cannot say that one classifier is always better than another. As these examples demonstrate, a classifier's performance depends on the data! If the class distributions are widely separated, most classifiers perform about the same. The maximum-likelihood classifier, however, must also have reasonably accurate covariance estimates, again previously demonstrated. Given that the Gaussian assumption is appropriate and given accurate parameter estimates, maximum-likelihood theoretically produces minimum class labeling error. Non-

FIGURE 9-28. DN histograms for the vegetation and initial water sites, with the corresponding Gaussian distribution model. The histograms and discrete probability functions are each normalized to unity. The model distribution for water only appears asymmetric because the mean is 6.4 and the values are calculated at integer DNs.

nearest-mean

maximum-likelihood

FIGURE 9-29. *Nearest-mean and maximum-likelihood classification maps of the image of Fig. 9-27 using one training site for water and one for vegetation.*

parametric algorithms that adapt to any class distributions, like the ANN, do not have a simple theoretical basis, but are generally easier to use since less care is required by the analyst in defining and validating the training data.

FIGURE 9-30. The effect of a priori probabilities on the goodness-of-fit between the class Gaussian models and the data. Only one training site is used for water. The two maximum-likelihood decision boundaries are for equal (right) and unequal (left) a priori probabilities.

FIGURE 9-31. Final classification training models and decision boundaries. The maximum-likelihood decision boundary has moved to the right by 6 DN from its location when only one water training site was used (Fig. 9-30), while the nearest-mean decision boundary has shifted by 1.5 DN to the right.

FIGURE 9-32. Maximum-likelihood classification map with pooled training statistics from two water sites and unequal a priori probabilities.

9.7   Spatial-Spectral Segmentation

So far, we have not used our knowledge that there is a significant statistical correlation among neighboring pixels (Chapter 4). The classification of pixels in a spectral feature space is independent of any spatial relationships. Another approach to classification, *spatial-spectral segmentation*, incorporates spatial and spectral information into unsupervised labeling. Spatial-spectral segmentation schemes tend to work best where the scene consists of relatively homogeneous objects of at least several pixels. Example algorithms are referenced in Table 9-5. Most employ either a local gradient operation (Chapter 6) to detect region boundaries (edge-based) or aggregate neighboring, similar pixels into larger regions (region-based). In the ECHO algorithm, for example, adjoining pixels are initially aggregated into small cells (e.g. 2-by-2 pixels) based on the similarity of their spectral vectors. Cells that cross spatial boundaries (edges) are detected by a threshold on the cell variance in each spectral band, and the pixels in the cell are not aggregated if an edge is detected within the cell. The homogeneous cells found at this stage are then aggregated further if they are spectrally similar to

TABLE 9-5. Some spatial-spectral segmentation algorithms.

name	data	type	reference
–	Landsat MSS	clustering	(Haralick and Dinstein, 1975)
ECHO	airborne MSS, Landsat MSS	edge detection and region growing	(Kettig and Landgrebe, 1976; Landgrebe, 1980)
BLOB	MSS	clustering	(Kauth *et al.*, 1977)
–	various	clustering	(Coleman and Andrews, 1979)
AMOEBA	MSS, multispec-tral aerial	clustering	(Bryant, 1979; Jenson *et al.*, 1982; Bryant, 1989; Bryant, 1990)
–	simulated	pyramid structure	(Kim and Crawford, 1991)
–	airborne MSS, SAR	hierarchical partitioning	(Benie and Thomson, 1992)
–	TM	iterative region growing	(Woodcock and Harward, 1992)
SMAP	simulated, SPOT	pyramid structure	(Bouman and Shapiro, 1994)
IPRG	TM	iterative parallel region growing	(Le Moigne and Tilton, 1995)

neighboring cells. The resulting spatially homogeneous areas are then classified in their entirety as *single* spectral samples, rather than pixel-by-pixel.

9.7.1 Region Growing

There are two ways in which a pixel can be allowed to connect to its immediate neighbors to form a multipixel region (Fig. 9-33). Generally, 8-connectedness is preferred because it is more general and includes 4-connected regions as a special case. Algorithms based on 4-connectedness are faster, however, since only four neighbors must be addressed.

Spatial connectivity is invoked to group similar pixels into an *image* region, presumed to represent a *physical* region of similar properties on the earth (Woodcock and Harward, 1992). A threshold must be specified for the similarity of a pixel and its neighbors; if their difference is within the threshold, they are

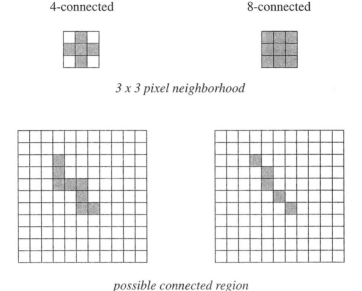

FIGURE 9-33. Two ways to define local spatial connections on a square pixel grid. Each small square represents a single pixel.

"grouped" and treated as a common entity in subsequent processing. Otherwise, they are maintained as individual pixels.

There are many techniques for spatial segmentation (Pal and Pal, 1993). A representative and robust *region-growing* algorithm will be described (Ballard and Brown, 1982; Ryan, 1985). Pixels in the image are processed in the normal order, left-to-right (along rows) and top-to-bottom (row-by-row), which is important for this algorithm. If the data were processed right-to-left, for example, a different (but equally valid) segmentation would result. Two parameters are specified by the user, a DN difference threshold, t, and a DN standard deviation threshold, σ.

A three pixel, four-connected neighborhood is used to make decisions on whether to add a current pixel to an adjoining region, or to initialize a new region (Fig. 9-34). The algorithm consists of several case rules applied to the DNs and labels of these three pixels to handle the various relations that can arise (Table 9-6). Case 4 provides a "filler" pixel between two diagonally-connected pixels with

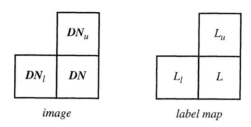

image   *label map*

FIGURE 9-34. *The spatial neighborhood used for the region-growing algorithm. The pixel being processed is the one in the lower right; it has an "upper" neighbor above and a "left" neighbor to the left.*

TABLE 9-6. *Rule system for spatial segmentation of a multispectral image.*

case	if	then
1. start new region	$\left\| DN - DN_u \right\| > t$ $\left\| DN - DN_l \right\| > t$	L = new label
2. merge with upper region	$\left\| DN - DN_u \right\| \leq t$ $\left\| DN - DN_l \right\| > t$	$L = L_u$
3. merge with left region	$\left\| DN - DN_u \right\| > t$ $\left\| DN - DN_l \right\| \leq t$	$L = L_l$
4. merge with upper and left region	$\left\| DN - DN_u \right\| \leq t$ $\left\| DN - DN_l \right\| \leq t$ $L_l = L_u$	$L = L_u$
5. relabel and merge	$\left\| DN - DN_u \right\| \leq t$ $\left\| DN - DN_l \right\| \leq t$ $L_l \neq L_u$	$L_l = L_u$ and $L = L_u$

the same label and case 5 provides for the connection of two previously separate regions, if the pixel values warrant the change. The label of the left region is changed to that of the upper region for all pixels in the left region. The algorithm requires dynamic lists of pixels and labels that must be continuously updated; a second pass through the labeled pixels is also needed to make the final labels sequential. When finished, the output labeled map consists of a region labeled "one" in the upper left (the first pixel), progressing to the "largest" label that represents the total number of regions in the image.

For reasonable amounts of smoothing, the dynamic range of the label image is large, so it cannot be displayed as a grey level image with full label resolution. A useful way to present the results is to replace the label of each region by the region's average DN in the original image (an example of the decoding discussed in relation to Fig. 9-2). This representation is used to display the segmentation results for the Marana image in Fig. 9-35. As the DN difference threshold, *t*, increases, pixels are more easily aggregated to existing regions, so there are fewer regions covering the full image.

The segmentation algorithm characteristics can also be understood in terms of magnitude error maps, as was done for clustering (Fig. 9-36). For a small DN threshold, the degree of aggregation into regions is small; as the threshold is increased, more aggregation occurs and fewer regions are formed. Therefore, the error map associated with approximating the original image data by the segmented map changes from local, noise-induced variations to larger areas. A plot of the number of regions and the average DN error in approximating the original image by the region mean DNs, for different DN difference thresholds, is shown in Fig. 9-37.

Another interpretation of the segmentation is possible by examining the scattergrams between bands 3 and 4 (Fig. 9-38). The segmentation algorithm results in a spatial-spectral clustering of the data because pixels in segmented spatial regions are spectrally similar and "clump" together into distinct spectogram clusters.

9.8   Subpixel Classification

We have pretended to this point that every pixel has only one labeled category, leading to a so-called *hard* classifiction. Since pixels represent a spatial average over the ground-projected spread function, it is inevitable that multiple spectral categories will be included in most of them (Chapter 4). Remote-sensing imagery is usually resampled at some point for geometric correction and registration (Chapter 7), and this resampling, if not nearest-neighbor, induces additional spatial mixing. If the bands of a multispectral image are not registered, mixing occurs even without resampling for registration (Billingsley, 1982; Townshend *et al.*, 1992). As the mixing proportions (class *fractions* or *abundances*, as they are sometimes called) change from pixel-to-pixel, the net spectral vector changes. This fact was recognized early on in the analysis of Landsat MSS data (Horwitz *et al.*, 1971; Nalepka and Hyde, 1972; Salvato, 1973). The advent of so-called

FIGURE 9-35. Segmentation results for two DN-difference thresholds.

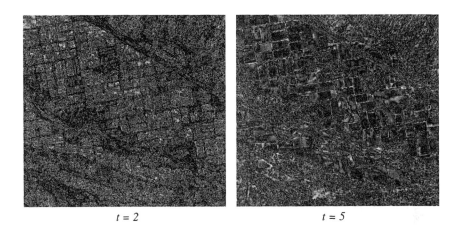

$t = 2$ $t = 5$

FIGURE 9-36. Residual error maps for segmentations with different DN thresholds. Notice how the narrow roads between the fields have a low error for the lower threshold and a high error for the higher threshold.

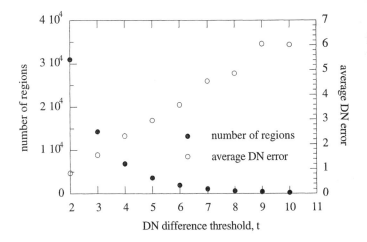

FIGURE 9-37. The convergence of the region growing segmentation algorithm and the associated average DN error as a function of the DN difference threshold. The general shape of these curves are typical for any image, but the convergence rate will depend on the spatial structure and contrast of the image.

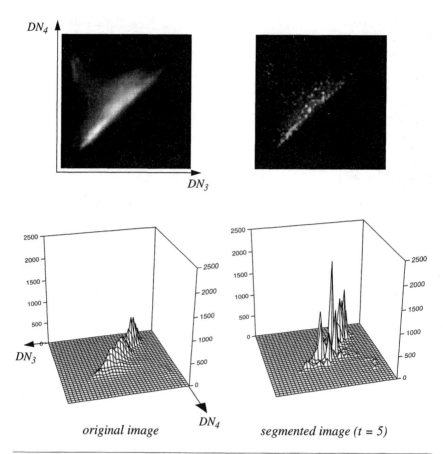

FIGURE 9-38. *Comparison of band 4 versus band 3 scattergrams before and after application of the segmentation algorithm to the imagery.*

hyperspectral sensors has prompted renewed interest in techniques for estimating mixture components because of the heritage of traditional spectroscopy (Adams *et al.*, 1993).

All natural, and most man-made surfaces, are nonuniform at some level of spatial resolution. Thus, signature mixing does not go away if one goes to higher resoluton imagery; it simply changes. However, there is a "natural" scale to many earth surface features (Markham and Townshend, 1981; Irons *et al.*, 1985). For example, paved roads range in width from about 3 m to 15 m or more, buildings have characteristic sizes according to their function, and shrub and tree canopy size varies characteristically by species. For specific classes, therefore, an increase in spatial resolution may reduce the percentage of mixed pixels, but mixed pixels will still occur at the boundary between objects regardless of their size or the sensor resolution. To see this, we create a simple synthetic scene consisting of buildings, streets and grass (Fig. 9-39). The three categories are assigned different DNs, and the "scene" is convolved with various *GIFOVs* of different sizes, simulating different sensor resolutions. The result shows the expected mixing at the boundaries between objects, and that it persists, even at relatively high resolution. This will *always* be true, because all real earth scenes have spatial detail with smaller dimensions than *any* given *GIFOV*.

9.8.1 The Linear Mixing Model

The idealized, pure signature for a class is called an *endmember*. Because of sensor noise and within-class signature variability, endmembers only exist as a conceptual convenience and as idealizations in real images. The spatial mixing of objects within a *GIFOV* and the resulting linear mixing of their spectral signatures are illustrated in Fig. 9-40. Linear mixture modeling assumes a single reflectance within the *GIFOV* and is an approximation to reality in many cases; *nonlinear* mixing occurs whenever there is radiation transmission through one of the materials (such as a vegetation canopy), followed by reflectance at a second material (such as soil); or if there are multiple reflections within a material or between objects within a *GIFOV* (Borel and Gerstl, 1994; Ray and Murray, 1996). We consider only linear mixing here.

The spatial area of integration for mixing analysis is usually assumed to be that of the *GIFOV* of the sensor, with a uniform weighting of radiances over that area. A better model is the total system spatial response (Chapter 3), which weights the individual object radiances according to their location (Fig. 9-41). It has been shown that spatial *restoration* of multispectral images, i.e. partial correction for the sensor spatial spread function, can improve subsequent unmixing analysis (Wu and Schowengerdt, 1993).

Another source of mixing is strictly spectral combination of components, which can occur, for example, with minerals or water solutions (Felzer *et al.*, 1994; Novo and Shimabukuro, 1994). These *intimate mixtures* are not distinguishable in the data from the spatial-spectral mixing induced by the sensor spatial response. The end result is the same in both cases, namely a mixed, integrated signal.

In two-dimensional feature space, we might have three endmembers as shown in Fig. 9-42. Now, if these endmembers are indeed "pure," and if they are an exhaustive basis for all spectral vectors in the image, the spectral vector for any pixel must lie within the *convex hull* defined by the envelope around the endmembers, as shown. The class fractions determine the location of the mixed-pixel

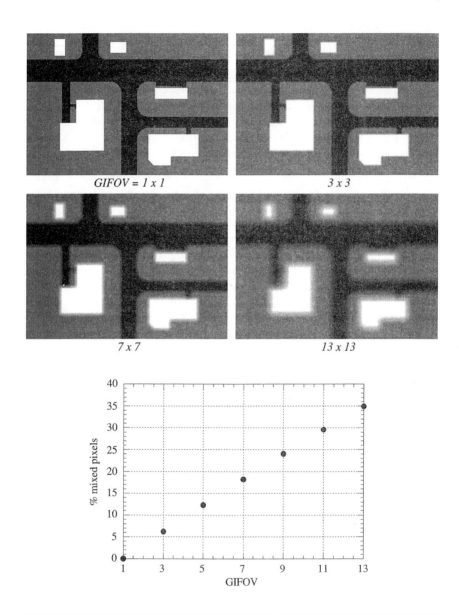

FIGURE 9-39. Simple example to illustrate spatial mixing. A synthetic scene consisting of three types of objects was created at a pixel size of one (upper left). Simulated images were then generated by spatial averaging over a range of GIFOVs. The percentage of mixed pixels for various GIFOVs is shown in the bottom graph.

one GIFOV:

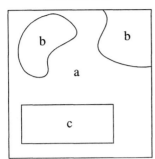

class a: 65% area, spectrum E_a

class b: 20% area, spectrum E_b

class c: 15% area, spectrum E_c

total spectrum at pixel: $\quad \boldsymbol{DN} = 0.65\boldsymbol{E}_a + 0.20\boldsymbol{E}_b + 0.15\boldsymbol{E}_c$

FIGURE 9-40. The linear mixing model for a single GIFOV. The boundaries between objects can be any shape and complexity; only the fractional coverages and individual spectral reflectances are important. An example using the spectral reflectance data of Chapter 1 is shown below. The mixed signature has lost most of the characteristic vegetation "edge" at 700nm.

scene   *detector GIFOV*   *effective net GIFOV*

FIGURE 9-41. The spatial integration involved in mixing of spectral signatures. An idealized scene radiance spatial distribution is shown on the left, the region of integration corresponding to the sensor GIFOV in the center, and the region of integration corresponding to the total sensor spatial response on the right (a response similar to that of TM is assumed). Since that response is not uniform, a weighting occurs as a function of distance from the center. Although the difference between the measured signals for the two cases on the right is only about 4% in this example, it is easy to conjure a spatial distribution of objects with different reflectances that would result in a large difference.

vector in feature space. The inversion problem, termed *unmixing*, is to estimate the fractions of each class from a given pixel vector.

Linear mixing is described mathematically as a linear vector-matrix equation,

$$DN_{ij} = Ef_{ij} + \varepsilon_{ij} \qquad (9-25)$$

where f_{ij} is the $L \times 1$ vector of L endmember fractions for the pixel at ij, and E is the $K \times L$ endmember signature matrix, with each column containing one of the endmember spectral vectors. The lefthand side DN_{ij} is the K-dimensional spectral vector at pixel ij. The added term ε_{ij} represents the residual error in the fitting of a given pixel's spectral vector by the sum of L endmember spectra and unknown noise. The relationship in Eq. $(9-25)$ is constrained by the assumption that an exhaustive set of endmembers (classes) has been defined, so that,

$$\sum_{l=1}^{L} f_l = 1 \qquad (9-26)$$

at each pixel. This last assumption is problematic in practice, since one is never sure that a sufficient number of endmembers has been defined for a given set of

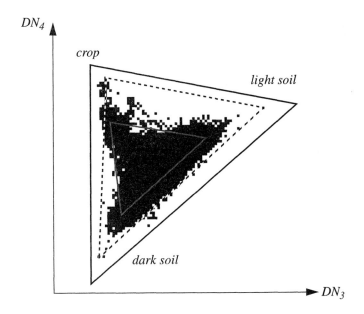

FIGURE 9-42. Three possible choices for endmembers for the classes "dark soil,"
"light soil," and "crop." The inner triangle is defined by the supervised class means
used earlier. The middle triangle is defined by extreme pixels along each DN axis. The
outer triangle contains all the pixels, but the endmembers are not actually present in
the image. Only the outer triangle is consistent with Eq. (9 – 26).

data. An additional constraint can also be included at each pixel, namely that each
fraction must be positive,

$$f_l \geq 0 . \tag{9 – 27}$$

Unmixing examples

A low-dimension example is the best way to introduce the calculation of class
fraction maps. The scatterplot in Fig. 9-42 is for TM bands 3 and 4 of the Marana
image. Three endmembers are defined, "dark soil," "light soil," and "crop," which
are associated with the tips of the scatterplot. We will use the model of Eq. (9 –
25) without the noise term, and write, for the 2-D multispectral vector at each
pixel,

$$\boldsymbol{DN}_{ij} = \boldsymbol{E}\boldsymbol{f}_{ij}, \tag{9 – 28}$$

or, writing out each element,

$$
\begin{bmatrix} DN_3 \\ DN_4 \end{bmatrix} = \begin{bmatrix} E_{dksoil3} & E_{ltsoil3} & E_{crop3} \\ E_{dksoil4} & E_{ltsoil4} & E_{crop4} \end{bmatrix} \begin{bmatrix} f_{dksoil} \\ f_{ltsoil} \\ f_{crop} \end{bmatrix}. \tag{9-29}
$$

This problem is underdetermined; we have three unknowns (the fraction components) but only two equations (for bands 3 and 4). The requirement that linear combinations of the three endmembers represent all spectral vectors in the image is expressed as,

$$
1 = f_{darksoil} + f_{lightsoil} + f_{crop} \tag{9-30}
$$

which applies to every pixel. This constraint can be combined with Eq. (9 – 29),

$$
\begin{bmatrix} DN_3 \\ DN_4 \\ 1 \end{bmatrix} = \begin{bmatrix} E_{dksoil3} & E_{ltsoil3} & E_{crop3} \\ E_{dksoil4} & E_{ltsoil4} & E_{crop4} \\ 1 & 1 & 1 \end{bmatrix} \begin{bmatrix} f_{dksoil} \\ f_{ltsoil} \\ f_{crop} \end{bmatrix}. \tag{9-31}
$$

We'll call this the *augmented* mixing equation. It can now be solved *exactly* for the fractions at each pixel,

$$
\begin{bmatrix} f_{dksoil} \\ f_{ltsoil} \\ f_{crop} \end{bmatrix} = \begin{bmatrix} E_{dksoil3} & E_{ltsoil3} & E_{crop3} \\ E_{dksoil4} & E_{ltsoil4} & E_{crop4} \\ 1 & 1 & 1 \end{bmatrix}^{-1} \begin{bmatrix} DN_3 \\ DN_4 \\ 1 \end{bmatrix}. \tag{9-32}
$$

For our example, the endmembers are defined in two ways, first as the extreme pixels of the scatterplot ("data-defined") and second, as "virtual" endmembers, which do not actually exist in the data, but whose convex hull will enclose all data in the scatterplot. These virtual endmembers are assumed to represent 100% pure pixels in their respective classes. The DN values are tabulated in Table 9-7, with the associated E and E^{-1} matrices in Table 9-8. With the E^{-1} matrix determined, Eq. (9 – 32) is applied to each multispectral pixel vector (augmented with the third, unit "band") to find the fractions of each endmember class. The results in Fig. 9-43 show little visual and numerical difference for the two endmember selection methods with these data.

In traditional hard classification, an "unknown" label is often included to accomodate everything not represented by the training sets. Any pixel that falls outside the decision threshold for any trained class is assigned to the "unknown" class. In the linear mixing model, however, we cannot define an "unknown" endmember, because such a spectral signature is not defined as a single spectral vector, or even a small cluster of spectral vectors. A "shade" endmember is data-defined endmembers   "virtual" endmembers

dark soil

light soil

crop

FIGURE 9-43. Class fraction maps produced by the data-defined endmembers (middle triangle, Fig. 9-42) and "virtual" endmembers (outer triangle, Fig. 9-42). All pictures have the same display LUT, so they are directly comparable.

TABLE 9-7. DN values for the 2-D unmixing example.

endmember type	band	dark soil	light soil	crop
data-defined	3	18	84	21
	4	14	72	84
virtual	3	15	14	93
	4	6	90	77

TABLE 9-8. Augmented matrices for unmixing.

endmember type	E	E^{-1}
data-defined	$\begin{bmatrix} 18 & 84 & 21 \\ 14 & 72 & 84 \\ 1 & 1 & 1 \end{bmatrix}$	$\begin{bmatrix} -0.0026991 & -0.01417 & +1.247 \\ +0.015744 & -0.00067476 & -0.27395 \\ -0.013045 & +0.014845 & +0.026991 \end{bmatrix}$
virtual	$\begin{bmatrix} 15 & 93 & 14 \\ 6 & 77 & 90 \\ 1 & 1 & 1 \end{bmatrix}$	$\begin{bmatrix} -0.0019629 & -0.011928 & +1.101 \\ +0.012683 & +0.00015099 & -0.19115 \\ -0.01072 & +0.011777 & +0.09014 \end{bmatrix}$

sometimes used to accomodate the variations in signatures arising from topography and subpixel surface texture (Adams *et al.*, 1993).

The constrained, linear unmixing example worked previously is simple and exact. More interesting unmixing challenges are presented when the number of spectral bands is greater than the number of classes, as in hyperspectral imagery. The linear mixing model then becomes an overdetermined set of K equations (K spectral bands) with L variables (L endmember fractions), where K is much greater than L. The endmember matrix is then $K \times L$ and cannot be inverted. A solution can be obtained via the pseudoinverse,

$$\hat{f}_{ij} = (E^T E)^{-1} E^T \cdot DN_{ij} , \qquad (9-33)$$

which minimizes the mean-squared error,

$$min[\varepsilon_{ij}^T \varepsilon_{ij}] = [(DN_{ij} - E\hat{f}_{ij})^T (DN_{ij} - E\hat{f}_{ij})] , \qquad (9-34)$$

in the fit of the estimated fraction mixture to the data. A map of $\|\varepsilon_{ij}\|$ at each pixel is called a *residual image*. This least-squares algorithm, Eq. (9 – 33) is quite intensive for high-dimensional hyperspectral data because the matrix-vector product in Eq. (9 – 33) must be calculated for every pixel. The matrix E is fixed for a given problem, so Eq. (9 – 33) is equivalent to a rotational transform of the *DN*, as in the PCT and TCT transforms (Chapter 5). If the additional constraints of Eq. (9 – 26) and Eq. (9 – 27) are included in the analysis, it is known as the *Constrained Least Squares (CLS)* method and requires additional mathematical techniques for solution (Shimabukuro and Smith, 1991).[14]

Specification of the endmembers can be a major problem in unmixing analysis. Clearly, accurate unmixing relies on accurate endmembers. It may be impossible to find completely pure pixels in the image data, i.e. pixels that contain only one endmember spectrum. Since we are considering subpixel content, how can we know if a pixel is "pure" or not? Even if such pixels exist, identifying them in high-dimensional hyperspectral imagery is especially difficult. The following methods for endmember specification have been proposed:

- Laboratory or field reflectance spectra (Boardman, 1990). In this case, the image data must be calibrated to reflectance (Chapter 7).

- Image pixels modeled as mixtures of library reflectance spectra (Smith *et al.*, 1990; Roberts *et al.*, 1993). Again, the data must be calibrated.

- Automated techniques based on image transforms (Full *et al.*, 1982; Boardman *et al.*, 1995)

- *K*-dimensional interactive visualization tools (Bateson and Curtiss, 1996)

Unmixing applications in remote sensing date to the earliest days of Landsat (Table 9-9). This intriguing topic, in which we attempt to make measurements beyond the "resolution limit" of the data, will continue to be important, particularly for EOS sensors such as MODIS.

9.8.2 Fuzzy Set Classification

The fuzzy set concept, in which an entity may have partial membership in more than one category, is a natural model for the unmixing problem. Two fuzzy algorithm tools for remote-sensing classification are described here: *Fuzzy C-Means (FCM)* clustering and fuzzy supervised classification.

---

[14] Note the similarity to the least-squares solution for the polynomial coefficients that best fit a set of GCPs (Chapter 7). We've assumed the noise is uncorrelated in feature space (for unmixing) and among GCPs (for warping), and that its covariance matrix is the diagonal identity matrix. If this assumption does not hold, the least-squares solution includes a dependence on C_ε (Settle and Drake, 1993).

TABLE 9-9. *Example applications of subpixel analyses.*

data	application	reference
simulated Landsat MSS from aerial MSS	crops and soil	(Nalepka and Hyde, 1972)
MSS	soil, rock and vegetation	(Marsh *et al.*, 1980)
AVHRR	temperature	(Dozier, 1981)
TM	desert vegetation	(Smith *et al.*, 1990)
AVHRR, TM	forest cover	(Cross *et al.*, 1991)
portable spectroradiometer	soil properties	(Huete and Escadafal, 1991)
MSS	forest, soil, shade	(Shimabukuro and Smith, 1991)
AVHRR, TM	vegetation/soil mapping	(Holben and Shimabukuro, 1993)
AVIRIS	vegetation, soil	(Roberts *et al.*, 1993)
simulated TM	crop types at field boundaries	(Wu and Schowengerdt, 1993)
AVHRR simulated from MSS	forest versus nonforest	(Foody and Cox, 1994)
AVHRR, TM	forest cover	(Hlavka and Spanner, 1995)
TM	tree, shrub cover	(Jasinski, 1996)

Fuzzy C-Means (FCM) clustering

This algorithm is much like the *K*-means unsupervised clustering algorithm described earlier. The essential difference is that the feature space is partitioned into fuzzy regions (Bezdek *et al.*, 1984; Cannon *et al.*, 1986). A *membership grade* matrix U is created for N pixels and L clusters,

$$U = \begin{bmatrix} u_{11} & \cdots & u_{1N} \\ \vdots & & \vdots \\ u_{L1} & \cdots & u_{LN} \end{bmatrix}. \qquad (9-35)$$

Each column of U represents the membership values of the image pixels in each of the C clusters. The following constraints also apply,

$$\sum_{n=1}^{N} u_{ln} > 0 \ , \ \sum_{l=1}^{C} u_{ln} = 1 \ , \ 0 \le u_{ln} \le 1 \ , \qquad (9-36)$$

which are similar to the constraints on fractions of endmembers in unmixing and on *a posteriori* probabilities in maximum-likelihood classification (Foody, 1992).

K-means hard clustering can be shown to minimize the squared-error function, for a fixed number of clusters *C*,

$$\varepsilon^2 = \sum_{n=1}^{N} \sum_{l=1}^{C} \left\| DN_n - \mu_l^* \right\|^2 \qquad (9-37)$$

where

$$\left\| DN_n - \mu_l^* \right\|^2 = (DN_n - \mu_l^*)^T (DN_n - \mu_l^*) \qquad (9-38)$$

is the square of the Euclidean distance L_2 from a pixel vector to the current *fuzzy mean vector* μ_l^* for cluster *l* (Jain and Dubes, 1988). To achieve a fuzzy partitioning of feature space, the function to be minimized incorporates the membership values,

$$J_m = \sum_{n=1}^{N} \sum_{l=1}^{C} u_{ln}^m \left\| DN_n - \mu_l^* \right\| , \ m \geq 1 . \qquad (9-39)$$

The parameter *m* determines the "fuzziness" of the partitioning; *m* equal to one results in a hard clustering, and values around two are typically used. The iterative adjustment process used for the *K*-means algorithm is followed (Sect. 9.4.2), with the following formulas to update the cluster means and membership values,

$$\mu_l^* = \left(\sum_{n=1}^{N} u_{ln}^m DN_n \right) \Big/ \left(\sum_{n=1}^{N} u_{ln}^m \right) \qquad (9-40)$$

$$u_{ln} = 1 \Big/ \sum_{j=1}^{C} \left[\left\| DN_n - \mu_l^* \right\| \Big/ \left\| DN_n - \mu_l^* \right\| \right]^{2/(m-1)} . \qquad (9-41)$$

These rather obscure equations are actually simple in concept. The fuzzy cluster means are simply calculated from the data samples weighted by their fuzzy membership values, and the membership values are updated by the normalized distance to the cluster means.

The Marana TM image is used to demonstrate the FCM algorithm (with an *m* value of two) in Fig. 9-44. Only three clusters are requested, so the crop areas are not distinct and there is considerable within-cluster data variation. The FCM algorithm shows similar overall cluster assignments as *K*-means, but also indicates the membership likelihood at each pixel (displayed on the same greyscale

K-means hard cluster maps   *Fuzzy C-means soft cluster maps*

cluster 1

cluster 2

cluster 3

FIGURE 9-44. Comparison of hard and fuzzy clustering results. (Software code and example courtesy of Te-shen Liang and Ho-yuen Pang, University of Arizona.)

so they can be compared). The active crop fields do not appear in any one fuzzy cluster, but have a nonzero likelihood of belonging to all three clusters. Areas of low vegetation density in the upper left appear with significant membership values in clusters 1 and 2.

Fuzzy supervised classification

This method is attributed to Wang (Wang, 1990a; Wang, 1990b). A membership matrix U is defined, as in FCM. The fuzzy mean is calculated from weighted training data as in Eq. (9 – 40) and the fuzzy covariance is calculated as,

$$C_l^* = \left[\sum_{n=1}^{N} u_{ln} (DN_n - \mu_l^*) (DN - \mu_l^*)^T \right] / \left(\sum_{n=1}^{N} u_{ln} \right) . \qquad (9-42)$$

These parameters are then used to define the fuzzy membership function as a modified Gaussian distribution,

$$U_l = P_i^*(DN) / \sum_{l=1}^{C} P_i^*(DN) \qquad (9-43)$$

where,

$$P_l^*(DN) = \frac{1}{|C_l^*|^{1/2} (2\pi)^{K/2}} e^{-(DN - \mu_l^*)^T C_l^{*-1} (DN - \mu_l^*)/2} , \qquad (9-44)$$

which is the normal distribution in K dimensions (Chapter 4).

Training data for a fuzzy supervised classification need not be homogeneous in a single class; if some pixels have known mixtures, the membership function can be used to calculate weighted fuzzy means and covariances. On the other hand, if the training data are pure in each class, the fuzzy means and covariances are the same as those in a conventional hard classification. The only difference then is that we interpret Eq. (9 – 43), given Eq. (9 – 44), as the fuzzy membership partitions of feature space. Fuzzy membership grades for each pixel can therefore be calculated from the *a posteriori* probabilities of a conventional hard maximum-likelihood classification.

9.9   Hyperspectral Image Analysis

While any of the multispectral classification techniques may be directly extended to hyperspectral imagery, that may not be desirable because

- computation costs in K dimensions are high

- more training data are required
- traditional classifiers do not necessarily exploit the greater information content of hyperspectral data

In this section we look at some of the unique tools that have been developed for hyperspectral image analysis.

9.9.1 Visualization of the Image Cube

Visualization is a persistent problem in the analysis of hyperspectral image data. The data are not only voluminous, but multidimensional (Chapter 1). One technique is to extract a line of pixels from a hyperspectral image and display the spectral values of each pixel as greylevels, resulting in an array where the x-coordinate is the pixel number within the line and the y-coordinate is the wavelength band (Fig. 9-45). This display is a *spatial spectrogram*. These at-sensor radiance data have been normalized to make the spectral information visible from 400 to 2400 nm (see Chapter 7).

Visualization of class spectral signatures is likewise difficult for hyperspectral data. An effective tool for second-order statistics is the *statistics image* (Benediktsson and Swain, 1992; Lee and Landgrebe, 1993). Examples from the Palo Alto AVIRIS data (Plate 1-2) are shown in Plate 9-2. The class covariance matrix is displayed as a pseudocolored map; the colors indicate the degree of covariance between bands, from negative to positive values. The mean signature of a set of training data can also be displayed, with a superimposed deviation in each band. The difference among the class means and covariances are clearly evident. The covariance matrix is required to distinguish among classes with similar mean signatures and some studies indicate that maximum-likelihood classifiers may be superior to nearest-mean classifiers for hyperspectral data because of information contained in the covariances (Lee and Landgrebe, 1993).

The large volume of data available permits unique dynamic visualizations, such as *spectral movies*, where the image bands are displayed sequentially in rapid succession. An eight–second "movie" results from 240 bands displayed at 30 bands/second (video frame rate). The main benefits from such a display are in visual detection of unique signatures and in rapid previewing of the data for bad bands.

9.9.2 Training for Classification

For any classifier, a sufficient number of training pixels must be used to estimate the class signature properties accurately. If a maximum-likelihood classifier is used and Gaussian class distributions are assumed, the class sample mean vectors and covariance matrices must be calculated. If K spectral or other features are

FIGURE 9-45. Visualization of hyperspectral image data by spatial spectrograms. A vertical profile of the display is the spectrum measured for the pixel in that column of the given line. These displays are equivalent to horizontal slices through the hyperspectral image cube (Fig. 1-7). Similar displays could be made along image columns or any arbitrary image profile. The horizontal noise lines are at atmospheric absorption bands. The display is much like the visualization of a single line of BIL-formatted data, shown in Fig. 1-12 of Chapter 1. The same format is also produced directly by a 2-D array imaging spectrometer.

used, the training set for each class must contain at least $K + 1$ pixels in order to calculate the sample covariance matrix. To obtain reliable class statistics, however, 10 to 100 training pixels *per class, per feature* are typically needed (Swain and Davis, 1978). The number of training pixels required for a given signature accuracy increases if there is large within-class variability.

This requirement is not too severe for multispectral imagery, where we have only a small number of bands. With hyperspectral imagery, however, it may be impossible to meet for some classes of limited spatial extent. Also, as the dimensionality of feature space increases, an accuracy degradation known as the *Hughes phenomenon* can require more training samples. Techniques for mitigating this situation have been proposed (Shahshahani and Landgrebe, 1994; Hoffbeck and Landgrebe, 1996).

9.9.3 Feature Extraction from Hyperspectral Data

The order of magnitude higher spectral resolution of hyperspectral imagery compared to multispectral imagery, creates opportunities for new feature-extraction techniques, some examples of which are described here.

Image residuals

One of the earliest feature-extraction specifications for hyperspectral data was the calculation of image "residual" spectra for mineral detection and identification (Marsh and McKeon, 1983). The intent is to remove the external factor of topographic shading and to emphasize the absorption bands of different minerals relative to an average signature without absorption features (see the discussion in Chapter 7). The process is:

1. *Divide the spectrum of each pixel by a "reference" band*, chosen to be relatively free of surface absorption features. This normalization insures that every pixel's spectral value in the reference band will be one. It also achieves topographic shading suppression (Chapter 5), particularly in the SWIR where atmospheric scattering is not a factor.
2. *Calculate the average normalized spectrum for the entire scene and subtract it from the normalized spectrum* produced in step 1. Surface absorption features that are below the average spectrum will now appear as negative "residuals." This step will be effective if the occurrence of absorption minerals is a relatively small proportion of the full scene. Otherwise, the absorption features would influence the mean spectrum, and not appear as residuals.

This feature-extraction technique is illustrated with AVIRIS data of Cuprite, Nevada (Plate 7-1). This area is highly mineralized as described in Chapter 7, containing several mineral types with distinguishing absorption bands at different wavelengths, particularly in the SWIR. For normalization, the band at 2.04 μm is used as in (Marsh and McKeon, 1983). The residual spectra were calculated for each pixel as described above. A color composite of the residual images at 2.3, 2.2 and 2.1 μm is shown in Plate 9-2. The different colors correspond well to known occurrences of alunite, buddingtonite and kaolinite (Kruse *et al.*, 1990; Abrams and Hook, 1995).

Spectral fingerprints

An interesting feature-extraction technique for hyperspectral data was developed in a series of papers by Piech and Piech (Piech and Piech, 1987; Piech and Piech, 1989). The basic concept is to locate the local points of inflection (maximum slope) in the spectral curve using scale-space filtering (Chapter 6). Absorption features lead to characteristic patterns in scale-space, termed "fingerprints." If the data are uncorrected for the atmosphere, then atmospheric absorption bands will be present in the fingerprints, along with bands due to material properties. If the data are corrected for atmospheric absorption, only material-related absorption features will be indicated.

A spectral fingerprint is calculated by convolving the spectrum with LoG filters of different widths and plotting the zero-crossings on a graph of the filter's sigma value (Eq. (6 – 38)) versus wavelength. As the scale (σ) increases, the number of zero-crossings cannot increase. This is a *scale-space* plot as originally defined in (Witkin, 1983); an example of zero-crossing determination is shown in Fig. 6-34. The logarithm of the scale at which a fingerprint loop closes is linear with the logarithm of the area of the absorption feature in the original spectrum, a relationship reminiscent of fractals (Piech and Piech, 1987).

Absorption-band parameters

The high-spectral resolution of imaging spectrometer systems creates the opportunity to identify some materials by their absorption-band characteristics. For single absorption bands, the parameters of depth, width and position can be defined (Fig. 9-48). The existence of an absorption band at a particular wavelength can be declared whenever the depth exceeds a certain threshold (Rubin, 1993); the width and position are then calculated, if desired. These features, as measured from hyperspectral imagery, can be compared to the same features derived from laboratory reflectance spectra for identification.

FIGURE 9-46. Spectral fingerprints for three mineral reflectance spectra (data from Fig. 1-6). Some characteristic absorption signatures are noted for each mineral. The closed loops indicate an absorption feature; loops that have not closed at a σ of 10 will close for a larger value of σ.

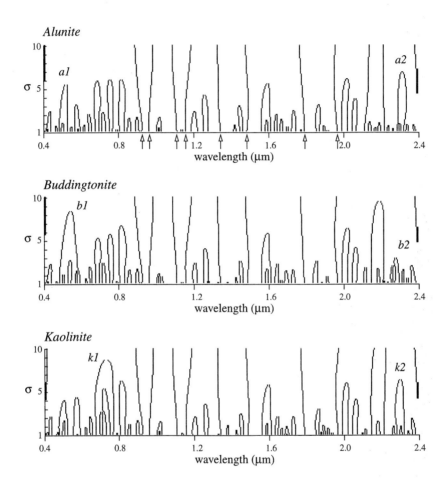

FIGURE 9-47. *Spectral fingerprints for three mineral radiance spectra (data from the AVIRIS Cuprite image in Plate 9-1). The zero-crossings arising from atmospheric absorption bands are clearly evident as similar features in each fingerprint (major bands are indicated by arrows; compare their wavelengths with those in Fig. 2-4). The distinguishing labeled features in Fig. 9-46 are seen also in these fingerprints.*

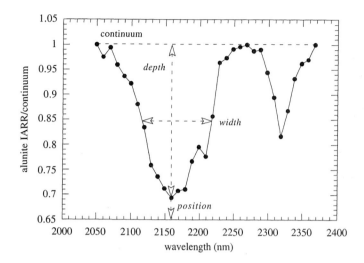

FIGURE 9-48. The definition of three absorption-band parameters. Depth is measured from continuum-corrected Internal Average Relative Reflectance (IARR) data (see Chapter 7). The width is measured at half the band depth, and the position is the wavelength at the band minimum (after (Kruse, 1988).)

Spectral derivative ratio

The derivative of a function tends to emphasize changes and suppress the mean level. This fact has been used to develop a technique that reduces the effects of atmospheric scattering and absorption on spectral signatures (Philpot, 1991). With the aid of a simple radiometric model for remote-sensing data, it can be shown that the ratio of any-order derivative of the at-sensor radiance data at two wavelengths approximately equals the ratio of the same-order derivative of the spectral reflectance,

$$\frac{d^n L/d\lambda^n\big|_{\lambda 1}}{d^n L/d\lambda^n\big|_{\lambda 2}} \approx \frac{d^n \rho/d\lambda^n\big|_{\lambda 1}}{d^n \rho/d\lambda^n\big|_{\lambda 2}}. \qquad (9-45)$$

For hyperspectral data collected in discrete wavelength bands, the first three continuous derivatives at wavelength λ can be approximated by the discrete derivatives,

$$\frac{dL}{d\lambda} \approx \frac{L(\lambda) - L(\lambda^+)}{\Delta\lambda}$$

$$\frac{d^2L}{d\lambda^2} \approx \frac{L(\lambda^-) - 2L(\lambda) + L(\lambda^+)}{(\Delta\lambda)^2}$$

$$\frac{d^3L}{d\lambda^3} \approx \frac{L(\lambda^-) - 3L(\lambda) + 3L(\lambda^+) - L(\lambda^{++})}{(\Delta\lambda)^3}$$

$(9-46)$

where λ^- is the nearest smaller wavelength, λ^+ is the nearest larger wavelength, λ^{++} is the next nearest larger wavelength, and $\Delta\lambda$ is the wavelength interval. Philpot determined the spectral ranges over which Eq. $(9-45)$ was valid for sample water and leaf spectra. The value of the approach is that radiance data, uncalibrated for the atmosphere, may be used within the spectral regions of validity to find spectral derivative signatures that will match those from reflectance data.

9.9.4 Classification Algorithms for Hyperspectral Data

In principle, any of the conventional algorithms described earlier for multispectral data can be directly applied to hyperspectral data. There are no theoretical limitations on the number of bands (or features) used by any of these algorithms. However, algorithms such as maximum-likelihood, even with efficiency improvements (Bolstad and Lillesand, 1991; Lee and Landgrebe, 1991; Jia and Richards, 1994), appear suddenly inefficient when applied to a 200-band hyperspectral image! A number of special classification tools have been developed specifically for hyperspectral imagery. Their structure is not only driven by a need for efficiency, but also by different types of pattern recognition made possible by the high-resolution spectral data.

Large, multiclass classifications, as commonly performed on TM or SPOT data, are computationally expensive for hyperspectral imagery. Therefore, three restricted classification modes have been described for imaging spectrometer data (Mazer *et al.*, 1988):

- Comparison of the spectrum of a single pixel, or the average of a group of pixels, to all pixels in the image. This is a traditional supervised classification.

- Match the spectra of all image pixels to a library reflectance spectrum. This mode is used to locate a particular material in the data and requires calibration, or at least partial normalization, to reflectance.

- Match a single spectrum from the remote-sensing data to the reflectrance library. In this case, we want to label a particular pixel or group of pixels. Remote-sensing data calibration is also required here.

It is also desirable to exclude the atmospheric water vapor absorption bands from spectral classifications. This will reduce computation time and avoid classification degradation from low *SNR* bands.

Binary encoding

The large size of hyperspectral images (about 140 MB for an AVIRIS image) and memory limits on computers prompted interest in techniques for simultaneous data reduction and pattern matching. Binary encoding of the spectral dimension was an early approach (Mazer *et al.*, 1988) that continues to be investigated and refined (Jia and Richards, 1993).

To code the spectral radiance, a single DN threshold is specified and values above the threshold are coded as one and values below as zero. A single bit can then be used in each band to code the spectrum. To illustrate, the radiance spectra from an AVIRIS image (Plate 1-2) are used in Fig. 9-49. A threshold of 700 captures the major absorption features, but soil and the bright roof of a building are indistinguishable. To aid discrimination, the derivative of the spectrum can also be encoded to zero or one, depending on whether it is negative or positive. The coded features can be made less sensitive to external factors of solar irradiance and the atmosphere if the coding is done with respect to the local spectral mean (Mazer *et al.*, 1988). The signature representation is also improved by using multiple thresholds, with however an increase in computation and the amount of data in the compressed features (Jia and Richards, 1993).

The coded spectra can be compared bit-by-bit using the *Hamming distance*, which is defined as the number of bits that are different in two binary numbers. The Hamming distances for our example are given in Table 9-10. We see that the building and soil signatures in fact differ by one bit, which is nevertheless probably below the noise level in the data. A minimum distance threshold is easily used to reduce noise effects in the classification.

Spectral-angle mapping

This classifier is like a nearest-mean classifier, using spectral-angle distance (the ANG distance measure defined in Table 9-3). Although originally developed for hyperspectral data (Kruse *et al.*, 1993), it does not use any of the special characteristics of that data and can be equally applied to multispectral data. The spectral-angle distance is independent of the magnitude of the spectral vectors, and therefore is insensitive to topographic variations (see the discussion in

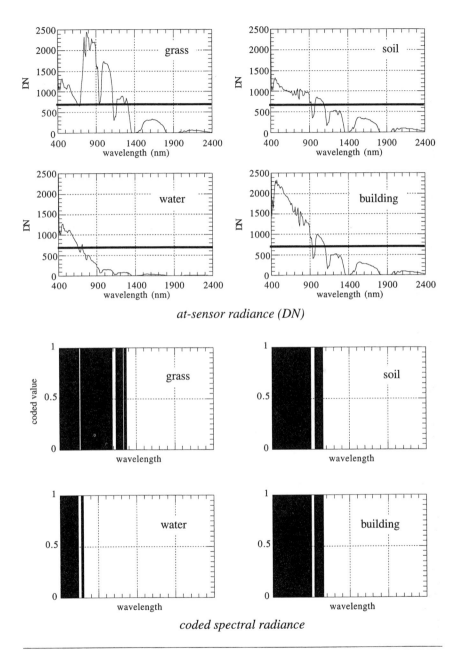

FIGURE 9-49. Binary encoding of the spectral radiance for four classes in the AVIRIS Palo Alto scene of Plate 1-2. The DN threshold is 700.

TABLE 9-10. Hamming distance table for the binary-coded spectral classes of Fig. 9-49. The matrix is symmetric.

	grass	soil	water	building
grass	0	20	58	19
soil		0	38	1
water			0	39
building				0

Sect. 4.7.1 and Fig. 4-38). The classifier can therefore be applied to remote-sensing data that have not been corrected for topography, and facilitates comparison of these data to laboratory reflectance spectra.

A general 2-D classification example is shown in Fig. 9-50. Classes b and c might represent dark and light soil, or the same soil with variations in topographic shading. In any event, they cannot be distinguished because their mean signatures lie very close to, and the class distributions are aligned with, the decision boundary.

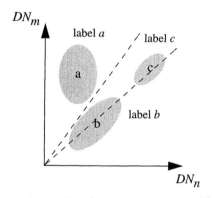

FIGURE 9-50. The spectral-angle classifier decision boundaries. Classes b and c cannot be distinguished. A threshold on the angular distance can be used to "tighten" the classification and exclude outliers, just as in the level-slice and nearest-mean classifiers.

Orthogonal subspace projection

This technique, originally derived in the context of maximizing signal-to-noise in detection of spectral signatures (Harsanyi and Chang, 1994), is equivalent to traditional spectral unmixing (Settle, 1996). The L spectral signatures of interest constitute a K x L matrix E, which may be thought of as the endmember matrix described earlier (Eq. (9 – 25)). E is assumed to consist of two parts, the first $L-1$ columns U containing $L-1$ endmember vectors and the last column containing a particular spectral signature of interest d. The optimal classification operator is then given by,

$$q^T = d^T(I - UU^\#)$$
(9 – 47)

where $U^\#$ is the pseudoinverse of U,

$$U^\# = (U^T U)^{-1} U^T ,$$
(9 – 48)

as used earlier for unmixing (Eq. (9 – 33)). The matrix $I\text{-}UU^\#$ is a *projection matrix*, P. The classifier is applied as a matrix-vector operator on an unknown pixel vector DN,

$$\alpha_P = \beta q^T DN .$$
(9 – 49)

The classifier can be viewed as "projecting" the unknown data vector onto the particular vector of interest, d, while simultaneously "nullifying" the other class signatures (Harsanyi and Chang, 1994). Larger values of α_P indicate a better match between d and DN. The scalar β is a normalizing factor equal to,

$$\beta = (d^T P d)^{-1} .$$
(9 – 50)

Although derived from a different viewpoint, this technique is mathematically equivalent to the unconstrained, least-squares fraction estimate, α_P, of the class whose endmember is d (Settle, 1996).

9.10 Summary

A wide assortment of techniques for thematic classification of remote-sensing imagery have been presented. They range from heuristic, intuitive approaches to theoretically well-founded, statistical pattern-recognition techniques. The important aspects of this chapter are:

- Thematic classification methods are either heuristic (nonparametric) or based on statistical distribution assumptions (parametric).

- It is not possible to declare one classifier better than another for all applications; their performance is strongly data-dependent if there is any overlap of class signatures. If the class distributions are Gaussian, the maximum-likelihood classifier results in minimum total misclassification error.

- Soft classifications have notable advantages over hard classifications, among them the ability to discriminate physical class continua and to provide estimates of subpixel class component fractions.

- Hyperspectral imagery from imaging spectrometers offers opportunities for new classification algorithms that can exploit high spectral resolution information.

In closing the discussion on thematic classification, we note that there are a number of other, "higher" level approaches. These methods may use external (to the imagery) knowledge to improve labeling of scene components and often include spatial context information (Tilton *et al.*, 1982; Wharton, 1982; Clément *et al.*, 1993). Approaches range from rule-based, "expert" systems (Nagao and Matsuyama, 1980; Wang *et al.*, 1983; McKeown *et al.*, 1985; Wharton, 1987; Goldberg *et al.*, 1988; Mehldau and Schowengerdt, 1990; Srinivasan and Richards, 1990; Wang, 1993) to evidential reasoning, such as the Dempster-Shafer theory (Lee *et al.*, 1987); a review of some of the work in these areas is provided in (Argialas and Harlow, 1990). Some algorithms can be applied to a label map produced by conventional classification. Such classification schemes, although they tend to produce large and complex analysis systems, appear promising for many applications, particularly those involving higher resolution imagery. They have not been discussed in detail here because they use high-level descriptive models for image content, rather than the physical models emphasized throughout this book.

9.11   Exercises

9–1. Select a Level II category from the Anderson classification scheme (Table 9-1) and expand it to Level III, using your own ideas of appropriate classes.

9–2. Suppose you have a three-band image and classify it into two classes using the nearest-mean algorithm with the Euclidean distance measure. Show mathematically that the intersection of the decision plane with the plane of any two bands is a line, but not necessarily the same line as the decision boundary obtained if only those two bands were used in the classification. What is required to make the two lines the same?

9–3. Calculate the DNs for the two decision boundaries in Fig. 9-22.

9–4. Derive the explicit mathematical form for the maximum-likelihood decision boundary between two classes, under the normal distribution assumption, in a two-dimensional feature space. What family of mathematical functions does the solution belong to? Specify some reasonable parameter values and plot the decision boundaries for the cases of:

   A. equal class means and unequal, diagonal class covariance matrices and
   B. unequal class means and equal class covariance matrices.

9–5. Explain how Eq. (9 – 8) and Eq. (9 – 9) are determined.

9–6. Explain how the derivative of the activation function (Eq. (9 – 4)) influences the iterative adjustment of weights in the back-propagation algorithm.

9–7. Develop a simple *nonlinear* mixing model similar to the linear model in Fig. 9-40. Assume each *GIFOV* consists of a partially-transmitting vegetation canopy and a soil background.

APPENDIX A

Sensor Acronyms

TABLE A–1. Some of the more common remote sensor system acronyms. The reference used is not necessarily the first for that sensor, nor the most complete, but is in either an archival journal or a readily available conference proceedings.

acronym	name	reference
ADEOS	ADvanced Earth Observing Satellite	(Kramer, 1996)
AIS	Airborne Imaging Spectrometer	(Vane *et al.*, 1984)
AOCI	Airborne Ocean Color Imager	(Wrigley *et al.*, 1992)
APT	Automatic Picture Transmission	(Bonner, 1969)
ASAS	Advanced Solid-State Array Spectroradiometer	(Irons *et al.*, 1991)
ASTER	Advanced Spaceborne Thermal Emission and Reflection Radiometer	(Asrar and Greenstone, 1995)
AVHRR	Advanced Very High Resolution Radiometer	(Kramer, 1996)
AVIRIS	Airborne Visible/InfraRed Imaging Spectrometer	(Porter and Enmark, 1987)
CZCS	Coastal Zone Color Scanner	(Kramer, 1996)
ETM	Enhanced Thematic Mapper	(Kramer, 1996)
GOES	Geostationary Operational Environmental Satellite	(Kramer, 1996)
HCMM	Heat Capacity Mapping Mission	(Short and Stuart, 1982)
HSI	HyperSpectral Imager	(Kramer, 1996)
HYDICE	HYperspectral Digital Imagery Collection Experiment	(Basedow *et al.*, 1995)
IRS	India Remote sensing Satellite	(Kramer, 1996)
JERS	Japanese Earth Resources Satellite (renamed Fuyo-1 after launch)	(Nishidai, 1993)
LISS	Linear Self Scanning Sensor	IRS
MAS	Modis Airborne Simulator	(Myers and Arvesen, 1995)
MISR	Muli-angle Imaging SpectroRadiometer	(Diner *et al.*, 1989)
MODIS	Moderate Resolution Imaging Spectroradiometer	(Salomonson *et al.*, 1989)
MSS	Multispectral Scanner System	(Lansing and Cline, 1975)

TABLE A–1. Some of the more common remote sensor system acronyms. The reference used is not necessarily the first for that sensor, nor the most complete, but is in either an archival journal or a readily available conference proceedings. (Continued)

acronym	name	reference
RBV	Return Beam Vidicon	(Kramer, 1996)
SPOT	Systeme Probatoire d'Observation	(Chevrel *et al.*, 1981)
SSM/I	Special Sensor Microwave/Imager	(Hollinger *et al.*, 1990)
TIMS	Thermal Infrared Multispectral Scanner	(Kahle and Goetz, 1983)
TM	Thematic Mapper	(Engel and Weinstein, 1983)
TMS	Thematic Mapper Simulator	(Myers and Arvesen, 1995)
VHRR	Very-High Resolution Radiometer	(Kramer, 1996)
WiFS	Wide Field Sensor	IRS-1C

APPENDIX B

Function Definitions

Several functions were used in Chapter 3 to model components of the Point Spread Function (PSF) of a sensor. In particular, the *rect* function is defined as follows,

$$rect(x/W) = \begin{array}{ll} 0 & |x/W| > 1/2 \\ 1/2 & |x/W| = 1/2 \\ 1 & |x/W| < 1/2 \end{array} \qquad (B-1)$$

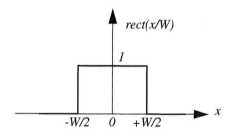

which is a square pulse of width W and amplitude one. The defined value of the function at the edges is not critical (Bracewell, 1995); we have used the definition in (Gaskill, 1978).

A separable function is defined as a two-dimensional function that is a product of two, one-dimensional functions in orthogonal directions,

$$f(x, y) = f_1(x)f_2(y) \qquad (B-2)$$

The 2-D, separable *rect* function is therefore,

$$rect(x/A, y/B) = rect(x/A)rect(y/B) \quad , \qquad (B-3)$$

which is a rectangular "box,"

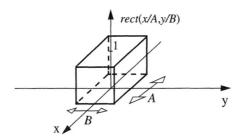

The 2-D *rect* function is a reasonable model for the spatial response of a single sensor detector element, excluding any nonuniform sensitivity across the detector or detector-to-detector "cross-talk" between elements.

Another useful function is the Gaussian ,

$$gaus(x/W) = e^{-x^2/2W^2} \qquad\qquad (B-4)$$

and its 2-D, separable form,

$$gaus(x/A, y/B) = e^{-x^2/2A^2} e^{-y^2/2B^2} \ . \qquad\qquad (B-5)$$

This function is particularly interesting because it is not only separable, but also rotationally symmetric (isotropic) if *A* equals *B*. The *gaus* function is commonly used to model the optical spatial response of a sensor. In that case, it should be normalized to unit volume under the function, as in Chapter 3.

Other examples of 2-D functions and their characteristics in separable and isotropic form are discussed in Chapter 4 in relation to modeling of spatial correlation. Finally, we note that the definition of a separable function is analogous to the definition of independent statistical variables, also discussed in Chapter 4. A multivariate distribution function is separable and equal to the product of the individual distribution functions, if the constituent variables are independent.

References

Image Gallery

Goode, J. P. (1925). "The Homolosine projection: a new device for portraying the Earth's surface entire." *Annals of the Association of American Geographers* **15**: 119-125.

Paola, J. D. and R. A. Schowengerdt (1995b). "A detailed comparison of backpropagation neural network and maximum-likelihood classifiers for urban land use classification." *IEEE Transactions on Geoscience and Remote Sensing* **33**(4): 981-996.

Short, N. M. and J. Locke M. Stuart (1982). *The Heat Capacity Mapping Mission (HCMM) Anthology*, Washington, D.C.: NASA, No. SP-465, 264 p.

Steinwand, D. R. (1994). "Mapping raster imagery to the Interrupted Goode Homolsine projection." *International Journal of Remote Sensing* **15**(17): 3463-3471.

CHAPTER 1 The Nature of Remote Sensing

Asrar, G. and R. Greenstone (1995). MTPE/EOS Reference Handbook. Greenbelt, MD, NASA/Goddard Space Flight Center, No. NP-215.

Avery, T. E. and G. L. Berlin (1992). *Fundamentals of Remote Sensing and Airphoto Interpretation*. 5th Ed. New York, NY: Macmillan Publishing Company, 472 p.

Badhwar, G. D., J. G. Carnes and W. W. Austin (1982). "Use of Landsat-derived temporal profiles for corn-soybean feature extraction and classification." *Remote Sensing of Environment* **12**(1): 57-79.

Bowker, D. E., R. E. Davis, D. L. Myrick, K. Stacy, and W. T. Jones (1985). Spectral Reflectances of Natural Targets for Use in Remote Sensing Studies, NASA, No. Reference Publication 1139.

Campbell, J. B. (1996). *Introduction to Remote Sensing*. Second Ed. New York, NY: The Guilford Press, 622 p.

Clark, B. P. (1990). "Landsat Thematic Mapper data production: a history of bulk image processing." *Photogrammetric Engineering and Remote Sensing* **56**(4): 447-451.

Clark, R. N., G. A. Swayze, A. J. Gallagher, T. V. V. King, and W. M. Calvin (1993). The U. S. Geological Survey Digital Spectral Library: Version 1: 0.2 to 3.0 microns. Denver, CO, U. S. Geological Survey, No. 93-592.

Colwell, R. N., Ed. (1983). *Manual of Remote Sensing.* Falls Church, VA: American Society for Photogrammetry and Remote Sensing.

Curlander, J. C. and R. N. McDonough (1991). *Synthetic Aperture Radar - Systems and Signal Processing.* New York, NY: John Wiley & Sons, 647 p.

Elachi, C. (1988). *Spaceborne Radar Remote Sensing: Applications and Techniques.* New York, NY: IEEE Press, 255 p.

EOSAT (1993). Fast Format Document. Lanham, Maryland: EOSAT.

Filiberti, D., S. Marsh and R. Schowengerdt (1994). "Synthesis of high spatial and spectral resolution imagery from multiple image sources." *Optical Engineering* **33**(8): 2520-2528.

Fritz, L. W. (1996). "The era of commercial earth observation satellites." *Photogrammetric Engineering and Remote Sensing* **62**(1): 39-45.

Goetz, A., G. Vane, J. E. Solomon, and B. N. Rock (1985). "Imaging spectrometry for Earth remote sensing." *Science* **228**(4704): 1147-1153.

Haralick, R. M., C. A. Hlavka, R. Yokoyama, and S. M. Carlyle (1980). "Spectral-temporal classification using vegetation phenology." *IEEE Transactions on Geoscience and Remote Sensing* **GE-18**: 167-174.

Haydn, R., G. W. Dalke, J. Henkel, and J. E. Bare (1982). "Application of the IHS color transform to the processing of multisensor data and image enhancement." In *International Symposium on Remote Sensing of Arid and Semi-Arid Lands,* Cairo, Egypt, Environmental Research Institute of Michigan: 599-616.

Hollinger, J. P., R. Lo, G. Poe, R. Savage and J. Peirce (1987). Special Sensor Microwave/Imager User's Guide. Washington, D.C., Naval Research Laboratory.

Hollinger, J. P., J. L. Peirce, and G. A. Poe (1990). "SSM/I instrument evaluation." *IEEE Transactions on Geoscience and Remote Sensing* **28**(5): 781-790.

Jensen, J. R. (1996). *Introductory Digital Image Processing - A Remote Sensing Perspective.* Second Ed. Upper Saddle River, New Jersey: Prentice Hall, 316 p.

Kramer, H. J. (1996). *Observation of the Earth and Its Environment: Survey of Missions and Sensors.* Third Ed. Berlin: Springer–Verlag, 960 p.

Landgrebe, D. A. (1978). The Quantitative Approach: Concept and Rationale. *Remote Sensing: The Quantitative Approach.* P. H. Swain and S. M. Davis, (Eds.). New York, NY: McGraw-Hill, 1-20.

Lillesand, T. M. and R. W. Kiefer (1987). *Remote Sensing and Image Interpretation.* 2nd Ed. New York, NY: John Wiley & Sons, 721 p.

Markham, B. L. and J. L. Barker, (Eds.) (1985). *Special LIDQA Issue.* Photogrammetric Engineering and Remote Sensing, American Society for Photogrammetry and Remote Sensing.

Marsh, S. E. and R. J. P. Lyon (1980). "Quantitative relationship of near-surface spectra to Landsat radiometric data." *Remote Sensing of Environment* **10**(4): 241-261.

Mather, P. M. (1987). *Computer Processing of Remotely–Sensed Images: An Introduction.* New York, NY: John Wiley & Sons, 352 p.

McDonald, R. A. (1995a). "Opening the Cold War sky to the public: declassifying satellite reconnaissance imagery." *Photogrammetric Engineering and Remote Sensing* LXI(4): 385-390.

McDonald, R. A. (1995b). "CORONA: Success for space reconnaissance, a look into the Cold War, and a revolution for intelligence." *Photogrammetric Engineering and Remote Sensing* LXI(6): 689-719.

Moik, J. G. (1980). *Digital Processing of Remotely Sensed Images.* Washington, D.C.: NASA, U.S. Government Printing Office, 330 p.

Niblack, W. (1986). *An Introduction to Digital Image Processing.* Prentice Hall International (UK) Ltd, 215 p.

Rast, M., S. J. Hook, C. D. Elvidge, and R. E. Alley (1991). "An evaluation of techniques for the extraction of mineral absorption features from high spectral resolution remote sensing data." *Photogrammetric Engineering and Remote Sensing* 57(10): 1303-1309.

Richards, J. A. (1993). *Remote Sensing Digital Image Analysis - An Introduction.* Second Ed. Berlin: Springer-Verlag, 334 p.

Rubin, T. D. (1993). "Spectral mapping with imaging spectrometers." *Photogrammetric Engineering and Remote Sensing* 59(2): 215-220.

Sabins, F. F. J. (1987). *Remote Sensing – Principles and Interpretation.* Second Ed. New York, NY: W. H. Freeman and Company, 449 p.

Salomonson, V. V., Ed. (1984). Special Issue on Landsat-4. *IEEE Transactions on Geoscience and Remote Sensing,* IEEE.

Salomonson, V. V., J. Barker, and E. Knight (1995). "Spectral characteristics of the Earth Observing System (EOS) Moderate Resolution Imaging Spectroradiometer (MODIS)." In *Imaging Spectrometry,* Orlando, FL, SPIE, vol. 2480: 142-152.

Schott, J. R. (1996). *Remote Sensing: The Image Chain Approach.* New York, NY: Oxford University Press, 394 p.

Schowengerdt, R. A. (1983). *Techniques for Image Processing and Classification in Remote Sensing.* Orlando, FL: Academic Press, 249 p.

Slater, P. N. (1980). *Remote Sensing - Optics and Optical Systems.* Reading, MA: Addison-Wesley, 575 p.

SPOTImage (1991). Reston, VA: SPOT Image Corporation.

Swain, P. H. and S. M. Davis, (Eds.) (1978). *Remote Sensing: The Quantitative Approach.* New York, NY: McGraw-Hill, 396 p.

Townshend, J., C. Justice, W. Li, C. Gurney, and J. McManus (1991). "Global land cover classification by remote sensing: present capabilities and future possibilities." *Remote Sensing of Environment* 35: 243-255.

Twomey, S., C. Bohren and J. Mergenthaler (1986). "Reflectance and albedo differences between wet and dry surfaces." *Applied Optics* 25(3): 431-437.

Vane, G. and A. F. H. Goetz (1988). "Terrestrial imaging spectroscopy." *Remote Sensing of Environment* 24: 1-29.

Vetter, R., M. Ali, M. Daily, J. Gabrynowicz, S. Narumalani, K. Nygard, W. Perrizo, P. Ram, S. Reichenbach, *et al.* (1995). "Accessing earth system science data and applications through high-bandwidth networks." *IEEE Journal on Selected Areas in Communications* **13**(5): 793-805.

Way, J. and E. A. Smith (1991). "The evolution of synthetic aperture radar systems and their progression to the EOS SAR." *IEEE Transactions on Geoscience and Remote Sensing* **29**(6): 962-985.

Welch, R. and M. Ehlers (1988). "Cartographic feature extraction with integrated SIR-B and Landsat TM images." *International Journal of Remote Sensing* **9**(5): 873-889.

Wong, F. H. and R. Orth (1980). "Registration of SEASAT-LANDSAT composite images to UTM coordinates." In *Proc. Sixth Canadian Symposium on Remote Sensing*, Halifax, Nova Scotia: 161-164.

CHAPTER 2   *Optical Radiation Models*

Berk, A., L. S. Bernstein and D. C. Robertson (1989). MODTRAN: A Moderate Resolution Model for LOWTRAN 7, U. S. Air Force Geophysics Laboratory, No. GL-TR-89-0122.

Chavez, P. S., Jr. (1988). "An improved dark-object subtraction technique for atmospheric scattering correction of multispectral data." *Remote Sensing of Environment* **24**: 459-479.

Curcio, J. A. (1961). "Evaluation of atmospheric aerosol particle size distribution from scattering measurement in the visible and infrared." *Journal of the Optical Society of America* **51**: 548-551.

Diner, D. J., C. J. Bruegge, J. V. Martonchik, T. P. Ackerman, R. Davies, S. A. W. Gerstl, H. R. Gordon, P. J. Sellers, J. Clark, *et al.* (1989). "MISR: A Multi-angle Imaging SpectroRadiometer for geophysical and climatological research from EOS." *IEEE Transactions on Geoscience and Remote Sensing* **27**(2): 200-214.

Dubayah, R. O. and J. Dozier (1986). "Orthographic terrain views using data derived from digital elevation models." *Photogrammetric Engineering and Remote Sensing* **52**(4): 509-518.

Gao, B.-C., A. F. H. Goetz and W. J. Wiscombe (1993). "Cirrus cloud detection from airborne imaging spectrometer data using the 1.38μm water vapor band." *Geophysical Research Letter* **20**(4): 301-304.

Giles, P. T., M. A. Chapman and S. E. Franklin (1994). "Incorporation of a digital elevation model derived from stereoscopic satellite imagery in automated terrain analysis." *Computers & Geosciences* **20**(4): 441-460.

Goel, N. S. (1988). "Models of vegetation canopy reflectance and their use in estimation of biophysical parameters from reflectance data." *Remote Sensing Reviews* **4**: 1-212.

Horn, B. K. P. (1981). "Hill shading and the reflectance map." *Proceedings of the IEEE* **69**(1): 14-47.

Proy, C., D. Tanre and P. Y. Deschamps (1989). "Evaluation of topographic effects in remotely sensed data." *Remote Sensing of Environment* **30**: 21-32.

Schott, J. R. (1996). *Remote Sensing: The Image Chain Approach.* New York, NY: Oxford University Press, 394 p.

Schowengerdt, R. A. (1982). "Enhanced thermal mapping with Landsat and HCMM digital data." In *Proc. Annual American Society of Photogrammetry Convention,* Denver, CO, American Society for Photogrammetry and Remote Sensing: 414-422.

Sjoberg, R. W. and B. K. P. Horn (1983). "Atmospheric effects in satellite imaging of mountainous terrain." *Applied Optics* 22(11): 1702-1716.

Slater, P. N. (1980). *Remote Sensing—Optics and Optical Systems.* Reading, MA: Addison-Wesley, 575 p.

Slater, P. N. (1996). personal communication.

Tanre, D., C. Deroo, P. Duhaut, M. Herman, J. J. Morcrette, J. Perbos, and P. Y. Deschamps (1990). "Description of a computer code to simulate the satellite signal in the solar spectrum: 5S code." *International Journal of Remote Sensing* 11: 659-668.

Teillet, P. M. and G. Fedosejevs (1995). "On the dark target approach to atmospheric correction of remotely sensed data." *Canadian Journal of Remote Sensing* 21(4): 374-387.

CHAPTER 3 Sensor Models

Anuta, P. E. (1973). Geometric correction of ERTS-1 digital multispectral scanner data, Laboratory for Applications of Remote Sensing, Purdue University, No. Information Note 103073.

Anuta, P. E., L. A. Bartolucci, M. E. Dean, D. F. Lozano, E. Malaret, C. D. McGillem, J. A. Valdes, and C. R. Valenzuela (1984). "LANDSAT-4 MSS and Thematic Mapper data quality and information content analysis." *IEEE Transactions on Geoscience and Remote Sensing* **GE-22**(3): 222-236.

Bachmann, M. and J. Bendix (1992). "An improved algorithm for NOAA-AVHRR image referencing." *International Journal of Remote Sensing* 13(16): 3205-3215.

Bernstein, R., J. B. Lospiech, H. J. Myers, H. G. Kolsky, and R. D. Lees (1984). "Analysis and processing of Landsat-4 sensor data using advanced image processing techniques and technologies." *IEEE Transactions on Geoscience and Remote Sensing* **GE-22**(3): 192-221.

Borgeson, W. T., R. M. Batson and H. H. Kieffer (1985). "Geometric accuracy of Landsat-4 and Landsat-5 Thematic Mapper images." *Photogrammetric Engineering and Remote Sensing* 51(12): 1893-1898.

Brush, R. J. H. (1985). "A method for real-time navigation of AVHRR imagery." *IEEE Transactions on Geoscience and Remote Sensing* **GE-23**(6): 876-887.

Brush, R. J. H. (1988). "The navigation of AVHRR imagery." *International Journal of Remote Sensing* 9(9): 1491-1502.

Bryant, N. A., A. L. Zobrist, R. E. Walker, and B. Gokhman (1985). "An analysis of Landsat Thematic Mapper P-product internal geometry and conformity to Earth surface geometry." *Photogrammetric Engineering and Remote Sensing* 51(9): 1435-1447.

Chavez, P. S., Jr. (1989). "Use of the variable gain settings on SPOT." *Photogrammetric Engineering and Remote Sensing* **55**(2): 195-201.

Chen, L. C. and L. H. Lee (1993). "Rigorous generation of digital orthophotos from SPOT images." *Photogrammetric Engineering and Remote Sensing* **59**(5): 655-661.

Dereniak, E. L. and G. D. Boreman (1996). *Infrared Detectors and Systems.* New York, NY, Wiley-Interscience, 560 p.

Desachy, J., G. Begni, B. Boissin, and J. Perbos (1985). "Investigation of Landsat-4 Thematic Mapper line-to-line and band-to-band registration and relative detector calibration." *Photogrammetric Engineering and Remote Sensing* **51**(9): 1291-1298.

Ehlers, M. and R. Welch (1987). "Stereocorrelation of Landsat TM images." *Photogrammetric Engineering and Remote Sensing* **53**: 1231–1237.

Emery, W. J., J. Brown and Z. P. Nowak (1989). "AVHRR image navigation: summary and review." *Photogrammetric Engineering and Remote Sensing* **55**(8): 1175-1183.

Emery, W. J. and M. Ikeda (1984). "A comparison of geometric correction nethods of AVHRR imagery." *Canadian Journal of Remote Sensing* **10**: 46-56.

Forrest, R. B. (1981). "Simulation of orbital image–sensor geometry." *Photogrammetric Engineering and Remote Sensing* **47**(8): 1187-1193.

Friedmann, D. E., J. P. Friedel, K. L. Magnussen, R. Kwok, and S. Richardson (1983). "Multiple scene precision rectification of spaceborne imagery with very few ground control points." *Photogrammetric Engineering and Remote Sensing* **49**(12): 1657-1667.

Fritz, L. W. (1996). "The era of commercial earth observation satellites." *Photogrammetric Engineering and Remote Sensing* **62**(1): 39-45.

Fusco, L., U. Frei and A. Hsu (1985). "Thematic Mapper: Operational activities and sensor performance at ESA/Earthnet." *Photogrammetric Engineering and Remote Sensing* **51**(9): 1299-1314.

Ho, D. and A. Asem (1986). "NOAA AVHRR image referencing." *International Journal of Remote Sensing* **7**: 895-904.

Justice, C. O., B. L. Markham, J. R. G. Townshend, and R. L. Kennard (1989). "Spatial degradation of satellite data." *International Journal of Remote Sensing* **10**(9): 1539-1561.

Krasnopolsky, V. M. and L. C. Breaker (1994). "The problem of AVHRR image navigation revisited." *International Journal of Remote Sensing* **15**(5): 979-1008.

Kratky, V. (1989). "On-line aspects of sterophotogrammetric processing of SPOT images." *Photogrammetric Engineering and Remote Sensing* **55**(3): 311-316.

Legeckis, R. and J. Pritchard (1976). Algorithm for correcting the VHRR imagery for geometric distortions due to the earth curvature, earth rotation and spacecraft roll attitude errors. Washington, D.C., NOAA, No. NESS 77.

Maling, D. H. (1992). *Coordinate Systems and Map Projections.* Second Ed. Oxford, England: Pergamon Press, 476 p.

Markham, B. L. (1985). "The Landsat sensors' spatial responses." *IEEE Transactions on Geoscience and Remote Sensing* **GE-23**(6): 864-875.

Markham, B. L. and J. L. Barker (1983). "Spectral characterization of the Landsat-4 MSS sensors." *Photogrammetric Engineering and Remote Sensing* **49**(6): 811-833.

Moik, J. G. (1980). *Digital Processing of Remotely Sensed Images.* Washington, D.C.: NASA, U.S. Government Printing Office, 330 p.

Moreno, J. F., S. Gandía, and J. Meliá (1992). "Geometric integration of NOAA AVHRR and SPOT data: low resolution effective parameters from high resolution data." *IEEE Transactions on Geoscience and Remote Sensing* **30**(5): 1006-1014.

Moreno, J. F. and J. Meliá (1993). "A method for accurate geometric correction of NOAA AVHRR HRPT data." *IEEE Transactions on Geoscience and Remote Sensing* **31**(1): 204-226.

Palmer, J. M. (1984). "Effective bandwidths for LANDSAT-4 and LANDSAT-D' multispectral scanner and Thematic Mapper subsystems." *IEEE Transactions on Geoscience and Remote Sensing* **GE-22**(3): 336-338.

Park, S. K., R. Schowengerdt, and M.-A. Kaczynski (1984). "Modulation-transfer-function analysis for sampled imaging systems." *Applied Optics* **23**(15): 2572-2582.

Park, S. K. and R. A. Schowengerdt (1982). "Image sampling, reconstruction, and the effect of sample–scene phasing." *Applied Optics* **21**(17): 3142-3151.

Puccinelli, E. F. (1976). "Ground location of satellite scanner data." *Photogrammetric Engineering and Remote Sensing* **42**(4): 537-543.

Richards, J. A. (1993). *Remote Sensing Digital Image Analysis - An Introduction.* Second Ed. Berlin: Springer-Verlag, 334 p.

Salamonowicz, P. H. (1986). "Satellite orientation and position for geometric correction of scanner imagery." *Photogrammetric Engineering and Remote Sensing* **52**(4): 491-499.

Sawada, N., M. Kidode, H. Shinoda, H. Asada, M. Iwanaga, S. Watanabe, K.-I. Mori, and M. Akiyama (1981). "An analytic correction method for satellite MSS geometric distortions." *Photogrammetric Engineering and Remote Sensing* **47**(8): 1195-1203.

SBRC (1984). Thematic Mapper - Design Through Flight Evaluation. Goleta, CA: Hughes Santa Barbara Research Center, No. NAS5-24200.

Schowengerdt, R. A., C. Archwamety, and R. C. Wrigley (1985). "Landsat Thematic Mapper image-derived MTF." *Photogrammetric Engineering and Remote Sensing* **51**(9): 1395-1406.

Schowengerdt, R. A., S. K. Park, and R. T. Gray (1984). "Topics in the two–dimensional sampling and reconstruction of images." *International Journal of Remote Sensing* **5**(2): 333-347.

Seto, Y. (1991). "Geometric correction algorithms for satellite imagery using a bi–directional scanning sensor." *IEEE Transactions on Geoscience and Remote Sensing*: 292-299.

Slater, P. N. (1979). "A re-examination of the Landsat MSS." *Photogrammetric Engineering and Remote Sensing* **45**(1): 1479-1485.

Slater, P. N. (1980). *Remote Sensing - Optics and Optical Systems.* Reading, MA: Addison-Wesley, 575 p.

Steiner, D. and M. E. Kirby (1976). "Geometrical referencing of Landsat images by affine transformation and overlaying of map data." *Photogrammetria* **33**: 41-75.

Tilton, J. C., B. L. Markham, and W. L. Alford (1985). "Landsat-4 and Landsat-5 MSS coherent noise: characterization and removal." *Photogrammetric Engineering and Remote Sensing* **51**(9): 1263-1279.

USGS/NOAA (1984). Landsat 4 Data Users Handbook. Alexandria, VA: U. S. Geological Survey and National Oceanic and Atmospheric Administration.

Verdebout, J., S. Jacquemoud, and G. Schmuck (1994). Optical properties of leaves: modelling and experimental studies. *Imaging Spectrometry - A Tool for Environmental Observations.* J. Hill and J. Megier, (Eds.). Dordrecht, The Netherlands: Kluwer Academic Publishers. 4: 335p.

Walker, R. E., A. L. Zobrist, N. A. Bryant, B. Gohkman, S. Z. Friedman, and T. L. Logan (1984). "An analysis of Landsat-4 Thematic Mapper geometric properties." *IEEE Transactions on Geoscience and Remote Sensing* **GE-22**(3): 288-293.

Welch, R., T. R. Jordan, and M. Ehlers (1985). "Comparative evaluations of the geodetic accuracy and cartographic potential of Landsat-4 and Landsat-5 Thematic Mapper image data." *Photogrammetric Engineering and Remote Sensing* **51**(11): 1799-1812.

Welch, R. and E. L. Usery (1984). "Cartographic accuracy of Landsat-4 MSS and TM image data." *IEEE Transactions on Geoscience and Remote Sensing* **GE-22**(3): 281-288.

Wessman, C. A. (1994). Estimating canopy biochemistry through imaging spectrometry. *Imaging Spectrometry - A Tool for Environmental Observations.* J. Hill and J. Megier, (Eds.). Dordrecht, The Netherlands: Kluwer Academic Publishers. 4: .

Westin, T. (1990). "Precision rectification of SPOT imagery." *Photogrammetric Engineering and Remote Sensing* **56**(2): 247-253.

Westin, T. (1992). "Inflight calibration of SPOT CCD detector geometry." *Photogrammetric Engineering and Remote Sensing* **58**(9): 1313-1319.

Wolf, P. R. (1983). *Elements of Photogrammetry.* Second Ed. , McGraw-Hill, 628 p.

Wong, K. W. (1975). "Geometric and cartographic accuracy of ERTS-1 imagery." *Photogrammetric Engineering and Remote Sensing* **41**: 621-635.

Wrigley, R. C., C. A. Hlavka, D. H. Card, and J. S. Buis (1985). "Evaluation of Thematic Mapper interband registration and noise characteristics." *Photogrammetric Engineering and Remote Sensing* **51**(9): 1417-1425.

CHAPTER 4 Data Models

Adams, J. B., M. O. Smith, and A. R. Gillespie (1993). "Imaging Spectroscopy: Interpretation Based on Spectral Mixture Analysis." *Remote Geochemical Analysis: Elemental and Mineralogical Composition.* C. M. Pieters and P. A. Englert, (Eds.) Cambridge: Cambridge University Press: 145-166.

Atkinson, P. M. (1993). "The effect of spatial resolution on the experimental variogram of airborne MSS imagery." *International Journal of Remote Sensing* **14**(5): 1005-1011.

Carr, J. R. (1995). *Numerical Analysis for the Geological Sciences.* Englewood Cliffs, NJ: Prentice Hall, 592 p.

Castleman, K. R. (1996). *Digital Image Processing.* Englewood Cliffs, NJ: Prentice Hall, 667 p.

Civco, D. L. (1989). "Topographic normalization of Landsat Thematic Mapper digital imagery." *Photogrammetric Engineering and Remote Sensing* **55**(9): 1303-1309.

Conese, C., M. A. Gilabert, F. Maselli, and L. Bottai (1993). "Topographic normalization of TM scenes through the use of an atmospheric correction method and digital terrain models." *Photogrammetric Engineering and Remote Sensing* **59**(12): 1745-1753.

Cornsweet, T. N. (1970). *Visual Perception.* New York, NY: Academic Press, 475 p.

Curran, P. J. (1988). "The semivariogram in remote sensing: an introduction." *Remote Sensing of Environment* **24**: 493-507.

Curran, P. J. and J. L. Dungan (1989). "Estimation of signal-to-noise: a new procedure applied to AVIRIS data." *IEEE Transactions on Geoscience and Remote Sensing* **27**(5): 620-628.

Djamdji, J.-P. and A. Bijaoui (1995). "Disparity analysis: A wavelet transform approach." *IEEE Transactions on Geoscience and Remote Sensing* **33**(1): 67-76.

Eliason, P. T., L. A. Soderblom and J. P. S. Chavez (1981). "Extraction of topographic and spectral albedo information from multispectral images." *Photogrammetric Engineering and Remote Sensing* **48**(11): 1571-1579.

Feder, J. (1988). *Fractals.* New York, NY: Plenum Press, 283 p.

Fukunaga, K. (1990). *Introduction to Statistical Pattern Recognition.* Second Ed. San Diego: Academic Press, 591 p.

Gaddis, L. R., L. A. Soderblom, H. H. Kieffer, K. J. Becker, J. Torson, and K. Mullins (1996). "Decomposition of AVIRIS spectra: Extraction of surface-reflectance, atmospheric, and instrumental components." *IEEE Transactions on Geoscience and Remote Sensing* **34**(1): 163-178.

Gohin, F. and G. Langlois (1993). "Using geostatistics to merge in situ measurements and remotely-sensed observations of sea surface temperature." *International Journal of Remote Sensing* **14**(1): 9-19.

Gonzalez, R. C. and R. E. Woods (1992). *Digital Image Processing.* Reading, MA: Addison-Wesley, 703 p.

Haralick, R. M., K. Shanmugan, and I. Dinstein (1973). "Textural features for image classification." *IEEE Transactions on Systems, Man, and Cybernetics* **SMC-3**: 610-621.

Holben, B. N. and C. O. Justice (1980). "The topographic effect on spectral response from nadir pointing sources." *Photogrammetric Engineering and Remote Sensing* **46**(9): 1191-1200.

Holben, B. N. and C. O. Justice (1981). "An examination of spectral band ratioing to reduce the topographic effect on remotely sensed data." *International Journal of Remote Sensing* **2**(2): 115-133.

Isaaks, E. H. and R. M. Srivastava (1989). *An Introduction to Applied Geostatistics.* New York, Oxford University Press, 561 p.

Itten, K. I. and P. Meyer (1993). "Geometric and radiometric correction of TM data of mountainous forested areas." *IEEE Transactions on Geoscience and Remote Sensing* **31**(4): 764-770.

Jain, A. K. (1989). *Fundamentals of Digital Image Processing.* Englewood Cliffs, NJ: Prentice Hall, 569 p.

Jasinski, M. F. and P. S. Eagleson (1989). "The structure of red-infrared scattergrams of semivegetated landscapes." *IEEE Transactions on Geoscience and Remote Sensing* **27**(4): 441-451.

Jasinski, M. F. and P. S. Eagleson (1990). "Estimation of subpixel vegetation cover using red-infrared scattergrams." *IEEE Transactions on Geoscience and Remote Sensing* **28**(2): 253-267.

Journel, A. G. and C. J. Huijbregts (1978). *Mining Geostatistics.* London: Academic Press.

Jupp, D. L. B., A. H. Strahler, and C. E. Woodcock (1989a). "Autocorrelation and regularization in digital images I. basic theory." *IEEE Transactions on Geoscience and Remote Sensing* **26**: 463-473.

Jupp, D. L. B., A. H. Strahler, and C. E. Woodcock (1989b). "Autocorrelation and regularization in digital images II. simple image models." *IEEE Transactions on Geoscience and Remote Sensing* **27**: 247-258.

Justice, C. O., S. W. Wharton, and B. N. Holben (1981). "Application of digital terrain data to quantify and reduce the topographic effect on Landsat data." *International Journal of Remote Sensing* **2**(3): 213-230.

Kawata, Y., S. Ueno, and T. Kusaka (1988). "Radiometric correction for atmospheric and topographic effects on Landsat MSS images." *International Journal of Remote Sensing* **9**(4): 729-748.

Lacaze, B., S. Rambal, and T. Winkel (1994). "Identifying spatial patterns of Mediterranean landscapes from geostatistical anlaysis of remotely-sensed data." *International Journal of Remote Sensing* **15**(12): 2437-2450.

Mandelbrot, B. B. (1967). "How long is the coast of Britain? Statistical self-similarity and fractal dimension." *Science* **155**: 636-638.

Mandelbrot, B. B. (1983). *The Fractal Geometry of Nature.* New York, NY: W. H. Freeman and Company, 468 p.

Peli, T. (1990). "Multiscale fractal theory and object characterization." *Journal of the Optical Society of America* **7**(6): 1101-1112.

Pelig, S., J. Naor, R. Hartley, and D. Avnir (1984). "Multiple resolution texture analysis and classification." *IEEE Transactions on Pattern Analysis and Machine Intelligence* **PAMI-6**(4): 518-523.

Pentland, A. P. (1984). "Fractal-based description of natural scenes." *IEEE Transactions on Pattern Analysis and Machine Intelligence* **PAMI-6**(6): 661-674.

Pratt, W. K. (1991). *Digital Image Processing.* Second Ed. New York, NY: John Wiley & Sons, 698 p.

Press, W. H., B. P. Flannery, S. A. Teukolsky, and W. T. Vetterling (1986). *Numerical Recipes–The Art of Scientific Computing.* First Ed. Cambridge: Cambridge University Press, 818 p.

Proy, C., D. Tanre, and P. Y. Deschamps (1989). "Evaluation of topographic effects in remotely sensed data." *Remote Sensing of Environment* **30**: 21-32.

Richards, J. A. (1993). *Remote Sensing Digital Image Analysis–An Introduction.* Second Ed. Berlin: Springer-Verlag, 334 p.

Rossi, R. E., J. L. Dungan, and L. R. Beck (1994). "Kriging in shadows: geostatistical interpolation for remote sensing." *Remote Sensing of Environment* **48**: 1-25.

Schott, J. R. (1996). *Remote Sensing: The Image Chain Approach.* New York, NY: Oxford University Press, 394 p.

Teillet, P. M., B. Guindon, and D. G. Goodenough (1982). "On the slope-aspect correction of multispectral scanner data." *Canadian Journal of Remote Sensing* **8**(2): 84-106.

Townshend, J. R. G., C. O. Justice, C. Gurney, and J. McManus (1992). "The impact of misregistration on change detection." *IEEE Transactions on Geoscience and Remote Sensing* **30**(5): 1054-1060.

Wald, L. (1989). "Some examples of the use of structure functions in the analysis of satellite images of the ocean." *Photogrammetric Engineering and Remote Sensing* **55**(10): 1487-1490.

Woodcock, C. E., A. H. Strahler, and D. L. B. Jupp (1988a). "The use of variograms in remote sensing: I scene models and simulated images." *Remote Sensing of Environment* **25**: 324-348.

Woodcock, C. E., A. H. Strahler, and D. L. B. Jupp (1988b). "The use of variograms in remote sensing: II real digital images." *Remote Sensing of Environment* **25**: 349-379.

CHAPTER 5 *Spectral Transforms*

Avery, T. E. and G. L. Berlin (1992). *Fundamentals of Remote Sensing and Airphoto Interpretation.* 5th Ed. New York, NY: Macmillan Publishing Company, 472 p.

Byrne, G. F., P. F. Crapper, and K. K. Mayo (1980). "Monitoring land-cover change by principal component analysis of multitemporal Landsat data." *Remote Sensing of Environment* **10**: 175-184.

Chavez, P. S., Jr., G. L. Berlin, and L. B. Sowers (1982). "Statistical method for selecting Landsat MSS ratios." *Journal of Applied Photographic Engineering* **8**(1): 23-30.

Crist, E. P. and R. C. Cicone (1984). "A physically-based transformation of Thematic Mapper data — the TM Tasseled Cap." *IEEE Transactions on Geoscience and Remote Sensing* **GE-22**: 256-263.

Crist, E. P., R. Laurin, and R. C. Cicone (1986). "Vegetation and soils information contained in transformed Thematic Mapper data." in: *IGARSS' 86*, Zurich, ESA Publications Division, vol. ESA SP-254: 1465-1472.

Crist, E. P. (1996). personal communication.

Dallas, W. J. and W. Mauser (1980). "Preparing pictures for visual comparison." *Applied Optics* **19**(21): 3586–3587.

Durand, J. M. and Y. H. Kerr (1989). "An improved decorrelation method for the efficient display of multispectral data." *IEEE Transactions on Geoscience and Remote Sensing* **27**(5): 611-619.

Eastman, J. R. and M. Fulk (1993). "Long sequence time series evaluation using standardized principal components." *Photogrammetric Engineering and Remote Sensing* **59**(6): 991-996.

Fahnestock, J. D. and R. A. Schowengerdt (1983). "Spatially-variant contrast enhancement using local range modification." *Optical Engineering* **22**(3): 378-381.

Fukunaga, K. (1990). *Introduction to Statistical Pattern Recognition.* Second Ed. San Diego: Academic Press, 591 p.

Fung, T. and E. LeDrew (1987). "Application of principal components analysis to change detection." *Photogrammetric Engineering and Remote Sensing* **53**(12): 1649-1658.

Gillespie, A. R., A. B. Kahle, and R. E. Walker (1986). "Color enhancement of highly correlated images. I. Decorrelation and HSI contrast stretches." *Remote Sensing of Environment* **20**: 209-235.

Gonzalez, R. C. and R. E. Woods (1992). *Digital Image Processing.* Reading, MA: Addison-Wesley, 703 p.

Green, A. A., M. Berman, P. Switzer, and M. D. Craig (1988). "A transformation for ordering multispectral data in terms of image quality with implications for noise removal." *IEEE Transactions on Geoscience and Remote Sensing* **26**(1): 65-74.

Haydn, R., G. W. Dalke, J. Henkel, and J. E. Bare (1982). "Application of the IHS color transform to the processing of multisensor data and image enhancement." in: *International Symposium on Remote Sensing of Arid and Semi-Arid Lands,* Cairo, Egypt: Environmental Research Institute of Michigan: 599-616.

Huete, A. R. (1988). "A Soil Adjusted Vegetation Index (SAVI)." *Remote Sensing of Environment* **25**: 295-309.

Huete, A. R. and R. D. Jackson (1987). "Suitability of spectral indices for evaluating vegetation characteristics on arid rangelands." *Remote Sensing of Environment* **23**: 213-232.

Ingebritsen, S. E. and R. J. P. Lyon (1985). "Principal components analysis of multitemporal image pairs." *International Journal of Remote Sensing* **6**: 687-696.

Jackson, R. D. (1983). "Spectral indices in N-space." *Remote Sensing of Environment* **13**(5): 409-421.

Justice, C. O., J. R. G. Townshend, and V. L. Kalb (1991). "Representation of vegetation by continental data sets derived from NOAA-AVHRR data." *International Journal of Remote Sensing* **12**(5): 999-1021.

Kauth, R. J. and G. S. Thomas (1976). "The Tasselled Cap – A graphic description of the spectral–temporal development of agricultural crops as seen by Landsat." in: *Symposium on Machine Processing of Remotely Sensed Data,* IEEE, vol. 76CH 1103–1MPRSD: 41-51.

Kruse, F. A., K. S. Kierein-Young, and J. W. Boardman (1990). "Mineral mapping at Cuprite, Nevada with a 63-channel imaging spectrometer." *Photogrammetric Engineering and Remote Sensing* **56**(1): 83-92.

Lavreau, J. (1991). "De-hazing Landsat Thematic Mapper images." *Photogrammetric Engineering and Remote Sensing* **57**(10): 1297-1302.

Lee, J. B., A. S. Woodyatt, and M. Berman (1990). "Enhancement of high spectral resolution remote-sensing data by a noise-adjusted principal components transform." *IEEE Transactions on Geoscience and Remote Sensing* **28**(3): 295-304.

Pizer, S. M., E. P. Amburn, J. D. Austin, R. Cromartie, A. Geselowitz, T. Greer, B. H. Romeny, J. B. Zimmerman, and K. Zuiderveld (1987). "Adaptive histogram equalization and its variations." *Computer Vision, Graphics and Image Processing* **39**: 355-368.

Pun, T. (1981). "Entropic thresholding, a new approach." *Computer Graphics and Image Processing* **16**: 210-239.

Ready, P. J. and P. A. Wintz (1973). "Information extraction, SNR improvement and data compression in multispectral imagery." *IEEE Transactions on Communications* **COM-21**(10): 1123-1131.

Richards, J. A. (1984). "Thematic mapping from multitemporal image data using the principal components transformation." *Remote Sensing of Environment* **16**: 35-46.

Richards, J. A. (1993). *Remote Sensing Digital Image Analysis–An Introduction.* Second Ed. Berlin: Springer-Verlag, 334 p.

Richardson, A. J. and C. L. Wiegand (1977). "Distinguishing vegetation from soil background information." *Photogrammetric Engineering and Remote Sensing* **43**: 1541-1552.

Rothery, D. A. and G. A. Hunt (1990). "A simple way to perform decorrelation stretching and related techniques on menu-drive image processing systems." *International Journal of Remote Sensing* **11**(1): 133-137.

Schowengerdt, R. A. (1983). *Techniques for Image Processing and Classification in Remote Sensing.* Orlando, FL: Academic Press, 249 p.

Singh, A. and A. Harrison (1985). "Standardized principal components." *International Journal of Remote Sensing* **6**(6): 883-896.

Smith, A. R. (1978). "Color gamut transform pairs." in: *ACM-SIGGRAPH*, ACM, vol. 12: 12-19.

Thompson, D. R. and O. A. Whemanen (1980). "Using Landsat digital data to detect moisture stress in corn-soybean growing regions." *Photogrammetric Engineering and Remote Sensing* **46**(8): 1087-1093.

Townshend, J. R. G. and C. O. Justice (1986). "Analysis of the dynamics of African vegetation using the normalized difference vegetation index." *International Journal of Remote Sensing* **7**(11): 1435-1445.

Tucker, C. J. and P. C. Sellers (1986). "Satellite remote sensing of primary production." *International Journal of Remote Sensing* **7**(11): 1395-1416.

CHAPTER 6 Spatial Transforms

Brigham, E. O. (1988). *The Fast Fourier Transform and Its Applications.* Englewood Cliffs, NJ: Prentice Hall, 448 p.

Burt, P. J. (1981). "Fast filter transforms for image processing." *Computer Graphics and Image Processing* **16**: 20-51.

Burt, P. J. and E. H. Adelson (1983). "The laplacian pyramid as a compact image code." *IEEE Transactions on Communications* **31**(4): 532-540.

Castleman, K. R. (1996). *Digital Image Processing*. Englewood Cliffs, NJ: Prentice Hall, 667 p.

Cohen, L. (1989). "Time-frequency distributions–A review." *Proceedings of the IEEE* **77**(7): 941-981.

Daubechies, I. (1988). "Orthonormal bases of compactly supported wavelets." *Communications on Pure and Applied Mathematics* **41**: 909-996.

Davis, L. S. (1975). "A survey of edge detection techniques." *Computer Graphics and Image Processing* **4**: 248-270.

Giardina, C. R. and E. R. Dougherty (1988). *Morphological Methods in Image and Signal Processing*. Englewood Cliffs, NJ: Prentice Hall, 321 p.

Grossman, A. and J. Morlet (1984). "Decomposition of Hardy functions into square integrable wavelets of constant shape." *SIAM Journal of Applied Mathematics* **15**: 723-736.

Hunt, B. R. and T. M. Cannon (1976). "Nonstationary assumptions for Gaussian models of images." *IEEE Transactions on Systems, Man & Cybernetics* **SMC-6**: 876-882.

Jensen, J. R. (1996). *Introductory Digital Image Processing - A Remote Sensing Perspective*. Second Ed. Upper Saddle River, New Jersey, Prentice Hall, 316 p.

Lee, J.-S. (1980). "Digital image enhancement and noise filtering by use of local statistics." *IEEE Transactions on Pattern Analysis and Machine Intelligence* **PAMI-2**(2): 165-168.

Levine, M. D. (1985). *Vision in Man and Machine*. New York, NY: McGraw-Hill.

Mallat, S. (1991). "Zero-crossings of a wavelet transform." *IEEE Transactions on Information Theory* **37**(4): 1019-1033.

Mallat, S. G. (1989). "A theory for multiresolution signal decomposition: the wavelet representation." *IEEE Transactions on Pattern Analysis and Machine Intelligence* **11**(7): 674-693.

Marr, D. (1982). *Vision*. New York, NY: W. H. Freeman and Company, 397 p.

Marr, D. and E. Hildreth (1980). "Theory of edge detection." *Proceedings of the Royal Society of London* **207**: 187-217.

McDonnell, M. J. (1981). "Box-filtering techniques." *Computer Graphics and Image Processing* **17**(1): 65-70.

Oppenheim, A. V. and J. S. Lim (1981). "The importance of phase in signals." *Proceedings of the IEEE* **69**(5): 529-541.

Pratt, W. K. (1991). *Digital Image Processing*. Second Ed. New York, NY: John Wiley & Sons, 698 p.

Press, W. H., B. P. Flannery, S. A. Teukolsky, and W. T. Vetterling (1986). *Numerical Recipes–The Art of Scientific Computing*. First Ed. Cambridge: Cambridge University Press, 818 p.

Robinson, G. S. (1977). "Detection and coding of edges using directional masks." *Optical Engineering* **16**(6).

Schalkoff, R. J. (1989). *Digital Image Processing and Computer Vision.* New York, NY: John Wiley & Sons, 489 p.

Schott, J. R. (1996). *Remote Sensing: The Image Chain Approach.* New York, NY: Oxford University Press, 394 p.

Serra, J. (1982). *Image Analysis and Mathematical Morphology.* New York, NY: Academic Press.

Shensa, M. J. (1992). "The discrete wavelet transform: Wedding the a trous and Mallat algorithms." *IEEE Transactions on Signal Processing* **40**: 2464-2482.

Wallis, R. (1976). "An approach to the space variant restoration and enhancement of images." in: *Proc. Symposium on Current Mathematical Problems in Image Science*, Monterey, CA: Naval Post-graduate School.

CHAPTER 7 Correction and Calibration

Abramson, S. B. and R. A. Schowengerdt (1993). "Evaluation of edge–preserving smoothing filters for digital image mapping." *ISPRS Journal of Photogrammetry and Remote Sensing* **48**(2): 2-17.

Ahern, F. J., D. G. Goodenough, S. C. Jain, V. R. Rao, and G. Rochon (1977). "Use of clear lakes as standard reflectors for atmospheric measurements." in: *Eleventh International Symposium on Remote Sensing of Environment*, Ann Arbor, MI: Environmental Research Institute of Michigan, : 583-594.

Algazi, V. R. and G. E. Ford (1981). "Radiometric equalization of nonperiodic striping in satellite data." *Computer Graphics and Image Processing* **16**: 287-295.

Anuta, P. E. (1973). Geometric correction of ERTS-1 digital multispectral scanner data, Laboratory for Applications of Remote Sensing, Purdue University, No. Information Note 103073.

Bernstein, R., J. B. Lospiech, H. J. Myers, H. G. Kolsky, and R. D. Lees (1984). "Analysis and processing of Landsat-4 sensor data using advanced image processing techniques and technologies." *IEEE Transactions on Geoscience and Remote Sensing* **GE-22**(3): 192-221.

Büttner, G. and A. Kapovits (1990). "Characterization and removal of horizontal striping from SPOT panchromatic imagery." *International Journal of Remote Sensing* **11**(2): 359-366.

Carrere, V. and J. E. Conel (1993). "Recovery of atmospheric water vapor total column abundance from imaging spectrometer data around 940 nm–sensitivity analysis and application to Airborne Visible/Infrared Imaging Spectrometer (AVIRIS) data." *Remote Sensing of Environment* **44**: 179-204.

Castleman, K. R. (1996). *Digital Image Processing.* Englewood Cliffs, NJ: Prentice Hall, 667 p.

Centeno, J. A. S. and V. Haertel (1995). "Adaptive low-pass fuzzy filter for noise removal." *Photogrammetric Engineering and Remote Sensing* **61**(10): 1267-1272.

Chavez, P. S., Jr. (1975). "Simple high-speed digital image processing to remove quasi-coherent noise patterns." in: *41st Annual Meeting*, Washington, D.C., American Society of Photogrammetry: 595-600.

Chavez, P. S., Jr. (1988). "An improved dark-object subtraction technique for atmospheric scattering correction of multispectral data." *Remote Sensing of Environment* **24**: 459-479.

Chavez, P. S., Jr. (1989). "Radiometric calibration of Landsat Thematic Mapper multispectral images." *Photogrammetric Engineering and Remote Sensing* **55**(9): 1285-1294.

Chavez, P. S., Jr. (1996). "Image-based atmospheric corrections–revisited and improved." *Photogrammetric Engineering and Remote Sensing* **62**(9): 1025-1036.

Chen, L.-C. and L.-H. Lee (1992). "Progressive generation of control frameworks for image registration." *Photogrammetric Engineering and Remote Sensing* **58**(9): 1321-1328.

Chin, R. T. and C.-L. Yeh (1983). "Quantitative evaluation of some edge-preserving noise-smoothing techniques." *Computer Vision, Graphics and Image Processing*(23): 67-91.

Clark, R. N. and T. L. Roush (1984). "Reflectance Spectroscopy: Quantitative Analysis Techniques for Remote Sensing Applications." *Journal of Geophysical Research* **89**: 6329-6340.

Colby, J. D. (1991). "Topographic normalization in rugged terrain." *Photogrammetric Engineering and Remote Sensing* **57**(5): 531-537.

Craig, M. D. and A. A. Green (1987). "Registration of distorted images from airborne scanners." *The Australian Computer Journal* **19**(3): 148-153.

Crippen, R. E. (1989). "A simple spatial filtering routine for the cosmetic removal of scan-line noise from Landsat TM P-tape imagery." *Photogrammetric Engineering and Remote Sensing* **55**(3): 327-331.

Curran, P. J. and J. L. Dungan (1989). "Estimation of signal-to-noise: a new procedure applied to AVIRIS data." *IEEE Transactions on Geoscience and Remote Sensing* **27**(5): 620-628.

Devereux, B. J., R. M. Fuller, L. Carter, and R. J. Parsell (1990). "Geometric correction of airborne scanner imagery by matching Delaunay triangles." *International Journal of Remote Sensing* **11**(12): 2237-2251.

Dikshit, O. and D. P. Roy (1996). "An empirical investigation of image resampling effects upon the spectral and textural supervised classification of a high spatial resolution mulitspectral image." *Photogrammetric Engineering and Remote Sensing* **62**(9): 1085-1092.

Eliason, E. and A. S. McEwen (1990). "Adaptive box filter for removal of random noise from digital images." *Photogrammetric Engineering and Remote Sensing* **56**(4): 453-458.

EOSAT (1993). Fast Format Document. Lanham, Maryland, EOSAT.

Filho, C. R. d. S., S. A. Drury, A. M. Denniss, R. W. T. Carlton, and D. A. Rothery (1996). "Restoration of corrupted optical Fuyo-1 (JERS-1) data using frequency domain techniqeus." *Photogrammetric Engineering and Remote Sensing* **62**(9): 1037-1047.

Fischel, D. (1984). "Validation of the Thematic Mapper radiometric and geometric correction algorithms." *IEEE Transactions on Geoscience and Remote Sensing* **GE-22**(3): 237-242.

Fusco, L., U. Frei, D. Trevese, P. N. Blonda, G. Pasquariello, and G. Milillo (1986). "Landsat TM image forward/reverse scan banding: characterization and correction." *International Journal of Remote Sensing* **7**(4): 557-575.

Gao, B.-C. and A. F. H. Goetz (1990). "Column atmospheric water vapor and vegetation liquid water retrievals from airborne imaging spectrometer data." *Journal of Geophysical Research* **95**: 3549-3564.

Gao, B.-C., K. B. Heidebrecht, and A. F. H. Goetz (1993). "Derivation of scaled surface reflectances from AVIRIS data." *Remote Sensing of Environment* **44**: 165-178.

Gilbert, E. N. (1974). "Distortion in maps." *SIAM Review* **16**(1): 47-62.

Helder, D. L., B. K. Quirk, and J. J. Hood (1992). "A technique for the reduction of banding in Landsat Thematic Mapper images." *Photogrammetric Engineering and Remote Sensing* **58**(10): 1425-1431.

Holben, B., E. Vermote, Y. J. Kaufman, D. Tanre, and V. Kalb (1992). "Aerosol retrieval over land from AVHRR data - application for atmospheric correction." *IEEE Transactions on Geoscience and Remote Sensing* **30**(2): 212-222.

Horn, B. K. P. and R. J. Woodham (1979). "Destriping Landsat MSS images by histogram modification." *Computer Graphics and Image Processing* **10**(1): 69-83.

Hummer-Miller, S. (1990). "Techniques for noise removal and registration of TIMS data." *Photogrammetric Engineering and Remote Sensing* **56**(1): 49-53.

Itten, K. I. and P. Meyer (1993). "Geometric and radiometric correction of TM data of mountainous forested areas." *IEEE Transactions on Geoscience and Remote Sensing* **31**(4): 764-770.

Jain, A. K. and R. C. Dubes (1988). *Algorithms for Clustering Data.* Englewood Cliffs, NJ: Prentice Hall, 320 p.

Jensen, J. R. (1996). *Introductory Digital Image Processing - A Remote Sensing Perspective.* Second Ed. Upper Saddle River, NJ: Prentice Hall, 316 p.

Keys, R. G. (1981). "Cubic convolution interpolation for digital image processing." *IEEE Transactions on Acoustics, Speech, and Signal Processing* **ASSP-29**: 1153-1160.

Khan, B., L. W. B. Hayes, and A. P. Cracknell (1995). "The effects of higher-order resampling on AVHRR data." *International Journal of Remote Sensing* **16**(1): 147-163.

Kruse, F. A. (1988). "Use of Airborne Imaging Spectrometer data to map minerals associated with hydrothermally altered rocks in the northern Grapevine Mountains, Nevada and California." *Remote Sensing of Environment* **24**(1): 31-51.

Kruse, F. A., K. S. Kierein-Young, and J. W. Boardman (1990). "Mineral mapping at Cuprite, Nevada with a 63-channel imaging spectrometer." *Photogrammetric Engineering and Remote Sensing* **56**(1): 83-92.

Lavreau, J. (1991). "De-hazing Landsat Thematic Mapper images." *Photogrammetric Engineering and Remote Sensing* **57**(10): 1297-1302.

Lee, J. (1983). "Digital image smoothing and the sigma filter." *Computer Vision, Graphics and Image Processing* **24**: 255-269.

Lee, J.-S. (1980). "Digital image enhancement and noise filtering by use of local statistics." *IEEE Transactions on Pattern Analysis and Machine Intelligence* **PAMI-2**(2): 165-168.

Lee, J.-S. (1981). "Speckle analysis and smoothing of synthetic aperture radar images." *Computer Graphics and Image Processing* **17**: 24-32.

Leprieur, C., V. Carrere, and X. F. Gu (1995). "Atmospheric corrections and ground reflectance recovery for Airborne Visible/Infrared Imaging Spectrometer (AVIRIS) data: MAC Europe'91." *Photogrammetric Engineering and Remote Sensing* **61**(10): 1233-1238.

Marsh, S. E. and J. B. McKeon (1983). "Integrated analysis of high-resolution field and airborne spectroradiometer data for alteration mapping." *Economic Geology* **78**(4): 618-632.

Mastin, G. A. (1985). "Adaptive filters for digital image noise smoothing: an evaluation." *Computer Vision, Graphics and Image Processing*(31): 103-121.

Moik, J. G. (1980). *Digital Processing of Remotely Sensed Images.* Washington, D.C.: NASA, U.S. Government Printing Office, 330 p.

Moran, M. S., R. D. Jackson, P. N. Slater, and P. M. Teillet (1992). "Evaluation of simplified procedures for retrieval of land surface reflectance factors from satellite sensor output." *Remote Sensing of Environment* **41**: 169-184.

Nagao, M. and T. Matsuyama (1979). "Edge preserving smoothing." *Computer Graphics and Image Processing* **9**: 394-407.

Pan, J.-J. (1989). "Spectral analysis and filtering techniques in digital spatial data processing." *Photogrammetric Engineering and Remote Sensing* **55**(8): 1203-1207.

Pan, J.-J. and C.-I. Chang (1992). "Destriping of Landsat MSS images by filtering techniques." *Photogrammetric Engineering and Remote Sensing* **58**(10): 1417-1423.

Park, S. K. and R. A. Schowengerdt (1983). "Image reconstruction by parametric cubic convolution." *Computer Vision, Graphics and Image Processing* **20**(3): 258-272.

Poros, D. J. and C. J. Peterson (1985). "Methods for destriping Landsat Thematic Mapper images - A feasibility study for an online destriping process in the Thematic Mapper Image Processing System (TIPS)." *Photogrammetric Engineering and Remote Sensing* **51**(9): 1371-1378.

Potter, J. F. (1984). "The channel correlation method for estimating aerosol levels from multispectral scanner data." *Photogrammetric Engineering and Remote Sensing* **50**: 43-52.

Potter, J. F. and M. Mendolowitz (1975). "On the determination of the haze levels from Landsat data." in: *10th International Symposium on Remote Sensing of Environment,* Ann Arbor, MI, Environmental Research Institute of Michigan, : 695-703.

Pratt, W. K. (1991). *Digital Image Processing.* Second Ed. New York, John Wiley & Sons, 698 p.

Quarmby, N. C. (1987). "Noise removal for SPOT imagery." *International Journal of Remote Sensing* **8**: 1229-1234.

Rao, C. R. N. and J. Chen (1994). Post-Launch Calibration of the Visible and Near Infrared Channels of the Advanced Very High Resolution Radiometer on NOAA-7, -9 and -11 Spacecraft. Washington, D.C.: National Oceanic and Atmospheric Administration, No. NESDIS 78.

Rast, M., S. J. Hook, C. D. Elvidge, and R. E. Alley (1991). "An evaluation of techniques for the extraction of mineral absorption features from high spectral resolution remote sensing data." *Photogrammetric Engineering and Remote Sensing* 57(10): 1303-1309.

Richards, J. A. (1993). *Remote Sensing Digital Image Analysis - An Introduction.* Second Ed. Berlin: Springer-Verlag, 334 p.

Richards, M. E. (1985). "A comparison of two nonlinear destriping procedures." in: *Architectures and Algorithms for Digital Image Processing II*, SPIE, vol. 534: 128-134.

Richter, R. (1996a). "A spatially adaptive fast atmospheric correction algorithm." *International Journal of Remote Sensing* 17(6): 1201-1214.

Richter, R. (1996b). "Atmospheric correction of satellite data with haze removal including a haze/clear transition region." *Computers & Geosciences* 22(6): 675-681.

Rindfleisch, T. C., J. A. Dunne, H. J. Frieden, W. D. Stromberg, and R. M. Ruiz (1971). "Digital processing of the Mariner 6 and 7 pictures." *Journal of Geophysical Research* 76(2): 394-417.

Ripley, B. D. (1981). *Spatial Statistics.* New York, NY: John Wiley & Sons, 252 p.

Rose, J. F. (1989). "Spatial interference in the AVIRIS imaging spectrometer." *Photogrammetric Engineering and Remote Sensing* 55(9): 1339-1346.

Roy, D. P. and O. Dikshit (1994). "Investigation of image resampling effects upon the textural information content of a high spatial resolution remotely sensed image." *International Journal of Remote Sensing* 15(5): 1123-1130.

Schürmann, J. (1996). *Pattern Classification–A Unified View of Statistical and Neural Approaches.* New York, NY: John Wiley & Sons, 373 p.

Simpson, J. J. and S. R. Yhann (1994). "Reduction of noise in AVHRR channel 3 data with minimum distortion." *IEEE Transactions on Geoscience and Remote Sensing* 32(2): 315-328.

Singh, S. M. and A. P. Cracknell (1986). "The estimation of atmospheric effects for SPOT using AVHRR channel-1 data." *International Journal of Remote Sensing* 7(3): 361-377.

Smith, J. A., T. L. Lin, and K. Ranson (1980). "The lambertian assumption and Landsat data." *Photogrammetric Engineering and Remote Sensing* 46(9): 1183-1189.

Snyder, J. P. (1987). *Map Projections–A Working Manual.* U.S. Geological Survey Professional Paper 1395, Washington, D.C.: U.S. Government Printing Office, 383 p.

Srinivasan, R., M. Cannon, and J. White (1988). "Landsat data destriping using power spectral filtering." *Optical Engineering* 27(11): 939-943.

Steiner, D. and M. E. Kirby (1976). "Geometrical referencing of Landsat images by affine transformation and overlaying of map data." *Photogrammetria* 33: 41-75.

Steinwand, D. R. (1994). "Mapping raster imagery to the Interrupted Goode Homolsine projection." *International Journal of Remote Sensing* 15(17): 3463-3471.

Steinwand, D. R., J. A. Hutchinson, and J. P. Snyder (1995). "Map projections for global and continental data sets and an analysis of pixel distortion caused by reprojection." *Photogrammetric Engineering and Remote Sensing* **61**(12): 1487-1497.

Swann, R., D. Hawkins, A. Westwell-Roper, and W. Johnstone (1988). "The potential for automated mapping from geocoded digital image data." *Photogrammetric Engineering and Remote Sensing* **54**(2): 187-193.

Switzer, P., W. S. Kowalik, and R. J. Lyon (1981). "Estimation of atmospheric path radiance by the covariance matrix method." *Photogrammetric Engineering and Remote Sensing* **47**: 1469-1476.

Tanre, D., C. Deroo, P. Duhaut, M. Herman, J. J. Morcrette, J. Perbos, and P. Y. Deschamps (1990). "Description of a computer code to simulate the satellite signal in the solar spectrum: 5S code." *International Journal of Remote Sensing* **11**: 659-668.

Teillet, P. M. and G. Fedosejevs (1995). "On the dark target approach to atmospheric correction of remotely sensed data." *Canadian Journal of Remote Sensing* **21**(4): 374-387.

Thome, K. J., S. F. Biggar, D. I. Gellman, and P. N. Slater (1994). "Absolute-radiometric calibration of Landsat-5 Thematic Mapper and the proposed calibration of the Advanced Spaceborne Thermal Emission and Reflection Radiometer." in: *IGARSS-94*, Pasadena, CA: IEEE, vol. 4: 2295-2297.

Tilton, J. C., B. L. Markham, and W. L. Alford (1985). "Landsat-4 and Landsat-5 MSS coherent noise: characterization and removal." *Photogrammetric Engineering and Remote Sensing* **51**(9): 1263-1279.

Wegener, M. (1990). "Destriping multiple sensor imagery by improved histogram matching." *International Journal of Remote Sensing* **11**(5): 859-875.

Weinreb, M. P., R. Xie, J. H. Lienesch, and D. S. Crosby (1989). "Destriping GOES images by matching empirical distribution functions." *Remote Sensing of Environment* **29**: 185-195.

Westin, T. (1990). "Filters for removing coherent noise of period 2 in SPOT imagery." *International Journal of Remote Sensing* **11**(2): 351-357.

Westin, T. (1990). "Precision rectification of SPOT imagery." *Photogrammetric Engineering and Remote Sensing* **56**(2): 247-253.

Wolberg, G. (1990). *Digital Image Warping*. Los Alamitos, CA: IEEE Computer Society Press, 318 p.

Wolf, P. R. (1983). *Elements of Photogrammetry*. Second International Student Ed., Singapore: McGraw-Hill, 628 p.

Wrigley, R. C., D. H. Card, C. A. Hlavka, J. R. Hall, F. C. Mertz, C. Archwamety, and R. A. Schowengerdt (1984). "Thematic Mapper image quality: Registration, noise, and resolution." *IEEE Transactions on Geoscience and Remote Sensing* **GE-22**(3): 263-271.

Wrigley, R. C., M. A. Spanner, R. E. Slye, R. F. Pueschel, and H. R. Aggarwal (1992). "Atmospheric correction of remotely sensed image data by a simplified model." *Journal of Geophysical Research* **97**(D17): 18797-18814.

Zagolski, F. and J. P. Gastellu-Etchegorry (1995). "Atmospheric corrections of AVIRIS images with a procedure based on the inversion of the 5S model." *International Journal of Remote Sensing* **16**(16): 3115-3146.

CHAPTER 8 *Image Registration and Fusion*

Anuta, P. (1970). "Spatial registration of multispectral and multitemporal digital imagery using Fast Fourier Transform techniques." *IEEE Transactions on Geoscience Electronics* **GE-8**(4): 353-368.

Barnea, D. I. and H. E. Silverman (1972). "A class of algorithms for fast digital image registration." *IEEE Transactions on Computers* **C-21**(2): 179-186.

Bernstein, R. (1976). "Digital image processing of earth observation sensor data." *IBM Journal of Research and Development* **20**(1): 40-57.

Brockelbank, D. C. and A. P. Tam (1991). "Stereo elevation determination techniques for SPOT imagery." *Photogrammetric Engineering and Remote Sensing* **57**(8): 1065-1073.

Brown, L. (1992). "A survey of image registration techniques." *ACM Computing Surveys* **24**(4).

Carper, W. J., T. M. Lillesand, and R. W. Kiefer (1990). "The use of intensity-hue-saturation transformations for merging SPOT panchromatic and multispectral image data." *Photogrammetric Engineering and Remote Sensing* **56**(4): 459-467.

Chavez, P. S., Jr. (1986). "Digital merging of Landsat TM and digitized NHAP data for 1:24,000-scale image mapping." *Photogrammetric Engineering and Remote Sensing* **52**(10): 1637-1646.

Chavez, P. S., Jr., S. C. Sides, and J. A. Anderson (1991). "Comparison of three different methods to merge multiresolution and multispectral data: Landsat TM and SPOT panchromatic." *Photogrammetric Engineering and Remote Sensing* **57**(3): 295-303.

Cliche, G., F. Bonn, and P. Teillet (1985). "Integration of the SPOT panchromatic channel into its multispectral mode for image sharpness enhancement." *Photogrammetric Engineering and Remote Sensing* **51**(3): 311-316.

Craig, M. D. and A. A. Green (1987). "Registration of distorted images from airborne scanners." *The Australian Computer Journal* **19**(3): 148-153.

Daily, M. I., T. Farr, C. Elachi, and G. Schaber (1979). "Geologic interpretation from composited radar and Landsat imagery." *Photogrammetric Engineering and Remote Sensing* **45**(8): 1109-1116.

Djamdji, J. P., A. Bijaoui, and R. Maniere (1993b). "Geometrical registration of images: the multiresolution approach." *Photogrammetric Engineering and Remote Sensing* **59**(5): 645-653.

Ehlers, M. (1991). "Multisensor image fusion techniques in remote sensing." *ISPRS Journal of Photogrammetry and Remote Sensing* **46**: 19-30.

Ehlers, M. and R. Welch (1987). "Stereocorrelation of Landsat TM images." *Photogrammetric Engineering and Remote Sensing* **53**: 1231–1237.

Filiberti, D., S. Marsh, and R. Schowengerdt (1994). "Synthesis of high spatial and spectral resolution imagery from multiple image sources." *Optical Engineering* **33**(8): 2520-2528.

Flusser, J. and T. Suk (1994). "A moment-based approach to registration of images with affine geometric distortion." *IEEE Transactions on Geoscience and Remote Sensing* **32**(2): 382-387.

Fonseca, L. M. G. and B. S. Manjunath (1996). "Registration techniques for multisensor remotely sensed imagery." *Photogrammetric Engineering and Remote Sensing* **62**(9): 1049-1056.

Garguet-Duport, B., J. Girel, J.-M. Chassery, and G. Pautou (1996). "The use of multiresolution analysis and wavelets transform for merging SPOT panchromatic and multispectral image data." *Photogrammetric Engineering and Remote Sensing* **62**(9): 1057-1066.

Gonzalez, R. C. and R. E. Woods (1992). *Digital Image Processing*. Reading, MA, Addison-Wesley, 703 p.

Goshtasby, A. (1988). "Registration of images with geometric distortions." *IEEE Transactions on Geoscience and Remote Sensing* **26**(1): 60-64.

Goshtasby, A. (1993). "Correction to "Registration of images with geometric distortions"." *IEEE Transactions on Geoscience and Remote Sensing* **31**(1): 307.

Goshtasby, A., G. Stockman, and C. Page (1986). "A region-based approach to digital image registration with subpixel accuracy." *IEEE Transactions on Geoscience and Remote Sensing* **GE-24**(3).

Greenfield, J. S. (1991). "An Operator–Based Matching System." *Photogrammetric Engineering and Remote Sensing* **57**(8): 1049-1055.

Hall, E. L. (1979). *Computer Image Processing and Recognition*. New York, NY: Academic Press, 584 p.

Harris, J. R., R. Murray, and T. Hirose (1990). "IHS transform for the integration of radar imagery with other remotely sensed data." *Photogrammetric Engineering and Remote Sensing* **56**(12): 1631-1641.

Haydn, R., G. W. Dalke, J. Henkel, and J. E. Bare (1982). "Application of the IHS color transform to the processing of multisensor data and image enhancement." in: *International Symposium on Remote Sensing of Arid and Semi-Arid Lands*, Cairo, Egypt: Environmental Research Institute of Michigan: 599-616.

Henderson, T. C., E. E. Triendl, and R. Winter (1985). "Edge- and shape-based geometric registration." *IEEE Transactions on Geoscience and Remote Sensing* **GE-23**(3): 334-341.

Hood, J., L. Ladner, and R. Champion (1989). "Image processing techniques for digital orthophotoquad production." *Photogrammetric Engineering and Remote Sensing* **55**(9): 1323-1329.

Konecny, G., P. Lohmann, H. Engel, and E. Kruck (1987). "Evaluation of SPOT imagery on analytical photogrammetric instruments." *Photogrammetric Engineering and Remote Sensing* **53**(9): 1223-1230.

Li, H., B. S. Manjunath, and S. K. Mitra (1995). "A contour-based approach to multisensor image registration." *IEEE Transactions on Image Processing* **4**(3): 320-334.

Liang, T. and C. Heipke (1996). "Automatic relative orientation of aerial images." *Photogrammetric Engineering and Remote Sensing* **62**(1): 47-55.

Moran, S. M. (1990). "A window-based technique for combining Landsat thematic mapper thermal data with higher-resolution multispectral data over agricultural lands." *Photogrammetric Engineering and Remote Sensing* **56**(3): 337-342.

Munechika, C. K., J. S. Warnick, C. Salvaggio, and J. R. Schott (1993). "Resolution enhancement of multispectral image data to improve classification accuracy." *Photogrammetric Engineering and Remote Sensing* **59**(1): 67-72.

Panton, D. J. (1978). "A flexible approach to digital stereo matching." *Photogrammetric Engineering and Remote Sensing* **44**(12): 1499-1512.

Pellemans, A. H. J. M., R. W. L. Jordans, and R. Allewijn (1993). "Merging multispectral and panchromatic SPOT images with respect to the radiometric properties of the sensor." *Photogrammetric Engineering and Remote Sensing* **59**(1): 81-87.

Pradines, D. (1986). "Improving SPOT images size and multispectral resolution." in: *Earth Remote Sensing using the Landsat Thematic Mapper and SPOT systems*, SPIE, vol. 660: 98-102.

Pratt, W. K. (1974). "Correlation techniques of image registration." *IEEE Transactions on Aerospace and Electronic Systems* **AES-10**(3): 353-358.

Pratt, W. K. (1991). *Digital Image Processing*. Second Ed. New York, NY: John Wiley & Sons, 698 p.

Price, J. C. (1987). "Combining panchromatic and multispectral imagery from dual resolution satellite instruments." *Remote Sensing of Environment* **21**: 119-128.

Quam, L. H. (1987). Hierarchical Warp Stereo. *Readings in Computer Vision: Issues, Problems, Principles, and Paradigms*. M. A. Fischler and O. Firschein, (Eds.). Los Altos, CA: Morgan Kaufmann Publishers, Inc.: 800.

Ramapriyan, H. K., J. P. Strong, Y. Hung, and J. Charles W. Murray (1986). "Automated matching of pairs of SIR-B images for elevation mapping." *IEEE Transactions on Geoscience and Remote Sensing* **GE-24**(4): 462-472.

Rao, T. C. M., K. V. Rao, A. R. Kumar, D. P. Rao, and B. L. Deekshatula (1996). "Digital Terrain Model (DTM) from Indian Remote Sensing (IRS) satellite data from the overlap area of two adjacent paths using digital photogrammetric techniques." *Photogrammetric Engineering and Remote Sensing* **62**(6): 727-731.

Reddy, B. S. and B. N. Chatterji (1996). "An FFT-based technique for translation, rotation, and scale-invariant image registration." *IEEE Transactions on Image Processing* **5**(8): 1266-1271.

Rodriguez, V., P. Gigord, A. C. d. Gaujac, P. Munier, and G. Begni (1988). "Evaluation of the stereoscopic accuracy of the SPOT satellite." *Photogrammetric Engineering and Remote Sensing* **54**(2).

Rosenfeld, A. and A. C. Kak (1982). *Digital Picture Processing*. Second Ed., Orlando, FL: Academic Press, 349 p.

Scambos, T. A., M. J. Dutkiewicz, J. C. Wilson, and R. A. Bindschadler (1992). "Application of image cross–correlation to the measurement of glacier velocity using satellite image data." *Remote Sensing of Environment* **42**: 177-186.

Schenk, T., J. C. Li, and C. Toth (1991). "Towards an Autonomous System for Orienting Digital Stereopairs." *Photogrammetric Engineering and Remote Sensing* **57**(8): 1057-1064.

Schowengerdt, R. A. (1980). "Reconstruction of multispatial, multispectral image data using spatial frequency content." *Photogrammetric Engineering and Remote Sensing* **46**(10): 1325-1334.

Schowengerdt, R. A. and D. Filiberti (1994). "Spatial frequency models for multispectral image sharpening." in: *Algorithms for Multispectral and Hyperspectral Imagery*, Orlando, Florida, SPIE, vol. 2231: 84-90.

Shettigara, V. K. (1992). "A generalized component substitution technique for spatial enhancement of multispectral images using a higher resolution data set." *Photogrammetric Engineering and Remote Sensing* **58**(5): 561-567.

SPOTImage, SPOTView Digital Ortho-Image Data Sampler CD-ROM, 1995.

Tateishi, R. and A. Akutsu (1992). "Relative DEM production from SPOT data without GCP." *International Journal of Remote Sensing* **13**(14): 2517-2530.

Tom, V. T., M. J. Carlotto, and D. K. Scholten (1985). "Spatial sharpening of Thematic Mapper data using a multiband approach." *Optical Engineering* **24**(6): 1026-1029.

Ton, J. and A. K. Jain (1989). "Registering Landsat images by point matching." *IEEE Transactions on Geoscience and Remote Sensing* **27**(5).

USGS (1995). Capital Cities of the United States. *Digital Raster Graphic Data CD-ROM*, U.S. Geological Survey.

Ventura, A. D., A. Rampini, and R. Schettini (1990). "Image registration by the recognition of corresponding structures." *IEEE Transactions on Geoscience and Remote Sensing* **28**(3): 305-387.

Welch, R. and M. Ehlers (1987). "Merging multiresolution SPOT HRV and Landsat TM data." *Photogrammetric Engineering and Remote Sensing* **53**(3): 301-303.

Welch, R. and M. Ehlers (1988). "Cartographic feature extraction with integrated SIR-B and Landsat TM images." *International Journal of Remote Sensing* **9**(5): 873-889.

Westin, T. (1990). "Precision rectification of SPOT imagery." *Photogrammetric Engineering and Remote Sensing* **56**(2): 247-253.

Wolf, P. R. (1983). *Elements of Photogrammetry*. Second Ed. , McGraw-Hill, 628 p.

Wong, F. H. and R. Orth (1980). "Registration of SEASAT-LANDSAT composite images to UTM coordinates." in: *Sixth Canadian Symposium on Remote Sensing*, Halifax, Nova Scotia: 161-164.

Yocky, D. A. (1996). "Multiresolution wavelet decomposition image merger of Landsat Thematic Mapper and SPOT panchromatic data." *Photogrammetric Engineering and Remote Sensing* **62**(9): 1067-1074.

Zheng, Q. and R. Chellappa (1993). "A computational vision approach to image registration." *IEEE Transactions on Image Processing* **2**: 311-326.

CHAPTER 9 *Thematic Classification*

Abrams, M. and S. J. Hook (1995). "Simulated Aster data for geologic studies." *IEEE Transactions on Geoscience and Remote Sensing* 33(3): 692-699.

Abuelgasim, A. A., S. Gopal, J. R. Irons, and A. H. Strahler (1996). "Classification of ASAS multiangle and multispectral measurements using artificial neural networks." *Remote Sensing of Environment* 57: 79-87.

Adams, J. B., M. O. Smith, and A. R. Gillespie (1993). Imaging Spectroscopy: Interpretation Based on Spectral Mixture Analysis. *Remote Geochemical Analysis: Elemental and Mineralogical Composition.* C. M. Pieters and P. A. Englert, (Eds.), Cambridge: Cambridge University Press: 145-166.

Anderberg, M. R. (1973). *Cluster Analysis for Applications.* New York, NY: Academic Press, 353 p.

Anderson, J. R., E. E. Hardy, J. T. Roach, and R. E. Witmer (1976). A Land Use and Land Cover Classification System for Use with Remote Sensor Data. Washington, D.C., U. S. Geological Survey, No. Professional Paper 964.

Arai, K. (1992). "A supervised Thematic Mapper classification with a purification of training samples." *International Journal of Remote Sensing* 13(11): 2039-2049.

Argialas, D. P. and C. A. Harlow (1990). "Computational image interpretation models: an overview and a perspective." *Photogrammetric Engineering and Remote Sensing* 56(6): 871-886.

Augusteijn, M. F., L. E. Clemens, and K. A. Shaw (1995). "Performance evaluation of texture measures for ground cover identification in satellite image by means of a neural network classifier." *IEEE Transactions on Geoscience and Remote Sensing* 33(3): 616-626.

Avery, T. E. and G. L. Berlin (1992). *Fundamentals of Remote Sensing and Airphoto Interpretation.* 5th Ed. New York, NY: Macmillan Publishing Company, 472 p.

Ball, G. and D. Hall (1967). "A clustering technique for summarizing multivariate data." *Behavioral Science* 12: 153-155.

Ballard, D. H. and C. M. Brown (1982). *Computer Vision.* Englewood Cliffs, NJ: Prentice Hall, 523 p.

Baraldi, A. and F. Parmiggiani (1995). "A neural network for unsupervised categorization of multivalued input patterns: an application to satellite image clustering." *IEEE Transactions on Geoscience and Remote Sensing* 33(2): 305-316.

Bateson, A. and B. Curtiss (1996). "A method for manual endmember selection and spectral unmixing." *Remote Sensing of Environment* 55: 229-243.

Benediktsson, J. A., J. R. Sveinsson, and K. Arnason (1995). "Classification and feature extraction of AVIRIS data." *IEEE Transactions on Geoscience and Remote Sensing* 33(5): 1194-1205.

Benediktsson, J. A. and P. H. Swain (1992). "Consensus theoretic classification methods." *IEEE Transactions on Systems, Man & Cybernetics* 22(4): 688-704.

Benediktsson, J. A., P. H. Swain, and O. K. Ersoy (1990a). "Neural network approaches versus statistical methods in classification of multisource remote sensing data." *IEEE Transactions on Geoscience and Remote Sensing* 28(4): 540-551.

Benie, G. B. and K. P. B. Thomson (1992). "Hierarchical image segmentation using local and adaptive similarity rules." *International Journal of Remote Sensing* **13**(8): 1559-1570.

Benjamin, S. and L. Gaydos (1990). "Spatial resolution requirements for automated cartographic road extraction." *Photogrammetric Engineering and Remote Sensing* **56**(1): 93-100.

Bezdek, J. C., R. Ehrlich, and W. Full (1984). "FCM: The fuzzy c-means clustering algorithm." *Computers and Geosciences* **10**(2-3): 191-203.

Billingsley, F. C. (1982). "Modeling misregistration and related effects on multispectral classification." *Photogrammetric Engineering and Remote Sensing* **48**(3): 421-430.

Bischof, H., W. Schneider, and A. J. Pinz (1992). "Multispectral classification of Landsat–images using neural networks." *IEEE Transactions on Geoscience and Remote Sensing* **30**(3): 482-490.

Boardman, J. (1990). "Inversion of high spectral resolution data." SPIE, vol. 1298 : 222-233.

Boardman, J. W., F. A. Kruse, and R. O. Green (1995). "Mapping target signatures via partial unmixing of AVIRIS data." in: *Fifth JPL Airborne Earth Science Workshop*, Pasadena, CA: JPL, vol. 95-1: 23-26.

Bolstad, P. V. and T. M. Lillesand (1991). "Rapid maximum likelihood classification." *Photogrammetric Engineering and Remote Sensing* **57**(1): 67-74.

Borel, C. C. and S. A. W. Gerstl (1994). "Nonlinear spectral mixing models for vegetative and soil surfaces." *Remote Sensing of Environment* **47**: 403-416.

Bouman, C. A. and M. Shapiro (1994). "A multiscale random field model for bayesian image segmentation." *IEEE Transactions on Image Processing* **3**(2): 162-177.

Bryant, J. (1979). "On the clustering of multidimensional pictorial data." *Pattern Recognition* **11**(2): 115-125.

Bryant, J. (1989). "A fast classifier for image data." *Pattern Recognition* **22**: 45-48.

Bryant, J. (1990). "AMOEBA clustering revisited." *Photogrammetric Engineering and Remote Sensing* **56**(1): 41-47.

Buchheim, M. P. and T. M. Lillesand (1989). "Semi-automated training field extraction and analysis for efficient digital image classification." *Photogrammetric Engineering and Remote Sensing* **55**(9): 1347-1355.

Cannon, R. L., J. V. Dave, J. C. Bezdek, and M. M. Trivedi (1986). "Segmentation of a Thematic Mapper image using the fuzzy c-means clustering algorithm." *IEEE Transactions on Geoscience and Remote Sensing* **GE-24**(3): 400-408.

Chen, K. S., Y. C. Tzeng, C. F. Chen, and W. L. Kao (1995). "Land-cover classification of multispectral imagery using a dynamic learning neural network." *Photogrammetric Engineering and Remote Sensing* **61**(4): 403-408.

Civco, D. L. (1993). "Artificial neural networks for land–cover classification and mapping." *International Journal of Geographical Information Systems* **7**(2): 173-186.

Clément, V., G. Giraudon, S. Houzelle, and F. Sandakly (1993). "Interpretation of remotely sensed images in a context of multisensor fusion using a multispecialist architecture." *IEEE Transactions on Geoscience and Remote Sensing* **31**(4): 779-791.

Coleman, G. B. and H. C. Andrews (1979). "Image segmentation by clustering." *Proceedings of the IEEE* **67**(5): 773-785.

Cross, A. M., J. J. Settle, N. A. Drake, and R. T. M. Paivinen (1991). "Subpixel measurment of tropical forest cover using AVHRR data." *International Journal of Remote Sensing* **12**(5): 1119-1129.

Dozier, J. (1981). "A method for satellite identification of surface temperature fields of subpixel resolution." *Remote Sensing of Environment* **11**: 221-229.

Dreyer, P. (1993). "Classification of land cover using optimized neural nets on SPOT data." *Photogrammetric Engineering and Remote Sensing* **59**(5): 617-621.

Duda, R. D. and P. E. Hart (1973). *Pattern Classification and Scene Analysis*. New York, NY: John Wiley & Sons, 482 p.

Dymond, J. R. (1993). "An improved Skidmore/Turner classifier." *Photogrammetric Engineering and Remote Sensing* **59**(5): 623-626.

Felzer, B., P. Hauff, and A. F. H. Goetz (1994). "Quantitative reflectance spectroscopy of buddingtonite from the Cuprite mining district, Nevada." *Journal of Geophysical Research* **99**(B2): 2887-2895.

Foody, G. M. (1992). "A fuzzy sets approach to the representation of vegetation continua from remotely sensed data: an example from lowland heath." *Photogrammetric Engineering and Remote Sensing* **58**(2): 221-225.

Foody, G. M. (1995). "Using prior knowledge in artificial neural network classification with a minimal training set." *International Journal of Remote Sensing* **16**(2): 301-312.

Foody, G. M. (1996). "Relating the land-cover composition of mixed pixels to artificial neural network classification output." *Photogrammetric Engineering and Remote Sensing* **62**(5): 491-499.

Foody, G. M. and D. P. Cox (1994). "Sub-pixel land cover composition estimation using a linear mixture model and fuzzy membership functions." *International Journal of Remote Sensing* **15**(3): 619-631.

Foody, G. M., M. B. McCulloch, and W. B. Yates (1995). "Classification of remotely sensed data by an artificial neural network: issues related to training data characteristics." *Photogrammetric Engineering and Remote Sensing* **61**(4): 391-401.

Fukunaga, K. (1990). *Introduction to Statistical Pattern Recognition*. Second Ed. San Diego: Academic Press, 591 p.

Full, W. E., R. Ehrlich, and J. C. Bezdek (1982). "Fuzzy Qmodel - a new model approach for linear unmixing." *Mathematical Geology* **14**(3): 259-270.

Goldberg, M., D. G. Goodenough, and G. Plunkett (1988). "A knowledge-based approach for evaluating forestry-map congruency with remotely sensed imagery." *Philosophical Transactions of the Royal Society of London* **324**: 447-456.

Gyer, M. S. (1992). "Adjuncts and alternatives to neural networks for supervised classification." *IEEE Transactions on Geoscience and Remote Sensing* **22**(1): 35-46.

Hara, Y., R. G. Atkins, R. T. Shin, J. A. Kong, S. H. Yueh, and R. Kwok (1995). "Application of neural networks for sea ice classification in polarimetric SAR images." *IEEE Transactions on Geoscience and Remote Sensing* **33**(3): 740-748.

Hara, Y., R. G. Atkins, S. H. Yueh, R. T. Shin, and J. A. Kong (1994). "Application of neural networks to radar image classification." *IEEE Transactions on Geoscience and Remote Sensing* **32**(1): 100-109.

Haralick, R. M. and I. h. Dinstein (1975). "A spatial clustering procedure for multi-image data." *IEEE Transactions on Circuits and Systems* **CAS-22**(5): 440-450.

Hardin, P. J. (1994). "Parametric and nearest-neighbor methods for hybrid classification: a comparison of pixel assignment accuracy." *Photogrammetric Engineering and Remote Sensing* **60**(12): 1439-1448.

Hardin, P. J. and C. N. Thomson (1992). "Fast nearest neighbor classification methods for multispectral imagery." *The Professional Geographer* **44**(2): 191-201.

Harsanyi, J. C. and C.-I. Chang (1994). "Hyperspectral image classification and dimensionality reduction: an orthogonal subspace projection approach." *IEEE Transactions on Geoscience and Remote Sensing* **32**(4): 779-785.

Hartigan, J. A. (1975). *Clustering Algorithms.* New York, NY: John Wiley & Sons, 351 p.

Heermann, P. D. and N. Khazenie (1992). "Classification of multispectral remote sensing data using a back–propagation neural network." *IEEE Transactions on Geoscience and Remote Sensing* **30**(1): 81-88.

Hlavka, C. A. and M. A. Spanner (1995). "Unmixing AVHRR imagery to assess clearcuts and forest regrowth in Oregon." *IEEE Transactions on Geoscience and Remote Sensing* **33**(3): 788-795.

Hoffbeck, J. P. and D. A. Landgrebe (1996). "Covariance matrix estimation and classification with limited training data." *IEEE Transactions on Pattern Analysis and Machine Intelligence* **18**(7): 763-767.

Holben, B. N. and Y. E. Shimabukuro (1993). "Linear mixing models applied to coarse spatial resolution data from multispectral satellite sensors." *International Journal of Remote Sensing* **14**(11): 2231-2240.

Horwitz, H. M., R. F. Nalepka, P. D. Hyde, and J. P. Morgenstern (1971). "Estimating the proportions of objects within a single resolution element of a multispectral scanner." in: *Seventh International Symposium on Remote Sensing of Environment,* Ann Arbor, Michigan, Environmental Research Institute of Michigan, vol. 2: 1307-1320.

Huete, A. R. and R. Escadafal (1991). "Assessment of biophysical soil properties through spectral decomposition techniques." *Remote Sensing of Environment* **35**: 149-159.

Irons, J. R., B. L. Markham, R. F. Nelson, D. L. Toll, D. L. Williams, R. S. Latty, and M. L. Stauffer (1985). "The effects of spatial resolution on the classification of Thematic Mapper data." *International Journal of Remote Sensing* **6**(8): 1385-1403.

Jain, A. K. and R. C. Dubes (1988). *Algorithms for Clustering Data.* Englewood Cliffs, NJ: Prentice Hall, 320 p.

James, M. (1985). *Classification Algorithms.* London: Collins, 209 p.

Jasinski, M. F. (1996). "Estimation of subpixel vegetation density of natural regions using satellite multispectral imagery." *IEEE Transactions on Geoscience and Remote Sensing* **34**(3): 804-813.

Jensen, J. R. (1996). *Introductory Digital Image Processing - A Remote Sensing Perspective.* Second Ed. Upper Saddle River, NJ: Prentice Hall, 316 p.

Jenson, S. K., T. R. Loveland, and J. Bryant (1982). "Evaluation of AMOEBA: a spectral-spatial classification method." *Journal of Applied Photographic Engineering* **8**(3): 159-162.

Jia, X. and J. A. Richards (1993). "Binary coding of imaging spectrometry data for fast spectral matching and classification." *Remote Sensing of Environment* **43**: 47-53.

Jia, X. and J. A. Richards (1994). "Efficient maximum likelihood classification for imaging spectrometer data sets." *IEEE Transactions on Geoscience and Remote Sensing* **32**: 274-281.

Kanellopoulos, I., A. Varfis, G. G. Wilkinson, and J. Mégier (1991). "Neural network classification of multi-date satellite imagery." in: *11th Annual International Geoscience and Remote Sensing Symposium,* Espoo, Finland: IEEE : 2215-2218.

Kanellopoulos, I., A. Varfis, G. G. Wilkinson, and J. Mégier (1992). "Land cover discrimination in SPOT imagery by artificial neural network–a twenty class experiment." *International Journal of Remote Sensing* **13**(5): 917-924.

Kauth, R. J., A. P. Pentland, and G. S. Thomas (1977). "BLOB, an unsupervised clustering approach to spatial preprocessing of MSS imagery." in: *Eleventh International Symposium on Remote Sensing of Environment,* Ann Arbor, MI: Environmental Research Institute of Michigan: 1309-1317.

Kepuska, V. Z. and S. O. Mason (1995). "A hierarchical neural network system for signalized point recognition in aerial photographs." *Photogrammetric Engineering and Remote Sensing* **61**(7): 917-925.

Kettig, R. L. and D. A. Landgrebe (1976). "Classification of multispectral image data by extraction and classification of homogeneous objects." *IEEE Transactions on Geoscience Electronics* **GE-4**(1): 19-26.

Key, J., J. A. Maslanik, and S. A. J. (1989). "Classification of merged AVHRR and SMMR arctic data with neural networks." *Photogrammetric Engineering and Remote Sensing* **55**(9): 1331-1338.

Kim, K. and M. M. Crawford (1991). "Adaptive parametric estimation and classification of remotely sensed imagery using a pyramid structure." *IEEE Transactions on Geoscience and Remote Sensing* **29**(4): 481-493.

Kruse, F. A. (1988). "Use of Airborne Imaging Spectrometer data to map minerals associated with hydrothermally altered rocks in the northern Grapevine Mountains, Nevada and California." *Remote Sensing of Environment* **24**(1): 31-51.

Kruse, F. A., K. S. Kierein-Young, and J. W. Boardman (1990). "Mineral mapping at Cuprite, Nevada with a 63-channel imaging spectrometer." *Photogrammetric Engineering and Remote Sensing* **56**(1): 83-92.

Kruse, F. A., A. B. Lefkoff, J. W. Boardman, K. B. Heidebrecht, A. T. Shapiro, P. J. Barloon, and A. F. H. Goetz (1993). "The Spectral Image Processing System (SIPS) - interactive viualization and analysis of imaging spectrometer data." *Remote Sensing of Environment* **44**: 145-163.

Landgrebe, D. A. (1980). "The development of a spectral-spatial classifier for earth observational data." *Pattern Recognition* **12**: 165-175.

Le Moigne, J. and J. C. Tilton (1995). "Refining image segmentation by integration of edge and region data." *IEEE Transactions on Geoscience and Remote Sensing* **33**(3): 605-615.

Lee, C. and D. A. Landgrebe (1991). "Fast likelihood classification." *IEEE Transactions on Geoscience and Remote Sensing* **29**(4): 509-517.

Lee, C. and D. A. Landgrebe (1993). "Analyzing high-dimensional multispectral data." *IEEE Transactions on Geoscience and Remote Sensing* **31**(4): 792-800.

Lee, T., J. A. Richards, and P. H. Swain (1987). "Probabilistic and evidential approaches for multisource data analysis." *IEEE Transactions on Geoscience and Remote Sensing* **GE-25**: 283-293.

Lillesand, T. M. and R. W. Kiefer (1987). *Remote Sensing and Image Interpretation.* 2nd Ed. New York, NY: John Wiley & Sons, 721 p.

Lippmann, R. P. (1987). "An introduction to computing with neural nets." *IEEE ASSP Magazine*(April): 4-22.

Liu, Z. K. and J. Y. Xiao (1991). "Classification of Remotely-Sensed Image Data Using Artificial Neural Networks." *International Journal of Remote Sensing* **12**(11): 2433-2438.

Markham, B. L. and J. R. G. Townshend (1981). "Land cover classification accuracy as a function of sensor spatial resolution." in: *Fifteenth International Symposium on Remote Sensing of Environment*, Ann Arbor, MI, Environmental Research Institute of Michigan, vol. III: 1075-1090.

Marsh, S. E. and J. B. McKeon (1983). "Integrated analysis of high-resolution field and airborne spectroradiometer data for alteration mapping." *Economic Geology* **78**(4): 618-632.

Marsh, S. E., P. Switzer, and R. J. P. Lyon (1980). "Resolving the percentage component terrains within single resolution elements." *Photogrammetric Engineering and Remote Sensing* **46**(8): 1079-1086.

Maxwell, E. L. (1976). "Multivariate systems analysis of multispectral imagery." *Photogrammetric Engineering and Remote Sensing* **42**(9): 1173-1186.

Mazer, A. S., M. Martin, M. Lee, and J. E. Solomon (1988). "Image processing software for imaging spectrometry data analysis." *Remote Sensing of Environment* **24**: 201-211.

McKeown, D. M., Jr., J. Wilson A. Harvey, and J. McDermott (1985). "Rule-based interpretation of aerial imagery." *IEEE Transactions on Pattern Analysis and Machine Intelligence* **PAMI-7**(5): 570-585.

Mehldau, G. and R. A. Schowengerdt (1990). "A C-extension for rule-based image classification systems." *Photogrammetric Engineering and Remote Sensing* **56**(6): 887-892.

Nagao, M. and T. Matsuyama (1980). *A Structural Analysis of Complex Aerial Photographs.* New York, NY: Plenum Press, 199 p.

Nalepka, R. P. and P. D. Hyde (1972). "Classifying unresolved objects from simulated space data." in: *Eighth International Symposium on Remote Sensing of Environment,* Ann Arbor, MI: Environmental Research Institute of Michigan, vol. 2: 935-949.

Novo, E. M. and Y. E. Shimabukuro (1994). "Spectral mixture analysis of inland tropical waters." *International Journal of Remote Sensing* **15**(6): 1351-1356.

Pal, N. R. and S. K. Pal (1993). "A review on image segmentation techniques." *Pattern Recognition* **26**(9): 1277-1294.

Pao, Y. H. (1989). *Adaptive Pattern Recognition and Neural Networks.* Reading, MA: Addison-Wesley, 299 p.

Paola, J. D. and R. A. Schowengerdt (1995a). "A review and analysis of backpropagation neural networks for classification of remotely-sensed multi-spectral imagery." *International Journal of Remote Sensing* **16**(16): 3033-3058.

Paola, J. D. and R. A. Schowengerdt (1995b). "A detailed comparison of backpropagation neural network and maximum-likelihood classifiers for urban land use classification." *IEEE Transactions on Geoscience and Remote Sensing* **33**(4): 981-996.

Paola, J. D. and R. A. Schowengerdt (1997). "The effect of neural network structure on a multispectral land-use/land-cover classification." *Photogrammetric Engineering and Remote Sensing.*

Philpot, W. D. (1991). "The derivative ratio algorithm: avoiding atmosheric effects in remote sensing." *IEEE Transactions on Geoscience and Remote Sensing* **29**(3): 350-357.

Piech, M. A. and K. R. Piech (1987). "Symbolic representation of hyperspectral data." *Applied Optics* **26**(18): 4018-4026.

Piech, M. A. and K. R. Piech (1989). "Hyperspectral interactions: invariance and scaling." *Applied Optics* **28**(3): 481-489.

Pratt, W. K. (1991). *Digital Image Processing.* Second Ed. New York, NY: John Wiley & Sons, 698 p.

Ray, T. W. and B. C. Murray (1996). "Nonlinear spectral mixing in desert vegetation." *Remote Sensing of Environment* **55**: 59-64.

Richards, J. A. (1993). *Remote Sensing Digital Image Analysis - An Introduction.* Second Ed. Berlin: Springer-Verlag, 334 p.

Richter, R. (1996a). "A spatially adaptive fast atmospheric correction algorithm." *International Journal of Remote Sensing* **17**(6): 1201-1214.

Ritter, N. D. and G. F. Hepner (1990). "Application of an artificial neural network to land–cover classification of Thematic Mapper imagery." *Computers & Geosciences* **16**(6): 873-880.

Roberts, D. A., M. O. Smith, and J. B. Adams (1993). "Green vegetation, nonphotosynthetic vegetation, and soils in AVIRIS data." *Remote Sensing of Environment* **44**: 255-269.

Rubin, T. D. (1993). "Spectral mapping with imaging spectrometers." *Photogrammetric Engineering and Remote Sensing* **59**(2): 215-220.

Rumelhart, D. E., G. E. Hinton, and R. J. Williams (1986). Learning internal representations by error propagation. *Parallel DIstributed Processing: Explorations in the Microstruction of Cognition.* D. E. Rumelhart and J. L. McClelland, (Eds.). Cambridge, MA, The MIT Press. **I:** 318-362.

Ryan, T. (1985). "Image segmentation algorithms." in: *Architectures and Algorithms for Digital Image Processing II*, SPIE, vol. 534: 172-178.

Salu, Y. and J. Tilton (1993). "Classification of multispectral image data by the binary neural network and by nonparametric, pixel-by-pixel methods." *IEEE Transactions on Geoscience and Remote Sensing* **31**(3): 606-617.

Salvato, P. J. (1973). "Iterative techniques to estimate signature vectors for mixture processing of multispectral data." in: *Symposium on Machine Processing of Remotely Sensed Data*, IEEE, vol. 73CH0834-2GE: 3B:48-62.

Schowengerdt, R. A. (1983). *Techniques for Image Processing and Classification in Remote Sensing*. Orlando, FL: Academic Press, 249 p.

Schowengerdt, R. A. (1996). "On the estimation of spatial-spectral mixing with classifier likelihood functions." *Pattern Recognition Letters,* **17**(13): 1379-1387.

Schürmann, J. (1996). *Pattern Classification–A Unified View of Statistical and Neural Approaches*. New York, NY: John Wiley & Sons, 373 p.

Serpico, S. B. and F. Roli (1995). "Classification of multisensor remote-sensing images by structured neural networks." *IEEE Transactions on Geoscience and Remote Sensing* **33**(3): 562-577.

Settle, J. J. (1996). "On the relationship between spectral unmixing and subspace projection." *IEEE Transactions on Geoscience and Remote Sensing* **34**(4): 1045-1046.

Settle, J. J. and N. A. Drake (1993). "Linear mixing and the estimation of ground cover proportions." *International Journal of Remote Sensing* **14**(6): 1159-1177.

Shahshahani, B. M. and D. A. Landgrebe (1994). "The effect of unlabeled samples in reducing the small sample size problem and mitigating the Hughes phenomenon." *IEEE Transactions on Geoscience and Remote Sensing* **32**(5): 1087-1095.

Shimabukuro, Y. E. and J. A. Smith (1991). "The least-squares mixing models to generate fraction images derived from remote sensing multispectral data." *IEEE Transactions on Geoscience and Remote Sensing* **29**(1): 16-20.

Skidmore, A. K. and B. J. Turner (1988). "Forest mapping accuracies are improved using a supervised nonparametric classifier with SPOT data." *Photogrammetric Engineering and Remote Sensing* **54**(10): 1415–1421.

Smith, M. O., S. L. Ustin, J. B. Adams, and A. R. Gillespie (1990). "Vegetation in deserts: I. A regional measure of abundance from multispectral images." *Remote Sensing of Environment* **31**: 1-26.

Srinivasan, A. and J. A. Richards (1990). "Knowledge-based techniques for multi-source classification." *International Journal of Remote Sensing* **11**: 505-525.

Strahler, A. H. (1980). "The use of prior probabilities in maximum likelihood classification of remotely sensed data." *Remote Sensing of Environment* **10**: 135-163.

Swain, P. H. and S. M. Davis, (Eds.) (1978). *Remote Sensing: The Quantitative Approach.* New York, NY: McGraw-Hill, 396 p.

Tilton, J. C., S. B. Vardeman, and P. H. Swain (1982). "Estimation of context for statistical classification of multispectral image data." *IEEE Transactions on Geoscience and Remote Sensing* **GE-20**(4): 445-452.

Townshend, J. R. G., C. O. Justice, C. Gurney, and J. McManus (1992). "The impact of misregistration on change detection." *IEEE Transactions on Geoscience and Remote Sensing* **30**(5): 1054-1060.

Wang, F. (1990a). "Improving remote sensing image analysis through fuzzy information representation." *Photogrammetric Engineering and Remote Sensing* **56**(8): 1163-1169.

Wang, F. (1990b). "Fuzzy supervised classification of remote sensing images." *IEEE Transactions on Geoscience and Remote Sensing* **28**(2): 194-201.

Wang, F. (1993). "A knowlege-based vision system for detecting land changes at urban fringes." *IEEE Transactions on Geoscience and Remote Sensing* **31**(1): 136-145.

Wang, S., D. B. Elliott, J. B. Campbell, R. W. Erich, and R. M. Haralick (1983). "Spatial reasoning in remotely sensed data." *IEEE Transactions on Geoscience and Remote Sensing* **GE-21**(1): 94-101.

Wharton, S. (1987). "A spectral-knowledge-based approach for urban and land-cover discrimination." *IEEE Transactions on Geoscience and Remote Sensing* **GE-25**(3): 272-282.

Wharton, S. W. (1982). "A contextural classification method for recognizing land use patterns in high resolution remotely sensed data." *Pattern Recognition* **15**(4): 317-324.

Witkin, A. P. (1983). "Scale-space filtering." in: *Ninth International Joint Conference on Artificial Intelligence*, Karlsruhe, West Germany: Morgan Kaufmann Publishers, 1019-1022.

Woodcock, C. and V. J. Harward (1992). "Nested-hierarchical scene models and image segmentation." *International Journal of Remote Sensing* **13**(16): 3167-3187.

Wu, H.-H. P. and R. A. Schowengerdt (1993). "Improved fraction image estimation using image restoration." *IEEE Transactions on Geoscience and Remote Sensing* **31**(4): 771-778.

Yhann, S. R. and J. J. Simpson (1995). "Application of neural networks to AVHRR cloud segmentation." *IEEE Transactions on Geoscience and Remote Sensing* **33**(3): 590-604.

Yoshida, T. and S. Omatu (1994). "Neural network approach to land cover mapping." *IEEE Transactions on Geoscience and Remote Sensing* **32**(5): 1103-1109.

Appendix A Sensor Acronyms

Asrar, G. and R. Greenstone (1995). MTPE/EOS Reference Handbook. Greenbelt, MD: NASA/Goddard Space Flight Center, No. NP-215.

Basedow, R. W., D. C. Carmer, and M. E. Anderson (1995). "HYDICE system, implementation and performance." in: *Imaging Spectrometry*, Orlando, FL: SPIE, vol. 2480: 258-267.

Bonner, W. D. (1969). "Gridding scheme for APT satellite pictures." *Journal of Geophysical Research* **74**(18): 4581-4587.

Chevrel, M., M. Courtois, and G. Weill (1981). "The SPOT satellite remote sensing mission." *Photogrammetric Engineering and Remote Sensing* **47**(8): 1163-1171.

Diner, D. J., C. J. Bruegge, J. V. Martonchik, T. P. Ackerman, R. Davies, S. A. W. Gerstl, H. R. Gordon, P. J. Sellers, J. Clark, et al. (1989). "MISR: A Multi-angle Imaging

SpectroRadiometer for geophysical and climatological research from EOS." *IEEE Transactions on Geoscience and Remote Sensing* **27**(2): 200-214.

Engel, J. L. and O. Weinstein (1983). "The Thematic Mapper - An overview." *IEEE Transactions on Geoscience and Remote Sensing* **GE-21**(3): 258-265.

Hollinger, J. P., J. L. Peirce, and G. A. Poe (1990). "SSM/I instrument evaluation." *IEEE Transactions on Geoscience and Remote Sensing* **28**(5): 781-790.

Irons, J. R., K. J. Ranson, D. L. Williams, R. R. Irish, and F. G. Huegel (1991). "An off-nadir-pointing imaging spectroradiometer for terrestrial ecosystem studies." *IEEE Transactions on Geoscience and Remote Sensing* **29**(1): 66-74.

Kahle, A. B. and A. F. H. Goetz (1983). "Mineralogic Information from a new airborne thermal infrared multispectral scanner." *Science* **222**(4619): 24-27.

Kramer, H. J. (1996). *Observation of the Earth and its Environment: Survey of Missions and Sensors.* Third Ed. Berlin: Springer–Verlag, 960 p.

Lansing, J. C., Jr. and R. W. Cline (1975). "The four- and five-band multispectral scanners for Landsat." *Optical Engineering* **14**: 312.

Myers, J. and J. Arvesen (1995). Sensor Systems of the NASA Airborne Science Program. Moffett Field, CA: NASA/Ames Research Center.

Nishidai, T. (1993). "Early results from 'Fuyo-1' Japan's Earth Resources Satellite (JERS-1)." *International Journal of Remote Sensing* **14**(9): 1825-1833.

Porter, W. M. and H. T. Enmark (1987). "A system overview of the Airborne Visible/Infrared Imaging Spectrometer (AVIRIS)." in: *31st Annual International Technical Symposium*, SPIE, vol. 834: 22-31.

Salomonson, V. V., W. L. Barnes, P. W. Maymon, H. E. Montgomery, and H. Ostrow (1989). "MODIS: Advanced facility instrument for studies of the earth as a system." *IEEE Transactions on Geoscience and Remote Sensing* **27**(2): 145-153.

Short, N. M. and J. Locke M. Stuart (1982). *The Heat Capacity Mapping Mission (HCMM) Anthology*, Washington, D.C.: NASA, No. SP-465, 264 p.

Vane, G., A. F. H. Goetz and J. B. Wellman (1984). "Airborne Imaging Spectrometer: a new tool for remote sensing." *IEEE Transactions on Geoscience and Remote Sensing* **GE-22**(6): 546-549.

Wrigley, R. C., R. E. Slye, S. A. Klooster, R. S. Freedman, M. Carle, and L. F. McGregor (1992). "The Airborne Ocean Color Imager: system description and image processing." *Journal of Imaging Science and Technology* **36**(5): 423-430.

Appendix B   Function Definitions

Bracewell, R. N. (1995). *Two-Dimensional Imaging.* Englewood Cliffs, NJ: Prentice Hall, 689 p.

Gaskill, J. D. (1978). *Linear Systems, Fourier Transforms, and Optics.* New York, NY: John Wiley & Sons, 554 p.

Index